THE COMPLETE GUIDE TO

COMPACT TREES AND SHRUBS FOR THE GARDEN

Duncan Coombs

THE COMPLETE GUIDE TO
COMPACT TREES AND SHRUBS FOR THE GARDEN

THE CROWOOD PRESS

CONTENTS

1 What Makes a Compact Tree or Shrub? 7

2 Trees and Shrubs in Garden Design 13

3 Directory of Suggested Trees 21

4 Directory of Suggested Shrubs 85

5 Directory of Suggested Conifers 239

6 Directory of Suggested *Calluna*, *Daboecia* and *Erica* 267

Appendix I: RHS Hardiness Ratings 278

Appendix II: USDA Plant Hardiness Zones Maps for the US and UK 279

Appendix III: Symbols and Abbreviations 280

Bibliography 280

General Index 281

Plant Index 281

Common Name Index 287

Acknowledgements 288

CHAPTER 1

WHAT MAKES A COMPACT TREE OR SHRUB?

The aim of this book is to produce a select list of trees and shrubs with all the necessary accompanying information to allow the would-be planter to choose with confidence 'the best plant for the intended space'.

A problem that confronts today's garden designer is often a lack of space.

Worldwide there is a lack of space. Increasing population means less space for each individual. In the developed world, increasing demand for land for building on has led to an increase in the value of land. Increased value has, in turn, led to increase in building density and increase in the use of vertical space, as so ably demonstrated by cities such as New York, Tokyo and, to a lesser but increasing extent, in London with skyscraper buildings. Roof gardens, green roofs and green walls have begun to proliferate, and many of these incorporate carefully selected compact trees and shrubs. In less pressurised situations we see greater use of 'brown field' sites and the increasing development of three-story or more town houses.

The traditional view has been that trees and shrubs are always large and can't be used in a small space, but many new cultivars are smaller, being specifically bred and selected to be more compact.

This book informs the reader of the large and often new range of compact trees and shrubs, which are so valuable when planting modern gardens, and how they may be used in modern garden design.

WHAT CONSTITUTES A COMPACT TREE OR SHRUB?

A modern approach would be to consider as a tree any plant that has a permanent branch structure, is large and perhaps dominant in a garden or planting. This idea is, however, dependent on context, since a plant such the slow-growing maple, *Acer palmatum* 'Dissectum', forms at maturity a mounded shape approximately 1.5m (5ft) tall and 2m (6ft 7in) across. In a small garden this would constitute a tree, but in a larger garden this would only be perceived as a shrub.

How a nurseryman (or nurserywoman) decides to produce and market a plant can alter its function in the garden. Fairly vigorous shrubs of normally flexuous, pendulous or creeping

Acer palmatum 'Dissectum' in all its autumn glory. A plant such as this would be regarded as a shrub in a large garden, but as a tree in a small garden.

Malus × robusta 'Red Sentinel' provides bright-red crab apples right through the seasons from autumn to the following early spring.

Cotoneaster 'Hybridus Pendulus' trained and then grown as a small tree.

habit can be grown as a single stem that is trained upwards to a height, usually about 1.8m (6ft) and then the leader allowed to fall over in a natural way. This, coupled with later formative pruning, transforms a shrub into a small, compact tree. Using this technique, plants such as *Cotoneaster* 'Hybridus Pendulus', often used for low-growing ground-cover, can be converted into a compact, pendulous tree.

THE KINDS OF TREES AND SHRUBS AVAILABLE

The Basic Material

As a result of the relatively recent last Ice Age and the slow recolonisation by plants, the UK has a rather small range of what are considered 'native' species. The paucity of the UK's native plant species and, more recently, decreasing garden size, led initially to a demand for a greater range of suitable plants. Plant hunters, most significantly our Victorian forebears, travelled to far-off lands to search for, collect and introduce new plant species, many of these being trees and shrubs. The exploits of these often intrepid and brave collectors are well documented and we, as gardeners, owe much to them for the very wide range of plants we can grow in our gardens. Some of this plant collection did produce bad consequences with habitat damage and, indeed, the possibility of plant extinction in the wild, in some cases.

Today, the introduction of new plants from the wild is very heavily regulated with international treaties banning the import and export of wild plants, except for scientific research.

The trade of plants across the world has expanded greatly in recent years. Driven by the demand for more and more plants to green our environment and inside our home, this trade has flourished. This movement of plants across international boundaries has, however, caused some problems, in particular the spread of plant pests and diseases. Plant phytosanitary regulations have been set up, but such pests and diseases are no respecters of regulations! Unfortunately, there have been several serious incidences of such introductions, perhaps the most infamous being of so-called Dutch elm disease and, currently, the threat of *Xylella*.

This has led to a demand for the practice of local sourcing of plant material from within a country's boundaries. In the UK, there is now a strong campaign for plant wholesalers, plant retailers and the plant-buying public to 'Buy British'. Imported trees and shrubs for sale should now carry a plant passport attached to them to indicate that they have passed phytosanitary regulations.

PLANT BREEDERS AND SELECTORS

The demand for new cultivars, often of a smaller size to suit smaller gardens, has stimulated the activity of plant breeders and selectors. A few nurseries and their propagators stand out in this respect, having produced many new cultivars, in an almost bewildering range, some notably smaller and more compact. Nowhere has this been developed and pursued more than with the conifers. This group has proved to be genetically very adept at producing new forms, both during their production from seed and in sports arising spontaneously from growing plants.

Elsewhere in shrubs, nurserymen have been able to spot any useful variation in their stocks. Over many centuries specific plants and genera have received enormous attention. In Japan and China, *Acer palmatum* has been cultivated and selected to a point where hundreds of different cultivars exist, many very well suited to the small garden. In China, an activity largely unknown in the West, the selection of different forms of the evergreen shrub *Osmanthus fragrans*, has also been practised for over a millennium. This shrub produces the most beautifully scented flowers in winter and spring, and for this the plant is greatly prized.

In Europe, selection has also been focused on evergreen shrubs, with mutations causing leaf variegation being a source of many new plants. Variegation is usually due to a fault in chlorophyll production, the green pigment present in leaves. Variegation causes reduced vigour in the plants and new leaf colour forms. Such new plants, often being small and compact and in numerous different colours, are very appealing to modern gardeners. Unfortunately, the commercial pressure to release these new, desirable cultivars to the buying public has resulted in the release of some poor plants. Ideally, these plants should be trialled over time for their hardiness and propensity to revert to the original green plant.

THE PRODUCTION AND SALE OF TREES

Trees may be sold as seeds, seedlings, whips and feathered whips. Seedlings have been produced for forestry plantings for many years. More recently, young seedlings of ornamental trees and shrubs, some of which are difficult to germinate, have appeared on the market.

Seedlings may be sold for several purposes, including:

- Use as hedging plants.
- As young trees to be grown on into larger trees for subsequent sale.
- For use as understocks, on to which other cultivars may be grafted.

The way the tree seedlings are cultivated in their initial years may affect their suitability for certain uses and their price. The method of tree production, their height and whether or not they have feathers, are usually mentioned when they are offered for sale.

Young seedlings in their seedbed can have several different treatments. They can just be left to grow, which is acceptable for a year, possibly two. After this the young trees may become excessively drawn by growing too close together and begin to develop a tap root (a root that just descends downwards without branching). A root system consisting of just a tap root is undesirable since such plants have been found not to transplant well. Excessive drawing upwards results in a tree that cannot support itself – very undesirable.

To overcome these problems, the seedlings can be dug up, called lifting, and then transplanted. The process of lifting the young tree breaks the developing tap root. This is desirable because this encourages a fibrous root system, which subsequently will transplant well. Upon replanting, each seedling is given more space in which to grow, thus reducing the problem of the plants being drawn upwards.

A cheaper way of preventing the development of a tap root is to undercut the root system of the seedlings. In the tree nursery this is done by a tractor blade, which is drawn through the soil below the seedlings. This blade cuts the developing tap root, stimulating the development of the desirable fibrous root system. This undercutting, whilst cheaper than transplanting the seedlings, does not, however, solve the problem of tree overcrowding.

During the young tree's growth, it may or may not develop side-shoots, called feathers. If the seedling is intended for use in forming hedging, then this is an advantage. Clearly, as the young trees grow, they become taller.

All these variables result in a grading system being employed to tell the potential purchaser how the tree has been produced, its suitability for various uses and it helps to justify different prices.

The Grading System for Seedlings and Very Young Trees

Trees are graded for height, usually in 10cm (4in) divisions.

They are graded for whether or not they are feathered. Feathered trees are better for hedging. Unfeathered trees, called whips, may be better for growing on to form larger trees or for use as understocks. Tree seedlings intended for use as understocks may also be graded for stem thickness.

An indication as to how the young plants have been produced is also usually given:

- 1+1 indicates that the seedling has been grown in the seedbed for one year and then transplanted.
- $\underline{1}$+1 indicates that the seedling has been grown in the seedbed for one year, undercut and then grown on in the seedbed for another year.

As the trees are grown on, this may become more complicated, but still understandable, and will affect the value of the young trees offered for sale, e.g. $\underline{1}$+2+1 indicates that the seedling has been grown in the seedbed for one year, undercut, then grown on in the seedbed for a further two years, then transplanted and grown on for a further year.

This is a somewhat complicated system, reflecting the initial complexity of producing young trees, but a system that the purchaser is advised to study and understand, so that he/she can purchase young trees best suited for their purpose.

Trees in the UK beyond the young stages detailed above, may be grown to conform to British Standard BS 3936. This standard is produced in ten sections, of which *BS 3936-1-1992 Nursery Stock: Specification for Trees and Shrubs* is of relevance here (*see* Table 1.1).

These specifications were developed in association with the Nursery Stock Industry and if adhered to by producers, and used by purchasers when specifying their requirements, should result in the production and supply of good-quality plant material.

Trees Outside the Standards

Some nurserymen specialize in the production of very small trees. In today's tiny gardens these can be very useful and can take the form of the following:

- **Top-worked trees.** Some nurserymen produce trees where a rootstock is used to produce the stem of the tree and several scions are grafted on to the top to form a 'top-worked tree', e.g. *Salix alba* 'Kilmarnock' and *Acer pseudoplatanus* 'Brilliantissimum'.

TABLE 1.1 NURSERY STOCK SPECIFICATIONS FOR FEATHERED AND STANDARD TREES

	Requirements		
Type of tree	Overall height from ground level	Clear stem height from ground level	Circumference of stem at 100cm (39in) above ground level
Light feathered	180–240cm (70–94in)	NS	NS
Heavy feathered	240–400cm (94–156in)	NS	NS
Light standard	250–275cm (98–107in)	120–150cm (47–59in)	NS
Standard	275–300cm (107–117in)	150–180cm (59–70in)	8–10cm (3–4in)
Selected standard	300–360cm (117–140in)	180–210cm (70–82in)	10–12cm (4–5in)
Heavy standard	360–425cm (117–166in)	NS	12–14cm (5–5½in)
Extra-heavy standard	450–600cm (176–234in)	NS	14–20cm (5½–8in)
Semi-mature	Min. 450cm (176in)	NS	25cm+ (10in+)

NS = not specified.

Garden centres and most amateur gardeners are concerned with standard trees and trees below this size. The form of the trees supplied must be typical of that species or cultivar. Where a single leader does not readily occur, this will not be required.

- **Shrubs on a leg.** A single stem of an often-vigorous, non-variegated form of a shrub can be taken up vertically and scions of an often special form of the same shrub grafted on top to form a very small tree, e.g. *Euonymus fortunei* 'Emerald 'n' Gold'.
- **Shrubs converted to small trees.** A single stem of a shrub with enough vigour can be grown vertically up a cane. Once enough height is reached, the growing point is pinched out and the resulting shoots used to form the head of the tree. Examples of shrubs where this is often done are *Buddleja alternifolia* and *Amelanchier lamarckii.*

In the USA the situation is different. The Tree Care Industry Association (TCIA) produces a set of ANSI A300 Standards. These are not intended to be used as standards but should be used to develop written specifications for work assignments leading to established objectives and measurable criteria.

Euonymus japonicus 'Bravo' grown as a standard on a UK nursery.

A young plant of *Amelanchier lamarckii* having been pruned to form a multi-stemmed tree. Several shoots at the base of the plant were removed when very young and the plant encouraged to grow with just a few upward growing shoots.

The Use of Rootstocks in Tree Production

Rootstocks can be important in the propagation of trees and can influence greatly the subsequent performance of the tree.

Some trees just cannot be grown from seed. Some species cannot be relied upon to produce viable seed, or they may be hybrids that will not 'come true' from seed. Rootstocks can, in many cases, enable the commercial production of uniform trees with a predictable performance. This they may do by controlling the vigour and ultimate size of trees. Or they may shorten the time period between planting, and flowering and fruiting. Such considerations are very important, especially when considering fruit trees.

'Treatment' of the Roots of Trees or Shrubs

Trees and shrubs can be supplied bare root, root wrapped, container grown or containerised. All these terms have specific meaning within the industry. Each method has its own advantages and disadvantages.

Bare Root

The plant is lifted from the ground, any remaining soil shaken off and supplied as such. This is acceptable for deciduous subjects during their dormant season, but usually only where transplantation is going to occur within the same garden. Bare roots are extremely susceptible to drying out. Once the roots have dried out, the plant is as good as dead. After lifting, during transportation and during planting, the roots must be protected from drying out. The use of some form of polyacrylamide gel has been advocated. The gel is used in the form of a root dip, the gel forming a thin film over the roots, protecting them to a limited extent from desiccation and is also said to lower transplant shock. The shock of transplanting is said to be lowered by the gel holding water at a low tension, so that plant roots can penetrate the gel and adsorb water from the gel. Such a root-dip treatment could be used in addition to wrapping the roots.

Root Wrapped

Immediately the plant is lifted from the ground, the roots are wrapped with some material, usually hessian or a polythene bag, to protect and protect the roots from desiccation. This is good nursery practice for deciduous plants during the dormant season.

Container Grown

When this system is employed, the young plant is transplanted and grown on in a container until it is well established. When ready for sale or planting out, the plant roots should have thoroughly permeated the compost to the extent that the root ball does not disintegrate when the plant is removed from its container. Conversely, the roots should not be seen to have become severely limited by the edge of the container with the roots running round the edge of the compost forming a solid ring of roots. Such plants are known as 'pot bound' and are not desirable. Where the plants are pot bound, but only to a limited degree, teasing out the roots prior to planting will help alleviate this problem.

Containerised

This is said to occur where a bare-root plant is placed into a container just before sale. The roots have no, or very little, time to permeate the compost before sale or planting out. This can be done to extend the life of an otherwise bare-root plant before sale or planting out. Less reputably, plants can be containerised just before sale or to make the plant more presentable and, therefore, of higher value. In either case, this should be made clear to the purchaser or planter. Unless great care is taken during planting in to open ground, the loose compost will fall off the roots and the planter left with an exposed bare-root plant.

THE PRODUCTION AND SALE OF SHRUBS

In addition to the British Standard 3936, the British Container Growers Group, to which a large number of prominent UK nurseries were signatories, published a *Specification of Standards for the Production of Hardy Container Grown Plants* several years ago (an undated publication). This contained a definition of standards to which a range of container plants should be grown. Included in this were criteria such as container size, number of breaks and overall height, plus numerous other points for a wide range of plants. If adhered to, or at least considered, good-quality plants should be produced and supplied. Today, in practice, most good suppliers meet, and indeed exceed, these standards.

CHAPTER 2

TREES AND SHRUBS IN GARDEN DESIGN

Trees, and to lesser degree shrubs, are fundamental to the design of a garden or successful planting. They function on two scales: the overall design of a garden or landscape and, when planted together, form plant associations, which the designer should consciously plan to form aesthetically pleasing plantings. These two scales are here considered separately.

One of the two Red Borders at Hidcote Manor Garden. The tall, coniferous hedge forms a perfect, uniform, dark-green backdrop to the magnificent mixed border in front.

THE OVERALL DESIGN

Provision of Shelter, Privacy and Enclosure

Starting with an open, new site, trees can be used to create shelter, necessary on any exposed or windswept site, privacy and to produce a sense of enclosure. Here, selection of the best plant is vital to fit the space available. Far better to plant *Thuja plicata* 'Atrovirens' (Western red cedar) that will produce a thin, uniform, green, fine-textured hedge, rather than the ubiquitous × *Cuprocyparis leylandii* (Leyland cypress) that, unless a considerable volume of space is available, will rapidly become far too large and problematic for the garden owner and neighbour alike!

Betula utilis subsp. *albosinensis* photographed in early winter. The pinkish-white stem is clearly shown against the dark-green background of an evergreen coniferous hedge.

Backdrop

Trees can provide an excellent uniform backdrop upon which to display your carefully composed plantings. This backdrop can remain largely the same throughout the year, as is the case with evergreens, such as many conifers, or can change with the seasons, as exemplified by deciduous trees. A further variation

Acer palmatum 'Wilson's Pink Dwarf' is a slow-growing plant that, when first planted, may function as a shrub, but eventually forms the dimensions of a tree in a small garden.

can be created by using hedges of trees such as *Fagus sylvatica* (beech) and *Carpinus betulus* (hornbeam) whose growth remains in a juvenile condition when clipped and does not drop in winter. Hence, the backdrop has three seasonal changes: green in summer, golden brown in winter and light-green and fine-textured in spring.

Division of Space

Within a garden or planting scheme, trees and shrubs can be used to block an overall view and divide the whole into a series of spaces. This can have several effects. The garden will feel larger than it actually is. By skillful positioning, the visitor can be guided around the garden and steered to see the best carefully contrived views. In the ancient scholar gardens of China, walls were often used to ensure that the visitor passed through a series of visual delights without appreciating how far he had walked or the overall garden size: such an effect can also be created by the use of trees and shrubs to obscure the view.

Framing of Views

Again, the ability of trees and large shrubs to limit and frame views, both within and outside a garden, can be employed by the designer. Here care must be taken to select the correct tree or shrubs, considering their ultimate size, so that the view is framed, rather than obscured in time.

Creating Vignettes

Small trees and shrubs can be used to create small spaces within a garden where a specially composed planting can be shown in partial isolation to the rest of the garden. This is exceptionally well demonstrated at Foggy Bottom the garden of Mr Adrian Bloom VMH.

Two plants of *Thuja occidentalis* 'Degroot's Spire' partially isolating a planting of *Crocosmia* 'Lucifer's Match?' in a mixed border at 'Foggy Bottom', the garden of Mr Adrian Bloom in Diss, Norfolk, England.

Mrs Winthrop's Garden at Hidcote Manor Garden. Here the tall hedge separates this garden from the rest of the larger garden, providing Mrs Winthrop with the privacy she desired whilst sunbathing. The garden is hedged on three sides. The south side is left open to allow sunlight to flood in and create a suntrap.

Three-Dimensional Planting

This is taking the idea of 'see through' planting to a higher level. It involves positioning a small tree or large shrub with a thin, largely transparent canopy near the front of a deep garden border, this having the visual effect of increasing the perceived depth of the planting.

Partial Views

Views within a garden can be partially or completely obscured by trees or hedges. When employed to partially block a view around a curving path, this can create a so-called 'come hither effect'. All humans, gardeners especially, are curious creatures and when presented with a hidden view will inevitably wish to walk around the deliberate obstruction and see what lies beyond. This is frequently seen at Hidcote Manor Garden, a National Trust property in the Cotswolds.

Similar effects and a general lightening beneath trees can be achieved by controlled thinning of a tree's canopy as the tree grows. This is well demonstrated at the famous woodland garden created by Princess Strada in Normandy.

Scale and Proportion

In this context, the word scale is taken to mean the relationship between a planting or garden and its overall situation. In the same context, the word proportion is taken to mean the relationship between the plants in a planting. In both comparisons, the concept of 'visual weight' is important, with a sense of balance wanting to be achieved in both.

The passing of time affects both these considerations. As time passes and as a planting develops, the relative sizes and visual weight of herbaceous plants, shrubs and trees alter. Herbaceous plants may be overgrown and disappear, and shrubs become proportionately smaller compared to the trees. Where one can watch a planting develop over years, this phenomenon may be observed. The originally well-designed planting developing into a pleasing, balanced composition.

PLANT ASSOCIATIONS

Here the designer and plantsman consider the fascinating area of plant associations. This being how one plant with its individual features relates visually with its neighbours. The designer needs to appreciate these individual plant visual features. These, plus

A small tree of a *Cytisus* cv. at the front of a deep border. The viewer can see beyond to the planting at the rear of the border, thus making the border appear much deeper than it actually is.

any tactile or olfactory features, are described in detail for each tree or shrub profiled in this book.

When one considers aesthetics, 'beauty lies in the eye of the beholder', but this can be analysed to a large extent by objectively considering the visual features of plants. Trees can provide visual beauty from their overall habit, meaning their branch pattern, form, their three-dimensional shape and texture, plus in closer inspection, visual delight from their foliage, flowers, fruits, stem and bark.

Our olfactory sense can also be stimulated by the scent released from trees, be it the smell of a magnolia in full bloom or the pineapple scent released from the pressed foliage of Western cedar (*Thuja plicata*).

Tactile qualities of trees can also be employed to add interest in the garden, e.g. the smooth and shiny bark of the Tibetan cherry (*Prunus serrula*) or the papery, flaking bark of the paper bark maple (*Acer griseum*).

This section briefly considers each of these features.

Size at Maturity

Size at maturity is a feature of vital importance, and one that has been the object of much work in breeding and selection but is unfortunately difficult to predict accurately, particularly in the case of trees. The growing conditions a plant experiences greatly affect both the growth rate and eventual size of a plant. In this book, a size within a range is given, based on knowledge of the plant and given average, not exceptional, conditions. Very large trees, often called 'champion trees' of exceptional size, are not considered.

In some cases, tree size can be controlled by choice of rootstock. In ornamental trees this can be important with tree genera, including *Malus* (ornamental apples) and *Prunus* (flowering cherries). With trees intended to produce edible fruit, choice of rootstock is very important, since this can affect not only eventual tree size, but other factors such as the length of time taken between planting and the regular production of fruit. Many of these compact or dwarf selections are listed and described in Chapter 3.

Trees and shrubs can be reduced in size, at least for a period, by pruning and, where appropriate, such techniques as head raising, pollarding and coppicing are discussed (*see* Chapter 3).

Shape (or Form) and Habit

These are two visual aspects that are best considered together. Shape (or form) is self-explanatory and refers to the three-dimensional space or volume of space occupied. Trees and shrubs vary considerably in their shape, and this is exemplified by conifers (*see* Chapter 5 for this book's selection).

How trees are propagated and produced in a plant nursery largely determines a tree's future development and final shape. The variously shaped trees available from nurseries and how these may be employed is detailed in Chapter 1.

Habit (the branching pattern of a plant) is an important characteristic when considering compact plants for limited space. Many of the plants, especially the trees, profiled in this book have been bred or selected for their compact, freely branching, upright or sometimes fastigiate habits; such trees do not use much floor area but give height to planting.

In garden design and the subject of plant association, the relationship between form and habit of deciduous plants changes with the seasons. In winter, habit is very evident; in the other seasons, because of the presence of leaves, individual plant form or overall shape assumes greater importance.

Colour

It is in the use of colour, with each plant making its contribution, that the artistic flair of the designer can come fully to the fore. He or she can employ the full range of the colour wheel to create the vivid contrasts or subtle harmonies that the client may desire. Colour may be seen as a pageant flowing through the seasons, with each plant making its seasonal contribution.

In the recent past, much of this colour was provided by seasonal bedding schemes, which reached a peak in the case of wealthy Edwardians who used to favour spring, summer, autumn and winter bedding displays. Such displays required the production of 'annuals' under heated glasshouse conditions and much labour, often in the form of paid gardeners. Very few of us have access to these facilities and the resources for this form of profligate horticulture.

Flowering trees and shrubs provide colour, but often only for a short period of time. An aim of the breeder has been to extend this period, trying to induce the plant to produce a succession of bloom, rather than just a short display. In some cases, this has been successful.

Colour in the form of leaf variegation, not just green, but also shades of lighter green and white, has been sought by breeders very successfully. Both deciduous and evergreen shrubs with coloured leaves are widely available. Certain colour combinations provide the most vivid contrasts.

A very strong colour contrast between the purple-red foliage of *Physocarpus opulifolius* 'Diabolo'PBR and *Choisya ternata* SUNDANCE 'Lich'.

Another exceptionally strong contrast of foliage colours between *Choisya ternata* SUNDANCE 'Lich' and *Veronica ochracea*. These are both small shrubs easily accommodated in today's small garden.

Shrubs such as this *Phormium* 'Sundowner' are now proving reliably hardy in many gardens and contribute vivid leaf colour and a dramatic habit all the year round.

The leaves of *Salvia argentea* just compel you to touch them.

The new leaves of *Mahonia aquifolium* (Oregon grape) really are as slippery as they look!

Some of these coloured-leaved shrubs provide even more opportunities by changing colour with the seasons. This might be both an opportunity and a problem for the designer!

In recent years, the length of winter has been more appreciated, perhaps because of its milder nature, and the need to provide colour and interest has been more fully appreciated. Here coloured-leaved evergreens have come to the fore. Approximately 20 per cent of evergreens was often recommended in the past as the best balance between evergreen and deciduous shrubs, a greater proportion being considered too 'heavy' and dreary. Today, with the advent of brightly coloured, compact evergreens, this percentage can be increased with great advantage. The fact that some of these coloured evergreens also have distinctive shapes, e.g. the numerous *Phormium* cultivars, adds to their versatility.

A further gain is that the increasing mildness of winters allows the wider planting of shrub genera, such as *Coprosma* and *Coryline*, previously considered too tender for outdoor planting, but whose members contain many coloured-leaf cultivars.

Texture

In this context, texture relates mainly to shrubs and their leaves, but can also be noticed in trees as the leaves expand and develop through spring into summer. Do you recall a treescape in late spring? That beautiful diverse appearance is due to shades of green presented by the developing leaves, but also to their developing and differing sizes.

Texture can be considered as have having two components: one tactile, the other visual.

To consider the tactile component first, try to remember the leaves of lamb's ears (*Stachys byzantina*) or *Salvia argentea* and compare those with the leaf of the Oregon grape (*Mahonia aquifolium*) when just emerged.

Clients love to touch the soft 'furry' leaves of the salvia and watch water run off the waxen surface of the mahonia. The opportunity to exploit this experience is limited, but several shrubs do have strong textural features. For example, *Viburnum rhytidophyllum*, with its large, leathery leaves, whose veins are deeply surface-indented, and *Rhododendron bureavii* with dark-green, glossy leaves whose undersides are covered with a rich, woolly, soft to the touch, indumentum.

The large, entire leaves of *Sorbus* 'John Mitchell' are visually coarse-textured.

The subdivided, pinnate leaves of *Sorbus aucuparia* are, when compared to those of *Sorbus* 'John Mitchell', visually more finely textured.

The second component of texture refers to the size and degree of division of leaves. A large-leaved shrub can be described as coarse-textured, e.g. *Aucuba japonica*. A shrub with many small leaves can be described as fine-textured, e.g. common box (*Buxus sempervirens*) or *Veronica rakaiensis*. The degree to which a leaf is subdivided will affect its appearance and texture: the broad, simple leaf of *Sorbus aria* (whitebeam) appearing coarser than the pinnate leaf of *Sorbus aucuparia* (mountain ash).

Leaf texture and colour tend to interact and the plain-green leaf of *Aucuba japonica* 'Hillieri', with plain, large, dark-green leaves, appears coarser than the yellow-spotted *Aucuba japonica* 'Variegata'.

A further way that texture can be employed by the designer is to partially obscure or highlight a feature by acting as a backdrop. Consider a fine, garden vase mounted on a plinth. Against a fine-textured background, the vase is highlighted, against a coarse-textured background the vase is less distinct. Theory has it that the eye is distracted by the relatively large components of the coarse background and hence the vase is less prominent. Behind a large, domestic gas-tank would a fine or coarse-textured backdrop tend to hide it?

Fragrance, Aromas and Odours

A wide range of trees and shrubs emit scents or odours and would-be planters would be wise to acknowledge this and use these plants accordingly.

Trees within the genera of *Magnolia*, *Malus*, *Prunus*, *Robinia* and *Tilia* are notable for their fragrance. Many of these are compact and suitable for restricted spaces, where their scent may be confined and concentrated.

Whilst the flowers of *Magnolia* FAIRY MAGNOLIA BLUSH ('Micjur01')PBR are beautiful to look at, they are also deliciously scented. This is another factor for the gardener to consider when selecting plants for his or her garden.

Roses are, of course, well known for their scent, but note that not all are scented. The new garden roses, first bred by David Austin, are an attempt to combine the scent of old roses with the recurrent flowering of modern garden roses and, in many cases, to produce a more compact, better branched shrub.

Other shrub genera with scented flowers to have received the breeder's attention and of note are the introduction of smaller, compact *Philadelphus* cultivars.

Various aromas and odours are released by many plants. *Aloysia citrodora*, lavenders and *Cistus* emit pleasant aromas when the leaves are crushed or, in some cases, just brushed by. Take care though – the author's experience has revealed that what is loved by one is disgusting to another! *Choisya ternata* (the Mexican orange) is a case in point: liked by some, abhorred by others. The genus *Salvia*, the sages, needs particular care. Many members

The foliage of *Choisya ternata* SUNDANCE ('Lich'PBR). The bright-yellow foliage colour is appreciated by all, but the smell of the foliage, once touched, is liked by some but hated by others – gardeners be aware!

Viburnum opulus (Guelder rose) photographed in early autumn when fruit colour, but leaf colour has yet to develop. This fine, deciduous shrub, native to the UK, is too large for today's small garden, but the cultivar *Viburnum opulus* 'Compactum' is a much smaller, compact shrub, which flowers and fruits just as freely. A good example of where a gardener with a small garden may wish to plant a native plant to encourage nature conservation and can do so, but by careful selection can choose a smaller more compact plant more suitable for where space is limited.

release a very bad smell when crushed (*Salvia turkestanica* being memorably bad) but *Salvia rosmarinus* (formerly *Rosmarinus officinalis*) is liked by all.

All interplay together, and the designer's role is to select and plant to create pleasing creations for their client. Good designers earn their fees!

THE RELEVANCE TO OUR ENVIRONMENT

Pollution and climate change are now acknowledged problems. How can gardeners and designers help with these problems, given declining space? By employing modern, compact trees and shrubs selected for this role and detailed in this book, they certainly can. (Where appropriate, species or selections of British native trees and shrubs have been chosen and included in the plant profiles contained in Chapters 3, 4 and 5.)

In recent years there has been much debate as to whether we, as a nation of gardeners, should plant 'native' or 'exotic' trees and shrubs. The term 'native' in this context means naturally occurring within the country in which it grows or is planted. This is laudable, in that such planting will also help to increase all the other native animal species dependent on it for food.

From an aesthetic point of view this could, however, be questioned. For example, in the case of the UK, the country has a somewhat depauperate flora, due to the relatively recent last Ice Age, with only relatively few species. Planting with only native species could, therefore, result in a somewhat uniform and perhaps visually boring planting.

The use of a much wider planting range, using plants from all over the world, could result in much more visually exciting and pleasing plantings. Would these mixed plantings be as good for encouraging other wildlife? Our native insects and birds would not have met these plants before; could they adapt to feed on them? The Royal Horticultural Society (RHS) has done some experimental work on this subject (RHSorg.uk/plants4bugs). Native plants were found to be most favoured by insects, but exotics were still visited and useful, but to a slightly lesser extent. Probably the answer for gardeners is to steal an analogy from another context – 'a little of what you fancy does you good'.

CHAPTER 3

DIRECTORY OF SUGGESTED TREES

ACER

A large genus composed of mainly deciduous trees. It ranges in size from large trees right through to shrubs and comes from all three northern continents.

Plants from the genus, known colloquially as 'maples', can be recognized collectively by useful common characteristics. They have buds arranged in opposite pairs that develop to give leaves also in opposite pairs. Their fruit is a winged samara (key) consisting of a nutlet or sometimes a pair and a large membranous wing. The wing is a mechanism for seed dispersal and in the autumn the whirling descent of the seeds delights children who frequently call them 'helicopters'.

The leaves are often palmate, but not always so. Cultivated forms of *Acer palmatum* have been selected to have a very elaborate and often dissected leaf shape.

As a genus they have several characteristics that commend them to the would-be cultivator. Their leaf shape and the colours developed in the autumn are a major attraction. In some, the bark may be very characteristic and attractive, as with *Acer griseum* (the paper bark maple). In others, the bark may feature elaborate patterns or striations, as in the snake bark maples. The flowers of maples are usually a visually insignificant feature but, in a few cases, they may be especially attractive, e.g. *Acer negundo* var. *violaceum* looks very attractive in spring when in flower with its pendant, long, reddish-pink flower tassels.

A mature tree of *Acer griseum* (paper bark maple).

The trifoliate leaf of *Acer griseum* showing autumn leaf colour.

Acer palmatum and its cultivars are considered collectively. The group of maples known as snake bark maples are also considered collectively. *Acer griseum,* the cultivars of *Acer japonicum* and *Acer negundo* 'Flamingo' are dealt with separately.

Acer griseum (paperbark maple)

This species was first collected and introduced by Ernest Wilson, whilst he was plant hunting for the famous plant nursery Messrs Veitch, from Central China in 1901.

It is a slow-growing deciduous tree of outstanding beauty, which may eventually reach a height of 14m (40ft).

Usually propagated from seed, which is generally only 1–2 per cent viable, it is best grown as a half-standard or standard tree to enable the bark, which is the main feature, to be seen and closely inspected. Some seed variability is present and if purchasing a tree for use as a specimen, inspection of individual trees before purchase is advised, as some from a young age produce masses of peeling bark that hangs on the stem in loose flakes to reveal the cinnamon-red young bark beneath. In others, the bark does not exfoliate so freely, and this very desirable feature is less prominent.

In mid-spring the leaves expand to reveal delicate trifoliate leaves. The central leaflet is slightly larger than the two laterals and each has three to five pairs of serrations.

Small, inconspicuous flowers are followed by winged seeds, known as winged samaras, and are typical of the maples. The paired seeds hang down on stalks, each seed with its wing being 7mm (1¼in) long.

The branches of the tree divide naturally to form a rounded or slightly taller than wide, somewhat irregular but very pleasing shape.

Growing Requirements

Hardiness: USDA 7b/8a. RHS H5.
Shelter: Avoid very exposed or very windy positions.
Sun: Best in full sun or part shade, but not heavy shade, to ensure that an evenly shaped specimen develops.

Prunus 'Snow Showers' forms a small, weeping tree and is ideal for where space is limited.

The cinnamon-red bark of *Acer griseum*, so clearly shown in winter.

Autumn leaf colour of *Acer griseum.*

Soil: The tree will grow in any soil providing that it is well drained in winter but avoid planting in very dry or excessively well-drained soils.
pH: The plant is not pH-sensitive, but avoid planting in very dry or excessively well-drained soils.

Garden Design Considerations

A superb, slow-growing, small specimen tree, providing fine leaf texture in spring and summer, plus bright late autumn leaf colour. Also very desirable is the distinct and attractive winter bark colour and texture.

It is often best planted near a path or at the front of a border to allow close inspection and appreciation of the bark, and is grown as a standard to allow underplanting for other seasonal interest.

When used in a planting intended mainly for autumn colour, *Acer griseum* reliably provides orange and red colours, which harmonize easily, but the colour develops late and may be employed to extend the display.

Acer griseum associates well with other acers, especially *Acer palmatum* and its many cultivars. With careful selection of *Acer palmatum* cultivars, interesting contrasts of habit, form and leaf texture can be produced.

Acer japonicum (Downy Japanese Maple or Fullmoon Maple)

This is a native of Japan and Korea and has been cultivated in Japan for many centuries. It is distinguished from *Acer palmatum* by its larger leaves, 8–15cm (3–6in) across, greater number of lobes to each leaf, having seven to ten lobes and with soft, hairy leaf petioles. *Acer japonicum* forms a bushy deciduous tree seldom reaching more than 6–9m (20–30ft) high, but often the same width. In cultivation it is generally larger than most forms of *Acer palmatum*.

Growing Requirements

These are the same as for *Acer palmatum* and its cultivars to which the reader is referred.

Hardiness: USDA 6b/7a. RHS H6.
The two cultivars most commonly grown are *Acer japonicum* 'Aconitifolium' AGM and *Acer japonicum* 'Vitifolium'.

Acer japonicum 'Aconitifolium' AGM

Height: 4–8m (13–26ft).
Spread: 4–8m (13–26ft).
The leaves of this cultivar are deeply lobed and cut, and as the specific epithet suggests, resemble those of many *Aconitum* species, commonly known as monkshoods or wolfsbane. Attractive, small, red flowers hang down beneath the branches in spring. Leaf colour is green during the growing season, but reliably develops rich autumn colours of reds and deep-purple.

Acer japonicum 'Vitifolium'

Height: 2.5–4m (8–13ft).
Spread: 2.5–4m (8–12ft).
The leaves of this cultivar are more divided than the species having ten to twelve lobes, the specific epithet suggesting a similarity to those of a vine. The leaves are green during spring and summer, but reliably develop rich autumn colours.

Garden Design Considerations

As these two cultivars are taller and wider than most *A. palmatum* cultivars, they can be usefully employed as backdrops to these and are also an excellent choice where a specimen bush-like small tree is required and adequate space is available to accommodate their spread.

Acer negundo 'Flamingo'

A variegated cultivar of *Acer negundo*, the box elder, which is native to North America.

Acer negundo 'Flamingo' is a fast-growing, medium-sized, deciduous tree capable of attaining a height of 12m (39ft) and a width of 8m (26ft). It is recommended here only for growing as a shrub, the growth being cut back hard every second year in late winter. When grown in this manner, the pinnate foliage is larger than normal and the variegated foliage more prominent. Each leaflet has a green centre, splashed with pink when young that later turns to cream. When cut back as recommended, the upright stems achieve a height of 2–3m (6½–10ft) and the brightly variegated foliage is well displayed.

In winter, once the foliage has fallen, the young, pale-grey stems add to a coloured winter stem display.

Leaves of *Acer japonicum* showing early autumn leaf colour, plus a pair of winged samaras (seeds).

The dissected leaves of *Acer japonicum* 'Aconitifolium' showing the development of the wonderful autumn leaf colour that is such a feature of this cultivar.

Growing Requirements

A plant of easy cultivation.
Hardiness: USDA 6b/7a. RHS H6.
Aspect: The growing situation can be of any orientation.
Shelter: No extra shelter is required.
Sun: Full sun or part shade.
Soil: A soil of any texture will suffice providing that it is well drained.
pH: It is not pH-sensitive, tolerating a wide range of soil acidity or alkalinity.

When cut down hard every other year it is recommended that the plant receives an application of balanced fertiliser and is mulched every spring to encourage the rapid new growth required.

Any sign of 'reverted' all-green growth should be removed as soon as seen. Reversion will occur, is much more vigorous than the desirable variegated growth and will soon dominate the growth unless rapidly removed.

Garden Design Considerations

This cultivar, when grown as a coppiced tree, provides light, bright foliage colour for the summer garden and pale-grey winter stems. Although of rapid growth, it does have a high-maintenance requirement if it is to be displayed to best advantage.

The brightly variegated foliage of *Acer negundo* 'Flamingo' in early summer.

Acer palmatum (Japanese Maple)

This species is native to Japan, China and Korea.

When grown from seed, this species can produce very variable seedlings. The trees can be 6m (20ft) high and wide at maturity; indeed, in wild populations they are often larger than this. At the other end of the range, true dwarf trees may occur that never exceed 1m (3ft) at maturity.

Features such as leaf colour and leaf shape tend to be characters that influence final size. Trees with the normal palmate five to seven lobes and plain-green leaves tend to be the most vigorous and attain the greatest size. Any change in leaf pigmentation – be it paler green, yellowed or reddish, or other variegation – tends to result in a reduction in vigour. The leaves can show a greater number of divisions than normal and the more the leaves are divided, generally, the greater reduction in vigour. The habit of the plants raised from seed can occasionally be very upright or pendulous.

This apparent natural variation has been selected for by growers, especially in Japan, over millennia. This has resulted in the naming of several hundred named cultivars. In an attempt to categorise, the cultivars are attributed to Group names following the scheme laid out in Vertrees, J.D. & Gregory P. (2010) *Japanese Maples: The Complete Guide to Selection and Cultivation.*

Acer negundo 'Winter Lightning' pollarded to develop and display the light-coloured stems clearly in winter.

The Groups are:

- **Amoenum Group:** leaf lobes shallowly to moderately divided, usually up to two-thirds of the way to the leaf base.
- **Palmatum Group:** leaf lobes moderately to deeply divided two-thirds to three-quarters of the way to the leaf base and, in some cases, almost to the base.
- **Matsumurae Group:** leaf lobes mostly very deeply divided more than three-quarters of the way to the leaf base.
- **Linearilobum Group:** lobes narrow, strap-like, divided to the leaf base.
- **Dissectum Group:** leaf lobes very deeply divided and deeply dissected into sub-lobes.
- **Dwarf Group:** cultivars whose mature height does not usually exceed 2m (6½ft).

Growing Requirements

Hardiness: Care needs to be taken when selecting cultivars. Most are USDA 6a/7a RHS H6, but some are USDA 7b/8a RHS H5. When grown in the ground, the top growth of *Acer palmatum*, if fully dormant, has been found to be hardy down to −18°C (0°F). In very cold climates, the roots and top growth of plants in containers must be given some cold protection. Wrapping the entire container and plant in several layers of bubble wrap should suffice. It is important to remember that *Acer palmatum* in the wild is often an understorey tree, growing beneath the canopy of larger trees.
Aspect: In cultivation, the aspect of the site can be south-, west- or east-facing, but preferably not north-facing.
Shelter: In all cases, some shelter will be required. The amount of shelter required by an individual plant can largely be predicted by the plant's leaf shape. The more the leaf is dissected, the more shelter will be required. Plants from the Linearilobum and Dissectum Groups have a particular need for shelter from wind. An east-facing site should be treated with caution. Acers are susceptible to late spring frosts, especially if the frost is followed by strong, early morning sunshine, as is often experienced in a site open to the east.
Sun: Providing that a regular water supply can be maintained throughout the growing season, then green-leaved cultivars will grow

well in full sun and purple-leaved cultivars will develop their characteristic coloration more clearly. Care must be taken because any interruption in water supply, or excessive drying winds, will result in scorching of the leaves. Pale and yellow-leaved cultivars require careful positioning since stronger light produces better coloration, but too much light will result in scorching. Growing the plants below taller deciduous trees that provide dappled shade is often optimal for nearly all cultivars.

Soil: An organic soil that is well drained but retains water in the summer and with an acid pH is ideal. Not many gardeners have this ideal, and this maple can be very successfully grown in soils that deviate from this. A sandy, clay or loamy soil is acceptable, providing that it is well drained and retains moisture in summer. The addition of organic matter will improve the structure and moisture retention. Mulching the ground with organic material after planting and thereafter on an annual basis will help the plants to establish and will gradually improve the soil.

pH: Soil pH should ideally be between 5.5 and 6.5 and not above pH 7. Thin, alkaline, chalky soils should be avoided.

Garden Design Considerations

Acer palmatum, including its numerous cultivars, has an exceptional range of applications in the garden and landscape.

From Table 3.1 it is clear to see that there is a considerable range in size of plant involved from moderate-sized trees to small, compact shrubs, and careful attention to this feature is necessary during the process of selection. Their need for at least some shelter could limit the use of these trees, although the larger cultivars with uncut leaves can be used to shelter those with more delicate, finely cut leaves.

In many gardens, the need to create the correct conditions for these trees to thrive can be clearly seen – larger trees being planted and left to develop a dappled, sheltered environment beneath them. This can be seen in both the Acer Glades at the Westonbirt Arboretum in Gloucestershire, England. The Old Glade is surrounded by a very diverse planting of trees and shrubs, all of which serve to provide the necessary environment. The Acer Loop in the younger part of the arboretum, known as Silk Wood, is a more recent creation with a range of quick-growing *Larix* (larch) employed to rapidly produce a lightly shaded woodland in which a collection of *Acer palmatum* cultivars grow to perfection.

Acer palmatum cultivars growing under the dappled deciduous shade of *Larix* sp. (larch) in Silk Wood, part of Westonbirt Arboretum Gloucestershire, England. This represents almost ideal growing conditions.

Within the correct environment, a range of garden design techniques can be employed to effectively display these choice plants. Taller plants can be used to act as a backdrop to their small companions and maples of differing habits placed adjacently to emphasise their differences. The height, but not width, of some cultivars can be increased to create the desired effect. *Acer palmatum* cultivars of the Dissectum Group usually are pendulous in habit, forming, when left to their own devices, a flowing mound of growth. If desired, a strong shoot can be trained up a cane and once the desired height is reached, the leader allowed to droop naturally. Followed by further careful formative pruning, this will result in a fountain, rather than a mound, of fine-textured pendulous growth.

The colour and texture of the maple's leaves are perhaps the most important visual features. Autumn leaf colour is the characteristic that first comes to mind, as some cultivars are legendary in their ability to produce amazing, vivid, autumnal displays. Many cultivars show changes in leaf colour from spring, through summer and into autumn. In addition, these colours can be readily predicted, especially with regard to final autumn leaf colour. The garden designer must, therefore, have a good 'eye for colour', considering the kaleidoscope of colour that can be presented to the garden viewer throughout the changing seasons. Fiery autumnal displays can be created with the strongest reds and yellows, although some of the most effective plant combinations employ the contrast between one blazing plant against the solid dark-green of a non-coloured cultivar.

In addition to leaf colour, the maples add leaf texture to the garden designer's palette. The basic five-lobed leaf of the original *Acer palmatum* is fine-textured compared with other trees and shrubs with much larger, simple leaves and coarser texture. Maples can be used against these coarser textures to create various degrees of textural contrast or, for a more subtle effect, planted as a small group with each *Acer palmatum* cultivar displaying a different leaf texture. The most accomplished designers enjoy a heady but planned mix of all the visual features maples display!

Late autumn and winter are other times when these maples can bring interest and, in some cases, drama to the garden. The black

TABLE 3.1 *ACER PALMATUM* CULTIVARS

Name	Group	Habit	Height	Spread	Leaf	Leaf colour			Remarks
						Spring	Summer	Autumn	
'Ariadne' AGM	Matsumurae	Spreading	3m (3ft)	3m (3ft)	5–7 lobes	Orange-pink	Purplish-red	Orange-pink	Red fruits
'Beni-shidare'	Dissectum	Pendulous	5m (16ft)	6m (19½ft)	7–9 lobes	Red	Bronzed	Orange-red	Name means red cascading.
'Burgundy Lace' AGM	Matsumurae	Bushy	6m (19½ft)	6m (19½ft)	7 lobes	Red	Wine-red	Green-red	Burns in full sun
'Dissectum Flavescens'	Dissectum	Pendulous	2m (6½ft)	3m (9ft)	5–7 lobes	Yellow-green	Green	Yellow tinged with orange	
'Eddisbury' AGM	Palmatum	Upright	4m (13ft)	3m (9ft)	7 lobes	Pale-yellow	Green	Golden with shades or red	Coral red young growth and leaf petioles.
'Ginko-san'	Palmatum	Small, rounded bush	1.5m (5ft)	1.5m (5ft)	5 lobes	Shrimp-pink	Reddish	Red	H5
'Inazuma'	Matsumurae	Tall, rounded tree	10m (33ft)	6m (20ft)	7 lobes	Purple-red	Green tinged purple	Orange and crimson	Excellent autumn leaf colour
'Koto-no-ito'	Linearilobum	Upright	2m (6½ft)	2m (6½ft)	5–7 lobes	Crimson tones	Green	Yellow and red	Note unusual leaf shape with long linear lobes
'Nuresagi'	Matsumurae	Upright	6–7m (20–23ft)	4m (13ft)	5–7 lobes	Light-purple	Dark purple-green	Red	Pale striations on bark
'Orange Dream'	Palmatum	Upright	3m (10ft)	2m (6½ft)	7 lobes	Orange	Lemon yellow	Yellow	Best in semi-shade
'Osakusuki' AGM	Amoenum	Rounded	7m (23ft)	5m (16ft)	7 lobes	Light-pink	Green	Red	Fiery autumn colour
'Sangu-kaku' AGM Syn. 'Senkaki'	Palmatum	Upright	8m (26ft)	6m (20ft)	5–7 lobes	Pale- green	Light-green	Yellow with tones of red	Coral bark maple. Young growth bright coral-red. Good winter bark colour
'Seiryu' AGM	Dissectum	Upright	5–7m (16–23ft)	3–4m (10–13ft)	7 lobes	Bright-green tinged red	Green	Orange, yellow and red	
'Skeeter's Broom'	Dwarf	Upright	2m (6½ft)	1.5m (5ft)	5–7 lobes	Red	Deep-purple-red	Red	

silhouette of sinuous acer branches against a reddening afternoon winter sky can certainly be stunning. A few Japanese maples have strongly coloured bark, e.g. *Acer palmatum* 'Sango-kaku' (the coral bark maple). In these cases, this colour is most apparent once the leaves have fallen and the tree should be positioned so that any direct winter sunlight falls upon and highlights this feature.

Acers create a strong oriental feel to a garden and can be used with other associated plants, such as conifers and bamboos, together with other appropriate features to create a Japanese- or Chinese-styled garden. In such gardens, the acers are often used as specimens to bring seasonal vivid colour to gardens otherwise noted for a very restrained use of colour.

As if this was not enough, maples are also widely used in container gardening where their slow growth, compact habit and small eventual size are all advantages. In the Japanese art form of bonsai (the word literally meaning plant in a pot), the beautiful features of acers already discussed are developed and cultivated on an individual and miniature scale. Many different types of tree and shrub can be used to form a specimen bonsai, but maples are very often chosen. In China, the ancient skill of penjing is still widely practiced, but here it is a miniature landscape that is created and maintained. Maples frequently feature in these miniature words.

Japanese maples with all these diverse attributes should surely form an important component in today's increasingly compact or small gardens.

A strong contrast of both leaf colour and texture. *Rheum palmatum,* an ornamental rhubarb, growing next to *Acer palmatum* 'Dissectum Atropurpureum'.

Acer palmatum 'Orange Dream'. Photographed in early spring just as the leaves were expanding.

Acer palmatum 'Orange Dream'. Photographed a little later that spring. Note the leaf expansion and change in leaf colour.

Acer palmatum 'Orange Dream' showing autumn foliage colour. These changes in leaf colour through the seasons is but one of the joys of growing acers.

Acer palmatum 'Skeeter's Broom'. A dwarf cultivar with excellent autumn leaf colour.

The red bark of the younger stems of *Acer palmatum* 'Sangu-kaku' (coral bark maple). This cultivar has light-yellow autumn foliage colour and can grow to make a tree 10m (30ft) tall.

Acer palmatum 'Koto-no-ito' of the Linearilobum Group. In cultivars such as this, the leaves are divided into fine, linear lobes. They need sheltered conditions if they are to grow well.

Snake Bark Maples

Acer davidii is perhaps the most widely cultivated and typical of this group of acer species. In the wild, *Acer davidii* is a deciduous small-to medium-sized, round-topped tree, still quite commonly found in central and western China. The species has been introduced into cultivation several times and has been found to be quite variable, making a tree between 9 and 15m (30–50ft) high. The leaves, although sometimes lobed when young, are normally undivided, 7.5–13.5cm (3–7in) long, heart-shaped at the base and unevenly toothed.

The main points giving this species great ornamental value are well-developed autumn leaf colour, plus green and white striated bark.

TABLE 3.2 SNAKE BARK MAPLES AND CULTIVARS

Name	Height	Spread	Habit	Leaf shape	Autumn leaf colour	Bark	Remarks
Acer capillipes (red snake bark maple)	8–12m (26–39ft)	4–8m (13–26ft)	Columnar and upright	3 lobes, 12cm (5in) long	Orange-red	White striations on green	
Acer × conspicuum 'Phoenix'	4–8m (13–26ft) as a small tree. 5m (16ft) when cut down hard every other year	4–8m (13–26ft) as a small tree. 2.5m (8ft) when cut down hard every other year	Upright	Broad ovate 5–20cm (2–8in) long 3–5 lobes	Yellow	Bright-red new growth. Older bark yellow	Can be cut down hard every early spring to form a coppiced plant with vivid winter-stem colour
Acer davidii	15m (49ft)	15m (49ft)		Ovate, unlobed or shallowly lobed, 15cm (6in) long	Orange-red	Green and white	
Acer davidii 'George Forrest' AGM	8–12m (26–39ft)	4–8m (13–26ft)	Upright arching	Ovate, unlobed or shallowly lobed, 15cm (6in) long	Orange-red	Green and white	Red young stems and red leaf petioles
Acer davidii VIPER 'Mindavi'	8–12m (26–39ft)	4–8m (13–26ft)	Upright, columnar	Ovate, unlobed or shallowly lobed, 15cm (6in) long	Orange-red	Strong green and white	Red young stems
Acer 'Serpentine' (syn. *Acer davidii* 'Serpentine')	8–12m (26–39ft)	4–8m (13–26ft)	Upright	Small, up to 10 cm (4in) long	Orange-yellow	Green and white.	Purple young stems
Acer grosseri	8–12m (26–39ft)	8–12m (26–39ft)	Broad-headed	Leaves triangular shallowly lobed 7cm (3in) long	Orange	Grey-green	
Acer grosseri var. *hersii* (Her's maple)	8–12m (26–39ft)	4–8m (13–26ft)	Broad-headed	Leaves broad, oval, 3 lobes	Orange	Green striped white	
Acer pensylvanicum 'Erythrocladum'	8–12m (26–39ft) as a single-stemmed tree, 3m (10ft) when cut down hard every other year	8m (26ft) as a single-stemmed tree, 2m (6½ft) when cut down hard every other year	Upright	Large leaves 15–25cm (6–10in) long	Yellow	Young growth bright-pink with white stripes. Older bark-reddish with white stripes	Not good as a single-stemmed tree. Better cut down hard every other year
Acer rufinerve (grey snake bark maple or red vein maple)	8–12m (26–39ft)	4–8m (13–26ft)	Broad-headed	Leaves oval with 3 lobes	Good	Green striped white	Young shoots glaucous

Growing Requirements

Hardiness: USDA 7b/8a RHS 5. Except *Acer capillipes* and *Acer pensylvanicum* 'Erythrocladum', which are USDA 6b/7a. RHS H6.
Aspect and shelter: For all species and cultivars mentioned in Table 3.2, the site can be facing any orientation, but it must be sheltered, especially from cold north or easterly winds.
Sun: Full sun or part shade is acceptable, and they can be used to provide the necessary dappled shade required by the less tolerant *Acer palmatum* and its cultivars.
Soil: The soil can be of any texture, but it needs to be well drained, fertile and moisture-retentive. A thin, excessively well-drained soil that dries out in summer should be avoided.
pH: All the snake bark maples mentioned in Table 3.2 are tolerant of a wide pH range, with the single exception of *Acer pensylvanicum* 'Erythrocladum', which requires a neutral or acidic soil.

Garden Design Considerations

All the trees in Table 3.2, except *Acer pensylvanicum* 'Erythrocladum', make superb specimen trees. Ideally, they should be positioned where the bark striation, young growth and buds can be inspected closely and appreciated, especially in the winter. They provide several seasons of interest, being shapely trees, having small, often pendant inflorescences in spring, attractive summer foliage and, in most years, very bright, autumn leaf colour.

Those grown as multi-stemmed trees or cut down hard every other year in late winter are good stand-alone plants contributing vivid autumn leaf and winter-stem colour. They are also good contributors to a winter garden or border, being a little taller than *Cornus* coppiced for winter-stem colour.

The fine, white striations on a green background make a clear pattern on the bark of *Acer capillipes*.

The younger red stems of *Acer × conspicuum* 'Phoenix' will gradually change to a yellow bark as it ages.

The young red stems of *Acer davidii* 'George Forrest'. As the branch ages and its bark develops, the pattern changes to white striations on a green background.

Acer pensylvanicum 'Erythrocladum'. Here a young branch shows the typical pink bark colour. The yellow autumn foliage colour is also shown.

Acer davidii VIPER ('Mindavi'). Here an older branch with white striations on a green background is shown.

ALBIZIA

A genus of about 130 species of deciduous trees, shrubs and climbers from Asia. They have pinnate foliage resembling mimosa and fluffy heads of flowers. Only *Abizia julibrissin* is cultivated in temperate areas. The species and its cultivars are widely cultivated in the USA, but in south-eastern parts it has become an invasive weed and should not be planted.

Growing Requirements

Hardiness: USDA 7b/8a. RHS H4.
Aspect: A south- or west-facing site should be selected.
Shelter: A warm, sheltered site is required, away from cold winds.
Sun: Full sun is required for best growth and flowering.
Soil: A soil of any texture is acceptable, but it must be well drained.
pH: A wide range of soil pH is tolerated.
In the UK, the planting of this tree is best restricted to warm gardens in southern England or sheltered gardens in warmer urban areas. During cold weather some damage to top growth may be expected, and cold protection may be required.

Albizia julibrissin (Persian Silk Tree)

A small- to medium-sized deciduous tree with arching branches 5–16m (16–50ft) tall.
Foliage: Each complete leaf is 20–40cm (8–16in) long and 12–25cm (5–10in) wide. The leaves are bipinnate having an acacia or fern-like appearance. Each leaf has six to twelve pairs of pinnae and each pinna is further subdivided into twenty to thirty pairs of leaflets.
Flowers: The flower consists of a small calyx and corolla with a cluster of very prominent stamens each 2–3cm (1–1.5in) long, giving the appearance of a mass of silken threads.

Garden Design Considerations

In the UK, the use of these small trees or large shrubs is best confined to relatively warm, sheltered gardens. Here they can be employed to introduce a subtropical effect along with other exotics, such as bananas (*Musa* and *Ensete*), ginger lilies (*Hedychium*) and the Indian shot plant (*Canna*). In such gardens, *Albizia* may be part of the permanent planting but may need winter protection.

Albizia can also be used as isolated specimen plants in containers or large pots and, as such, they are ideal for patios or small gardens. Here their fine-textured, coloured leaves and unusual flowers can be appreciated in fine detail when viewed close up. When used in this way, they can easily be given the extra winter protection they may need or they can be moved to a cold greenhouse or conservatory for the winter.

Albizia julibrissin f. *rosea.* The head of the small tree is typically rather broad, and the beautiful pinnate leaves can be clearly seen.

Albizia julibrissin f. *rosea.* The flowers can be seen to visually consist of the very prominent pink stamens. This feature with the foliage gives a subtropical feel to any planting where it is included.

TABLE 3.3 *ALBIZIA JULIBRISSIN*, FORMA *ROSEA* AND CULTIVARS

Name	Habit	Height	Spread	Foliage	Flowers	Remarks
A. julibrissin	Small- to medium-sized tree with an open arching habit	5–16m (16–50ft)	4m (13ft)	Leaves green	Greenish-yellow, followed by long, narrow, brown pods	May make a larger tree in USA
A. julibrissin CHOCOLATE FOUNTAIN ('Ncaj1')	Large shrub or small tree with a slightly weeping habit	6m (20ft)	4m (13ft)	Emerges chocolate-purple, later greyish-green	Rich purple flower heads	
A. julibrissin OMBRELLE ('Boubri'PBR)	Large shrub, small tree	4m (13ft)	3m (10ft)	Leaves green	Deep-pink flower heads	
A. julibrissin f. *rosea* AGM (pink silk tree)	Large shrub, or broad-headed small tree	4–8m (13–26ft)	2.5–4m (6.5–13ft)	Leaves green	Rose-pink flower heads	Said to be more cold-tolerant than *A. julibrissin*
A. julibrissin f. *rosea* 'Summer Chocolate'PBR AGM	Large shrub, small tree	4–8m (13–26ft)	2.5–4m (6.5–13ft)	Pinkish-purple in early summer	Pink flowers in summer	

AMELANCHIER

A genus of about twenty species found mainly in North America, but also Europe and Asia. The number of *Amelanchier* species still confuses botanists due to hybridization, polyploidy and apomixis. They are all deciduous shrubs and trees. Their flowers appear in spring and are followed by the fruit, which is a berry-like pome.

Growing Requirements

Hardiness: USDA 6a-1. RHS H7.
Aspect: A site facing any direction is suitable.
Shelter: A sheltered or exposed site is suitable.
Sun: Full sun or part shade is required. A situation in full sun will result in the best autumn leaf colour.
Soil: This must be well drained. A clay or loam-based soil is suitable but avoid a chalky soil.
pH: An acid or neutral pH is suitable. Avoid a soil with a high, alkaline pH.

Garden Design Considerations

The genus *Amelanchier* contains an extremely versatile and useful range of small trees and shrubs. The only limitation is their dislike of alkaline soils, with the sole exception of *A. ovalis* and its cultivars, which are lime-tolerant.

Amelanchier lamarckii is perhaps the most useful species to gardeners and particularly garden designers, since it provides several seasons of interest in the garden. In the spring, the profuse, white flowers are well displayed against the young, bronzed foliage. During the summer, it provides light dappled shade from its canopy of leaves and finally reliable, vivid autumn leaf colour. It can be employed as a small, single-stemmed tree or a multi-stemmed tree as a large shrub and can even be clipped to a regular formal shape, if required. *Amelanchier canadensis* is very similar and in the nursery industry the two are frequently confused.

The *Amelanchier* × *grandiflora* cultivars provide a slightly larger tree and in the case of *A.* × *grandiflora* 'Robin Hill' AGM, a little pink colour instead of white.

Upright shrubs and trees are very useful where space is limited, *A. alnifolia* 'Obelisk'PBR is a very upright-growing form and *A. canadensis* 'Glenn Form' (syn. *A. canadensis* Rainbow Pillar) is a slightly larger, upright cultivar.

In addition to all this versatility, they provide a crop of small fruits. These are mainly initially red but darken later. This is if they get the chance, since they are eagerly and rapidly eaten by birds.

Amelanchier ovalis and its cultivars are similar, if a little smaller, than the above but have the added benefit of tolerating soils with a high pH.

Amelanchier canadensis 'Glenn Form' (syn. *Amelanchier canadensis* Rainbow Pillar), showing the very upright habit of this cultivar.

Amelanchier canadensis 'Glenn Form' syn. *Amelanchier canadensis* Rainbow Pillar, showing the upright habit, shape of the leaves and the beginning of autumn leaf colour.

A small, multi-stemmed tree of *Amelanchier lamarckii* in full flower in spring.

Amelanchier lamarckii in flower, the contrasting colour of the young foliage is very evident.

Amelanchier lamarckii in fruit in early summer. The photographer did well to get this picture before the bird ate the fruit!

The autumn leaf colour of *Amelanchier lamarckii* is reliably produced every autumn.

TABLE 3.4 SPECIES AND CULTIVARS OF *AMELANCHIER*

Name	Habit	Height	Spread	Foliage	Flower	Fruit	Remarks
A. alnifolia 'Obelisk'[PBR]	Very slender tree or large shrub. Upswept stems	4–8m (13–26ft)	1–1.5m (3–5ft)	Young leaves bronzed, later green, then autumn colour of orange and red	Dense, terminal clusters of white flowers with green centres, early to mid-spring	Scarlet becoming darker	Fruit soon eaten by birds
A. canadensis	Large, erect shrub or small tree	4–8m (13–26ft)	4–8m (13–26ft)	Young leaves bronzed, later green, then autumn colour of orange and red	White flowers	Dark-purple becoming black	Fruit soon eaten by birds
A. canadensis 'Glenn Form' (syn. *A. canadensis* RAINBOW PILLAR)	Upright suckering shrub	4m (13ft)	3m (10ft)	Oblong, mid-green, autumn colours green, yellow, orange, red	Upright sprays of small, white flowers	Blue becoming black	Fruit soon eaten by birds
A. × *grandiflora* 'Ballerina'	Small tree or large shrub	4–8m (13–26ft)	4–8m (13–26ft)	Young leaves bronzed, later green, then autumn colour of orange and red	White in short nodding racemes	Red becoming purple	Fruit soon eaten by birds
A. × *grandiflora* 'Robin Hill' AGM	Conical, less loose-growing than *A. lamarckii*, upright small tree	5–10m (16–32ft)	5m (16ft)	Young leaves bronzed, later green, then autumn colour of orange and red	Pink buds open pale-pink, in short nodding racemes	Red becoming purple	Fruit soon eaten by birds
A. 'La Paloma' AGM	Small tree or large shrub	2.5–4m (8–15ft)	2.5–4m (8–15ft)	Young leaves bronzed, later green, then autumn colour of orange and red	Drooping sprays of white flowers in mid-spring	Red becoming purple	Fruit soon eaten by birds
A. laevis 'R. J. Hilton' AGM (smooth serviceberry)	Small tree or large shrub	2.5–4m (8–15ft)	2.5–4m 8–15ft)	Young leaves bronzed, later green, then autumn colour of orange and red	Pink buds opening to pink-tinged flowers, in nodding racemes, mid-spring	Red becoming purple	Fruit soon eaten by birds
A. laevis 'Snowflakes' (smooth serviceberry)	Small tree or large shrub	2.5–4m (8–15ft)	2.5–4m (8–15ft)	Young leaves bronzed, later green, then autumn colour of orange and red	Single, white flowers in nodding racemes, mid-spring	Red becoming purple	Fruit soon eaten by birds
A. lamarckii AGM (snowy mespilus, juneberry)	A shrub with an open habit or a small tree	8–12m (26–40ft)	4–8m (13–26ft)	Young leaves bronzed, later green, then autumn colour of orange and red	White in a lax raceme	Red becoming dark-purple	Fruit soon eaten by birds. Can be grown as a multi-stemmed tree
A. ovalis 'Edelweiss' (smooth service tree)	A medium-sized shrub with an upright habit	2.5–4m (8–15ft)	2.5–4m (8–15ft)	Young leaves bronzed, later green, then autumn colour of orange and red	Short erect racemes of relatively large white flowers.	Red becoming black	Will tolerate some lime in the soil

Amelanchier ovalis, the inflorescence is typically more upright than that of *A. lamarckii*.

Amelanchier ovalis in fruit. Note also the more rounded shape of the leaf.

ARBUTUS

A genus of about fourteen species of evergreen, shrubby trees found in North and South America, across Europe to the Mediterranean. Belonging to the family Ericaceae they are unusual in that some species are tolerant of lime in the soil. The flowers are urn-shaped and followed by strawberry-like fruits.

Arbutus unedo, the Killarney strawberry tree, is a member of the Artic-alpine flora of the British Isles, having survived the last Ice Age in the south-west extremity, which escaped glaciation.

Arbutus × andrachnoides is a naturally occurring hybrid between *Arbutus andrachne* and *Arbutus unedo*.

Growing Requirements

Hardiness: Variable (see individual entries).
Aspect: A south- or west-facing site is required.
Shelter: A sheltered site is required, especially from cold, desiccating easterly or northerly winds.
Sun: Full sun is necessary.
Soil: A soil with high organic matter content, but of any texture is preferred. The soil must be well drained.
pH: In contrast to most other ericaceous plants, a wide pH range is accepted, including alkaline soils. This is with the sole exception of *Arbutus menziesii*, which requires an acid or neutral soil.

Arbutus unedo in flower. These are normally produced in the late autumn.

Arbutus unedo in fruit. Fruit from the previous year ripens and colours, usually at the same time as the plant flowers. Overall, this can give a very decorative display, the flowers and fruit showing up well against the dark-green foliage.

Arbutus Species

Flowers: The panicles of white, urn-shaped flowers are produced in the autumn. In *Arbutus unedo* f. *rubra* the flowers are pink.
Fruit: The fruit starts to develop after fertilisation of the flowers. The following year, at flowering time, last year's fruit ripens and becomes red-coloured. Once ripe, the fruit becomes sweeter and is edible.

Garden Design Considerations

Arbutus forms a relatively unusual group of evergreen shrubs and trees. Those that flower and fruit in autumn and winter are especially valuable in producing a degree of interest in what can be an otherwise rather dull period in the garden.

They all require shelter from cold, desiccating winds, and inland they all need sheltered sites in warm, sunny, south-facing positions. *Arbutus* can, however, withstand the warmer, moist, often strong winds found in coastal gardens. In such gardens, *Arbutus* grows well and can function as wind protection for other more susceptible plants.

Arbutus menziesii is the largest of those considered and is too large for small or compact gardens. In larger coastal gardens it makes a fine but large, spreading tree.

Arbutus unedo and its forms and cultivars make either a spreading, evergreen shrub or, if trained as a single-stemmed specimen, a small, evergreen tree. In either form it adds colour and interest, provides structure and helps in the balance between evergreen and deciduous subjects in the winter garden. For flower and fruit production, *Arbutus unedo* f. *rubra* or *Arbutus unedo* 'Atlantic' are the best selections, but where a more compact shrub or small, compact tree is required, then plant *Arbutus unedo* ROSELILY ('Minilily'[PBR]).

Arbutus unedo is well suited to city and courtyard gardens where space is restricted and there is possibly only enough room for one tree. *Arbutus unedo*, or perhaps one of its more floriferous selections, is a good choice since they have several features giving the plant character and interest, including broad, evergreen foliage, attractive bark, flowers and fruit in autumn and winter, plus the provision of shade in the summer.

Arbutus is also very appropriate for the cottage or less formal gardens. Left largely to its own devices, it naturally has an interesting form and habit. Often, it forms an irregular, multi-stemmed, small tree or large shrub whose informal shape associates well with the overall loose design of the garden.

These plants come from an essentially Mediterranean type of climate with hot, dry summers and mild, wet winters. As such they are very appropriate in a Mediterranean garden or where a Mediterranean style of garden is desired.

AZARA MICROPHYLLA 'VARIEGATA'

This evergreen species, from Chile and Argentina, was introduced in 1861 by Richard Pearce and it is the variegated cultivar that is here recommended because of its brighter and lighter effect.

The plant forms a shrub or an exceptionally elegant small tree when kept to a single stem. Growth is slow, but the ultimate height is 4–8m (13–26ft) with a spread of 2.5–4m (8–13ft), taking twenty to fifty years to achieve.

Branch development takes on an almost frond-like pattern, forming attractive sprays. The small, alternate leaves, the largest being about 2.5cm (1in) long, appear to be in opposite pairs, one leaf often being almost three times the size of the other. The leaves are unevenly edged with creamy-white. Overall, the leaves give a fine-textured effect.

Tiny, yellow flowers consisting mainly of anthers appear on the underside of branches in late winter or early spring and have a very delicate scent of vanilla. Only very occasionally are these followed by white or purple berries.

Growing Requirements

Hardiness: USDA8b/9a. RHS H4.
Shelter: Avoid cold, wind-swept sites. In very cold winters, some limited damage may occur, but regrowth will soon take place.
Sun: Best placed in sun or part shade. Avoid heavy shade.
Soil: A soil that is well drained with some organic matter is required.
pH: A wide pH range is accepted.

TABLE 3.5 *ARBUTUS* SPECIES AND CULTIVARS

Name	Habit	Height	Spread	Foliage	Flowers	Fruit	Bark	Remarks
A. andrachne (Grecian strawberry tree)	Small tree	6m (20ft)	6m (20ft)	Leaves oblong 5–10cm (2–4in) long	Panicles each 10cm (4in) long with white flowers, produced in spring	Spherical 1–2cm (½–¾in) across. Last year's fruit ripen and become red at the same time as flowering occurs	Cinnamon-brown smooth, peeling to reveal green layer that later becomes orange, brown	USDA RHS H3. Becomes hardier as it matures
A. × *andrachnoides* AGM	Bushy evergreen tree	4–8m (13–26ft)	4–8m (13–26ft)	Leaves ovate 10cm (4in) long	Panicles with urn-shaped, white flowers, produced from autumn into spring	Small orange-red strawberry-like, rarely produced	Cinnamon red branches	USDA RHS H4
A menziesii (Pacific madrone, madrone)	Upright columnar tree	12m+ (40ft+)	8m+ (26ft+)	Leathery	Upright panicles of white flowers in early summer	Orange-red fruits ripen the following year	Smooth reddish-brown bark peels to reveal green young bark	USDA RHS H4. Requires acid or neutral soil
A. unedo (Killarney strawberry tree)	Spreading shrubby tree	4–8m (13–26ft)	4–8m (13–26ft)	Glossy, bright-green, leathery, 10cm (4in) long	Panicles with urn-shaped, white flowers, produced in late autumn	Spherical, strawberry-like. Last year's fruit become red and ripen at same time as flowering occurs	Deep-brown and shedding	USDA RHS H5
A. unedo 'Atlantic' AGM	Spreading shrubby tree	4–8m (13–26ft)	4–8m (13–26ft)	Glossy, bright-green, leathery 10cm (4in) long	Panicles with urn-shaped white flowers. Produced profusely in late autumn	Profuse spherical, strawberry-like. Last year's fruit become red and ripen at same time as flowering occurs	Deep brown and shedding	USDA RHS H5
A. unedo ROSELILY ('Minilily'[PBR])	Small, shrubby evergreen	2.5–4m (8–13ft)	2.5–4m (8–13ft)	Large, glossy, dark-green leaves	Drooping panicles of whiteish-pink flowers in autumn	Yellow-red fruits in following year	Red-brown peeling	USDA RHS H5
A. unedo f. *rubra* AGM		4–8m (13–26ft)	4–8m (13–26ft)		Panicles with urn-shaped, pink flowers, produced in autumn	Profuse spherical, strawberry-like. Last year's fruit become red and ripen at same time as flowering occurs	Red-brown bark	USDA RHS H5

Garden and Landscape Design Considerations

An excellent small tree in a slightly sheltered position. Place near a path or where close access is easy so that the fine growth habit, variegation, flowers and scent may be easily appreciated.

It will produce a fine wall shrub when grown against a tall south-, west- or east-facing wall.

Good for introducing textural leaf contrasts when paired against coarse-textured shrubs, such as *Viburnum davidii* or *Camellia japonica* cultivars.

Well suited for inclusion in a border or garden designed for winter interest, where both its foliage and flowers will contribute.

Azara microphylla 'Variegata'. The whole tree has a very fine textured almost mist-like effect and stands out against a dark background.

Azara microphylla 'Variegata'. When the plant flowers in late winter, the variegated foliage largely conceals the tiny, pale-yellow blooms, but their vanilla-like fragrance reveals their presence.

BETULA

The birches comprise a group of about forty-five species all found in the northern temperate region. All are hardy, deciduous trees and their main decorative attributes are their bark, which may be white, and yellow and golden autumn leaf colour. As a group, many are too large for modern, small gardens, but those mentioned below have been selected for moderate size and exceptional garden merit.

Growing Requirements

A tough genus of trees tolerant of a wide range of conditions. Normal length of life should be considered, since birches are not naturally long-lived trees. They take perhaps only twenty years to reach their maximum size and gradually decline after this; a fifty-year-old tree is past its best and many trees die before this age is reached.

Hardiness: USDA 6a-1. RHS H7.

Sun: In the natural environment, *Betula* is a pioneer genus, occupying previously bare soil and being eventually superseded by slower growing, but eventually more persistent tree genera. As such, birches are adapted to, and grow best in, full sun. They will grow in part shade, but growth will be slower and often rather thin.

Soil: Soil type may be almost any, but a deep, fertile, well-drained soil will produce the largest trees. Poor drainage is tolerated by some species and light soil is acceptable but produces a smaller tree.

pH: A wide pH range is tolerated, but highly alkaline soils will produce slower growth and smaller trees.

Planting: Birches are essentially shallow-rooted and efficient in collecting moisture and nutrients from the soil. This has consequences that should be taken into account by those considering planting these trees. Under the tree canopy the soil will be dry, which limits the cultivation of many shade-loving but moisture-requiring plants. Main roots will be near or will break the surface of the soil. This may be a problem where they are grown in fine grass and a grass-free circle out to at least the outer edge of the tree canopy should be maintained. When planting birch trees it is essential that the roots are positioned to be at the same depth as they were before being transplanted. As mentioned above, birches are naturally shallow-rooted trees. If the roots are planted too deeply, this is a serious mistake and frequently leads to the death of the tree.

Pruning: If pruning is required, this should only be done in late summer or early autumn. Pruning at any other time will result in serious loss of sap, which is injurious to the tree. Never top or lop birch, it completely ruins the habit of the tree. If a tree is too broad, reduction can be achieved by complete removal of one or several side-branches, removing the branch just above the branch bark ridge. If the tree is too tall or just too big, it is better to remove the tree.

Garden Design Considerations

Some trees detailed shown above are too large for small gardens, but small and very small alternatives are given. The garden designer and gardener are earnestly encouraged to think of any birch tree's eventual size and thereby avoid the need for any size reduction by pruning, which is very undesirable.

In the landscape and garden, birches find a very wide range of applications. After planting, especially when planted as very small, young trees, they will grow fast and rapidly, providing an attractive and effective screen. Recently, this approach was taken at RHS Garden Wisley where birches were planted to form a screen after severe damage to trees due to the 'Great Storm' of 1987. The trees themselves looked attractive and rapidly closed gaping gaps caused by this destructive event. Several years later, when less urgent planting of exotic trees could be considered, the birches were removed and replaced by other trees.

Examples of the natural variation found Betula bark colour.

Betula 'Fascination' AGM. A cultivar with a very dark, shiny bark and very different from the white-barked birch more commonly seen. Planted with other white-barked birch it would contrast and stand out clearly.

Betula nigra (river birch) with a very shaggy bark.

Betula pendula. The true species has this characteristic black and white pattern to its bark. This makes it distinct from similar but hybrid birch that do not have this pattern.

Betula utilis subsp. *jacquemontii* 'Jermyns' AGM. One of the brightest white-stemmed birches.

Such use of birches, especially native species, is of great environmental value, the trees providing a very valuable food source for many other native species and encouraging much needed species' diversity.

A single tree in a small garden, or several in a larger garden, makes a beautiful, fine-textured background. This then acts as a fine backdrop to set off and display other smaller plantings. This is a technique widely employed in show gardens at spring garden shows, such as Chelsea Flower Show. In late spring, the emerging small leaves of birches deliver a perfect background and the trees' light shade allows plantings beneath them. Here a word of caution is appropriate for, although the shade is light, in 'the real garden' the tree roots are hungry for water and nutrients; this, unfortunately, restricts the range of ground-layer plants that can be successfully grown.

Birches change greatly in their appearance with the seasons. In summer, they form a background. In autumn, the vivid brown and gold of the leaves before they fall is a glorious, if transient, sight. In winter, again their presence becomes very important due to their bark colour. In the garden or bed intended mainly for winter interest, this is the main reason for their inclusion. *Betula pendula*, whose common name of silver birch recalls the tree's prominent white bark, is a good example to start with. Later, as the tree matures, the base of the trunk goes on to develop its characteristic black and white fissured bark.

If a more striking and vivid display of bark colour is required, the genus contains a range of colours. Perhaps most familiar to gardeners is the vivid, often startling white bark of *Betula utilis* subsp. *jacquemontii*, whose visual effect can be increased by planting a multi-stemmed tree, thereby multiplying the number of white stems apparent within a small area.

Multi-stemmed trees may be produced by cutting out the leader of a young tree near ground level. This stimulates the growth of several branches, which eventually form a multi-stemmed tree. This copies the effect often seen in nature, where a young tree is damaged by a grazing rabbit. A multi-stemmed tree can also be effectively produced by planting several small trees in one pot or several young trees very close together. This can be seen to very good effect in the garden of The Garden House in Devon.

TABLE 3.6 *BETULA* SPECIES AND CULTIVARS

Name	Common name	Height	Spread	Bark	Remarks
B. gmelinii (syn. *B. apoiensis*) 'Mount Apoi' AGM		4m (15½ft)	3m (10ft)	Creamy-white	A slow-growing tree, ideal for a small garden, forming a neat, pyramidal head
B. 'Hergest'		6m (20ft)	6m (20ft)		Light copper-brown, shiny bark
B. nigra	River birch or red birch	8m (31ft)	6m (20ft)	Dark very flaky bark.	Likes extra moisture in the soil and does well on clay
B. nigra 'Little King'		5m (16ft)	4m (15½ft)		Like a small version of the species
B. nigra 'Siloh Splash'		3m (10ft)	3m (10ft)		Leaves with a white, irregular variegation
B. pendula	Silver birch or lady of the woods	12m (46ft)	8m (31ft)	White when young, but becoming fissured at the base giving a black and white irregular pattern	Native in Europe and very good for wildlife. When grown from seed the degree to which the trees are pendulous is very variable
B. pendula subsp. pendula FASTIGIATA JOES ('Jolep1')		6m (20ft)	3m (10ft)	Bark will whiten with age	Narrow, pyramidal head. Good where space is limited
B. pendula subsp. pendula 'Schneverdinger Goldbirke' (syn. *Betula pendula* 'Golden Cloud')		6m (20ft)	4m (15½ft)		Yellow-leaved selection. Popular when first introduced, but later leaves found to burn in the sun and susceptible to fungal infection
B. pendula 'Laciniata' (syn. 'Dalecarlica') AGM	Swedish birch	8m	6m (20ft)		Pendulous habit. Leaves oval and cut by narrow lobes
B. pendula 'Youngii'	Young's weeping birch	4m (15½ft)	4m (15½ft)		A small, weeping tree, eventually forming a dome-shaped head with branches arching down to the ground
B. pubescens	Downy birch	6m (20ft)	4m (15½ft)		Native to Europe, but generally found on damper soil or higher elevation than *B. pendula*
B. utilis subsp. *jacquemontii*	Himalayan birch	6m (20ft)	4m (15½ft)	Variable, can be white or shades of brown	Upright, vigorous habit. Some wonderful plants are contained within this subspecies and can be obtained by individual plant selection*
B. utilis subsp. *albosinensis*	Himalayan birch	6m (20ft)	4m (15½ft)	White when new, later turning pinkish	A variable subsp. (*see* selections in Table 3.8)

* *Betula utilis* subsp. *jacquemontii* is itself a white-stemmed selection of *Betula utilis*. Those cultivars shown below are themselves further selections from the subsp. and they should, therefore, be considered the very best of white-stemmed birch and should only be propagated by grafting. As such they will commend a much higher price than seed raised plants.

Betula pendula and some of its cultivars

Betula pendula. The basic species.

Betula pendula subsp. *pendula* 'Schneverdinger Goldbirke' (syn. *Betula pendula* 'Golden Cloud'). Distinct with its yellow foliage but, after introduction, found to be a poor-growing plant.

Betula pendula 'Tristis'. A very beautiful, even more pendulous cultivar.

Betula pendula subsp. *pendula* FASTIGIATA JOES ('Jolep1'). A comparatively new cultivar that is hoped to be a smaller and more upright form of the original species.

Betula pendula 'Youngii'. A very popular, small,weeping tree, whose height may be determined by deciding at what height the leader is allowed to weep and removing lateral stems, as required.

TABLE 3.7 *BETULA UTILIS* SUBSP. *JACQUEMONTII* SELECTIONS

Name	Height	Remarks
'Dorenbos' (syn. 'Snow Queen') AGM	6m (20ft)	Bark pale-orange when young, turning a striking white when older
'Jermyn's' AGM	6m (20ft)	A very vigorous white-barked form
'McBeath'	4m (15½ft)	A markedly less vigorous and ultimately smaller tree than other selections, but still with a wonderfully white stem
'Silver Shadow' AGM	6m (20ft)	Dazzling white stems. Large drooping leaves

TABLE 3.8 *BETULA UTILIS* SUBSP. *ALBOSINENSIS* SELECTIONS

Name	Height	Bark
var. *septentrionalis*	6m (20ft)	Pinkish to coppery-red
'Bowling Green' AGM	6m (20ft)	An excellent selection with dark-red peeling bark
'China Rose'	6m (20ft)	Solidly red bark
'Kansu' AGM	6m (20ft)	Multicoloured copper and pink bark
'Red Panda' AGM	6m (20ft)	Copper-pink bark

When grown from seed, *Betula utilis* subsp. *jacquemontii* produces seedlings with a range of bark colours, some shades of brown, others white. This may be what is required to give what might be considered a more 'naturalistic effect'. If the one tree, or trees, within a garden are required to be all dazzling white, then this can be assured by the purchaser individually selecting white-stemmed individuals from the tree nursery or by purchasing trees of a named clone of this subspecies that has itself been selected for outstanding stem colour.

Other colours are available, ranging from the coppery-red of *Betula utilis* subsp. *albosinensis* 'Bowling Green' or the more subtle pink coloration of other selections of *Betula utilis* subsp. *albosinensis*.

As a further inducement to the would-be cultivator, the trees are essentially very low maintenance. Removal of low lateral branches to reveal the stem may be required in the young, developing tree, but then they are best left alone and just enjoyed!

Betula utilis subsp. *albosinensis* 'Bowling Green' AGM.

Betula utilis subsp. *albosinensis* 'Bowling Green' AGM.

Betula utilis subsp. *jacquemontii*. A multi-stemmed tree, giving a real lift to this small garden in the depths of winter.

Betula utilis subsp. *jacquemontii*. A fine little group setting the garden off well both in winter and summer.

CARAGANA

A genus of leguminous, small, deciduous shrubs distributed over Central Asia. Only one species, *Caragana arborescens*, is capable of forming a small tree.

Caragana arborescens (Siberian Pea Tree)

A deciduous shrub or small tree.
Height: 4.5–6m (15–20ft).
Leaves: Pinnate, bright-green, 4–7.5cm (1.5–3in) long with four to six pairs of leaflets.
Flowers: Produced in late spring, pea-like and pale yellow.
Fruit: A pod 4–7.5cm (1.5–3in) long.

Growing Requirements

Hardiness: USDA 6a-1, RHS H7
Aspect: A site that is facing south or west is preferred.
Shelter: Some shelter, especially as a top-worked tree is necessary.
Sun: Full sun will produce the most prolific flowering.
Soil: The soil can be of any texture, as long as it is well drained.
pH: A very wide pH range is accepted. As long as good drainage can be provided, almost any soil is acceptable; the plant roots obtain nitrogen from nitrogen-fixing bacteria growing in nodules on the plant roots and it is tolerant of low levels of other nutrients.

Garden Design Considerations

When grown as trees, these are ideal for the small garden or where space is limited. They combine and contribute to the garden, both delicate foliage and attractive flowers.

Caragana arborescens 'Lorbergii' is larger than the top-worked trees. It has a characteristic graceful, arching habit; the leaves are very finely textured and, overall, it makes a small- to medium-sized, very attractive tree.

Caragana arborescens 'Pendula' and *C. arborescens* 'Walker' are both similar very small trees with a very distinctive pendulous habit, bright-green, pinnate foliage and a good display of yellow flowers in early summer. Their use would be appropriate in a large planter on a patio, making an interesting and unusual specimen or as feature and distinctive plants in a larger mixed planting.

TABLE 3.9 *CARAGANA ARBORESCENS* AND CULTIVARS

Name	Habit	Height	Spread	Leaves	Flowers	Remarks
C. arborescens	Bushy	3m (10ft)	3m (10ft)	Pinnate	Pale-yellow, pea-like	Dimensions given as a shrub
C. arborescens 'Lorbergii'	Arching	4–6m (15–20ft)	3–4m (10–13ft)	Narrow linear leaflets, giving a very fine texture	Yellow, smaller than *C. arborescens*	
C. arborescens 'Pendula'	Stiffly pendulous	1–1.5m (3–5ft)	1.1.5m (3–5ft)	Bright-green, pinnate	Pale-yellow, pea-like	Top-worked on to *C. arborescens*
C. arborescens 'Walker'	Pendulous	1.–1.5m (3–5ft)	1–1.5m (3–5ft)	Bright-green, pinnate	Pale-yellow pea-like	Top-worked on to *C. arborescens*

Caragana arborescens 'Walker'. A young, top-worked tree on a nursery prior to planting out.

Caragana arborescens 'Walker'. Photographed in late spring with the pea-like flowers about to open and the finely divided foliage developing.

Caragana arborescens 'Pendular' in flower and the less finely divided foliage.

CLERODENDRUM

A genus of mainly tropical trees, shrubs and climbers that is frequently used for ornamental purposes in tropical gardens and in heated glasshouses and conservatories in colder climbers.

Three species are hardy, and one is recommended and detailed below.

Clerodendrum trichotomum var. *fargesii* AGM (Harlequin Glory Bower)

A vigorous, deciduous, large shrub from western China. When constrained to a single stem, it forms a very attractive small tree.
Height: 5–6m (16–20ft).
Spread: 5–6m (16–20ft).
Leaves: Large opposite, ovate leaves 10–22cm (4–9in) long and 5–12cm (2–5in) wide emerge in late spring.
Flowers and fruit: From pink buds, very fragrant flowers appear in late summer. The flower has a white corolla, green sepals and a red calyx. This is followed by a vivid, metallic-blue berry, which forms a unique colourful combination with the persistent red, five-pointed calyx. Flowers and berries are freely produced. This display continues into the autumn. Unfortunately, the foliage has a most unpleasant smell, but only when touched or brushed against.

Clerodendrum trichotomum var. *fargesii* 'Carnival' AGM

A less vigorous, variegated cultivar.
Height: 2.5–4m (7–13ft).
Spread: 1.5–2.5m (5–8ft); perhaps better treated as a large shrub rather than a very small tree.
Leaves: Green shade by darker green and margined with yellow. Later the margins mature to a creamy-yellow.
Flowers and fruit: A little less free-flowering and fruiting, but still a colourful addition.

Growing Requirements

Hardiness: USDA 7b/8a. RHS H5.
Shelter and sun: Both require a sheltered, preferably south-facing position in full sun or part shade.
Soil: Soil can be any texture, but well drained and preferably with high organic content.
pH: Tolerant of a wide pH range.

Garden Design Considerations

An unusual, distinctive, small garden tree providing late season scent, flower and eye-catching fruit colour. Good as an isolated specimen or in an informal cottage-style garden.

Clerodendrum trichotomum var. *fargesii* 'Carnival' provides a lot of summer and early autumn interest, adding a subtropical feel to a bed or border designed for a late season colourful display.

Clerodendrum trichotomum var. *fargesii*. The small tree accommodated in a large border where the plant is adding to flower colour and scent in late summer.

Clerodendrum trichotomum var. *fargesii*. Here the detail of the white, scented flowers is clearly seen. After the flowers are finished, they will be followed by metallic-blue, spherical fruits, each coupled with a persistent red calyx.

Clerodendrum trichotomum var. fargesii 'Carnival'. The picture shows the smaller, more compact nature of this variegated cultivar.

The variegated foliage and flowers of *Clerodendrum trichotomum* var. fargesii 'Carnival' contributes colour from both foliage and flowers to the late summer scene.

CRATAEGUS

Known collectively as the thorns, this genus is found in Europe, western Asia, the Himalayas, China, Japan and west Africa, but principally in North America. The size of the genus has yet to be determined, as it is now thought that many of the species previously described are, in fact, hybrids of, or merely variants of, other species.

The plants within the genus are nearly all small, deciduous trees or shrubs armed with thorns. The flowers are usually white and produced in early summer and followed by fruits, called pomes or haws, in various colours, including black, yellow and blue, but red is by far the most common.

Growing Requirements

Hardiness: USDA 6a-1. RHS H7, except *C. orientalis*, which is USDA 6a/7a. RHS H6. They are tough plants, tolerant of windy, coastal and polluted conditions.
Aspect: A site of any orientation is acceptable.
Shelter: The planting position may be sheltered or exposed.
Sun: Sun or part shade will produce good growth, although full sun will result in more flowers followed by more fruit.
Soil: A soil of almost any texture – be it clay, sand, chalky or largely organic is acceptable, but good drainage is required, and a waterlogged site is not tolerated.
pH: A pH covering the full range normally found is accepted.

Garden Design Considerations

As a genus, there is some justification in saying it is under-used. In new gardens, a specific problem is often lack of shelter, there being no established trees or shrubs to provide this. *Crataegus* has traditionally been known to be excellent for windswept and coastal plantings, with its ability to withstand exposure, and could have an important role here.

Soil, the lack of or quality of, is again sometimes a problem, especially in new gardens. Thorns with their exceptional tolerance to almost any soil can be very helpful. Their aversion to waterlogged conditions should, however, be noted.

The fact that, as a group, they have potentially a long season of interest is valuable. Early summer is welcomed by profuse flowering, followed by plentiful coloured fruits and often vivid autumn leaf colour, which, in some cases, extends into early winter. *Crateagus orientalis* has the further advantage of distinctive light-grey leaf coloration, something that is pleasingly distinctive when viewed in early summer against the foliage of other trees.

Gardening to encourage wildlife is, correctly, very much in vogue. In this respect, the planting of *Crataegus monogyna* is entirely appropriate. Where native, this species supports a very wide range of insects, including the larvae of many moths. It is also an excellent source of pollen and nectar for bees. Several different species of birds are encouraged by its presence. The dense nature of its growth – whether as a hedge or a free-standing tree – is ideal for cover and nesting for birds, and the haws provide a valuable late winter source of food. Coupled with this, the white flowers associate so well with the other native flora, so many also producing a mass of white flowers at this time of year.

Hawthorn, of course, excels as a hedge. On top of the benefits mentioned above, when well-maintained it is stockproof, something much appreciated by the farming community. Where intrusion by people may be a potential problem, a hawthorn hedge is invaluable by being virtually impenetrable due to thick sturdy growth and thorns! A possible disadvantage is hawthorn's susceptibility to fire blight, a disease caused by the bacterium *Erwinia amylovora*, formerly but no longer a notifiable disease in the United Kingdom. This is a weakness it shares with other members of the Rosaceae plant family, and its planting may not be encouraged in areas where apple and pear orchards are prevalent.

Finally, it should be remembered that double-flowered cultivars cannot set fruit. This is a severe disadvantage where nature conservation is paramount but may be a boon in urban areas where fallen fruit could conceivably be a nuisance, perhaps a hazard to foot traffic.

Crataegus orientalis as a small tree in a mixed border.

Crataegus orientalis showing the cut foliage, still with a greyed surface and large, red haws in autumn.

TABLE 3.10 *CRATAEGUS* SPECIES AND CULTIVARS

Name	Habit	Height	Spread	Leaves	Flowers	Fruits	Remarks
C. monogyna (common hawthorn, single-seeded hawthorn, May, quick thorn)	Rounded head, very dense and thorny	4–8m (14–26ft)	4–8m (14–26ft)	1.5–3.5cm (½–1¼in) long. Deeply lobed with 3–7 lobes. Glossy	In early summer, usually white, but may be cream or pink with one stigma. Scented in flat-topped clusters	Haws with a single seed, produced in autumn, red lasting into winter	Native and the most common species in UK. Trunk deeply grooved and gnarled
C. laevigata (syn. *C. oxycantha*, *C. oxycanthoides*) (midland hawthorn)	Rounded head, very dense and thorny	4–8m (14–26ft)	4–8m (14–26ft)	1.5–5cm (½–2in) long. Less deeply cut than *C. monogyna* with 3–5 lobes	In early summer, white, with two stigmas. Flowers scented	Haws with two seeds, produced in autumn, red, lasting into winter	Native to UK, but much less common than *C. monogyna*. Hybrids between *C. monogyna* and *C. laevigata* do exist
C. laevigata 'Crimson Cloud' AGM (syn. *C. laevigata* 'Punicea')	Rounded head, very dense and thorny	4–8m (14–26ft)	4–8m (14–26ft)	As above	As above, but flowers pink with a white centre	As above	
C. laevigata 'Paul's Scarlet' AGM	Rounded head, dense and very thorny	4–8m (14–26ft)	4–8m (14–26ft)	As above	In early summer. Light red, fully double, no stigmas	No fruits produced	
C. laevigata 'Rosea Flore Pleno' AGM	Rounded head, dense and very thorny	4–8m (14–26ft)	4–8m 14–26ft)	As above	In early summer. Pink, fully double, no stigmas	No fruits produced	
C. × *lavalleei* 'Carrierei 'AGM	Small, dense headed	4–8m (14–26ft)	8m+ (26ft+)	Long, glossy dark-green leaves, few thorns	Early summer, large clusters of white flowers	Orange-red fruits. Persistent into early winter	Fruits well displayed against dark-green leaves. Leaves turn orange-red before falling
C. persimilis 'Prunifolia' AGM	Small, broad-headed	4–8m (14–26ft)	8m+ (26ft+)	Glossy, broad oval. Has long thorns	Early summer, white	Rounded, showy, red, persistent fruits	Brilliant orange and red autumn foliage colour
C. orientalis AGM (syn. *C. laciniata*) (eastern thorn)	Small, bushy, spreading	4–8m (14–26ft)	4–8m (14–26ft)	Deeply cut with 5–9 lobes. At first silvery-grey overall, becoming dark-green on upper surface, but remaining silver-grey below	Early summer. Flowers 1.5cm across, scented, in sprays of 10 or more	Globose ¾in diameter, yellow and red	Leaves yellow before falling. Flaking, silvery bark on main stem. Overall light colour of foliage very noticeable and attractive in early summer. Fading a little later but followed by fruit colour

A young tree of *Crateagus monogyna* (common hawthorn) flowering in early summer. This is very common in the UK and may also be seen as a multi-stemmed tree or as a component of a hedgerow.

An unkempt hawthorn hedge. Depending on your point of view, a wonderful feast for pollinating insects or a very untidy hedge.

The profuse flowers of *Crataegus monogyna*. An excellent source of nectar for pollinating insects.

The fruits (haws) of *Crataegus monogyna* are a valuable source of food for birds and small mammals in autumn and winter.

A young tree of *Crataegus laevigata* 'Paul's Scarlet' in a small garden.

The double flowers of *Crataegus laevigata* 'Paul's Scarlet' will provide some nectar for pollinators but will produce no fruit. In some situations, this may be regarded as an advantage, e.g. by a path where fruit may represent a slip hazard and be 'messy', but in a more natural habitat, a loss of environmental value.

Crataegus persimilis 'Prunifolia'. The picture shows what a compact, bushy, small tree this cultivar grows into.

Crataegus persimilis 'Prunifolia' in full autumnal glory.

Crataegus persimilis 'Prunifolia'. Here, details of the autumn leaf colour and the bright-red fruit can be clearly seen. Although the leaves will drop, the fruit will persist well into the winter.

EUONYMUS

A large genus of plants coming from both the northern and southern hemispheres. The plants may be deciduous or evergreen and can be trees, shrubs, climbers or low-growing, ground-covering plants. The flowers are always small and inconspicuous, but the fruits may be coloured, as may the seeds.

Many species are too large to consider for this book and only those of suitable dimensions are included.

For clarity and ease of use, the genus is here divided into deciduous and evergreen plants.

Growing Requirements

Speaking generally, the genus is easy to cultivate, having requirements that are not demanding. Occasional exceptions are mentioned under the plant concerned.
Aspect: A site of any orientation may be employed.
Shelter: Shelter is not generally a requirement, but there are exceptions, especially with evergreen cultivars; this is noted where appropriate.
Sun: Sun or shade is tolerated. Those with variegated foliage need higher light levels and their use is better restricted to full sun or part shade only. Those grown for autumn leaf colour or colourful autumn fruit production are also much better grown in full sun or only light shade.
Soil: As a genus it is not demanding regarding soil, growing well in soil of any texture, providing that it is well drained.
pH: The plants grow well in any pH but prefer soils with a high pH, including those both thin and chalky.
Caution: All species and cultivars are prone to attack by vine weevil (*Otiorhynchus sulcatus*), particularly when grown in containers. This should be checked for before planting. Symptoms include semi-circles cut out of the leaves, caused by the feeding of the adult weevils, and cream-coloured, crescent-shaped larvae present and visible in the compost in the containers. The larvae eat the plant roots and, in severe cases, the plants may be found to be almost without roots! Once in the ground the weevil will spread, damaging the appearance of the foliage and reducing plant growth. It is far better to avoid introducing this pest rather than trying to control it once it is established in the ground.

Toxicity: All parts of all *Euonymus* may cause severe discomfort if ingested. The coloured fruit of some species and cultivars may be attractive to young children.

Evergreen Species and Cultivars of *Euonymus*

Hardiness: Generally, USDA 7b/8a RHS H5. Where there are exceptions, these are mentioned specifically in Table 3.11.

TABLE 3.11 EVERGREEN SPECIES AND CULTIVARS OF *EUONYMUS*

Name	Habit	Height	Spread	Leaves	Remarks
E. fortunei 'Emerald Gaiety' AGM	A small compact, bushy shrub	0.5–1m (1½–3ft)	1–1.5m (3–5ft)	Green with an irregular white margin	Will climb against a wall. Leaf edges become pink due to winter cold
E. fortunei 'Emerald 'n' Gold'	A dense-growing dwarf shrub	0.5–1m (1½–3ft)	1–1.5m (3–5ft)	In spring, deep-green with a golden central splash, this becoming cream later	Will climb against a wall. Leaves become pink due to winter cold
E. fortunei 'Silver Queen' AGM	A small, compact shrub	0.5–1m (1½–3ft)	1–1.5m (3–5ft)	Leaves ovate and a little larger than previous two cultivars. Creamy-yellow in spring. Later green with a cream margin	Will climb 2–3m (6½–9¾ft) against a wall
E. japonicus 'Bravo'	Bushy shrub	2.5–4m (8–13ft)	1.5–2.5m (5–8ft)	Leaves ovate, green with irregular green margin. Where they overlap, grey-green	
E. japonicus EL DORADO (Br2012)	Bushy, compact shrub	0.6m (2ft)	1m (3ft)	Young leaves bright-yellow, later fading to light-green	Best in full sun
E. japonicus 'Green Rocket'	Upright and compact	0.5–1m (1.6–3ft)	0.1–0.5m (0.3–1½ft)	Green, glossy, leathery, oval leaves	
E. japonicus 'Green Spire'	Upright and compact	0.5–1m (1½–3ft)	0.1–0.5m (0.3–1½ft)		6b/7a
E. japonicus 'Kathy'PBR	Bushy shrub	0.5–1m (1½–3ft)	0.5–1m (1½–3ft)	Oval dark-green leaves, fading to light-green with white margins	Full sun required
E. japonicus 'Microphyllus Aureovariegatus' (syn. *E. jap.* 'Microphyllus Aureus Pulchellus', syn. *E. jap.* 'Pulchellus Aureovariegatus')	Upright and bushy	0.5–1m (1½–3ft)	0.5–1m (1½–3ft)	Small, oval, dark-green leaves edged with yellow	
Euonymus japonicus PALOMA BLANCA ('Lankveld03'PBR)	Compact, dwarf	0.6m (2ft)	1m (3ft)	Young foliage greenish-white, held against older deep-green foliage	
E. japonicus 'Ovatus Aureus' AGM (syn. *E. jap.* 'Marieke', syn. *E. Jap.* 'Aureovariegatus'	Bushy medium-sized shrub	2.5–4m (8–13ft)	1.5–2.5m (5–8ft)	Oval leaves bright-yellow when young, later green-yellow with a yellow margin	May produce a few flowers and a few pink fruits

Euonymus fortunei 'Emerald Gaiety'. Here the plants have been clipped into low mounds to form a ground cover. The plants could now be allowed to coalesce to produce an almost sea-like effect.

Euonymus fortunei 'Emerald 'n' Gold'. A much brighter variegation.

Euonymus fortunei GOLDEN HARLEQUIN ('Hoogi'[PBR]). When photographed not showing a very golden variegation, but the colour of these variegations does seem to vary considerably during the growing season.

Euonymus fortunei 'Silver Queen'. Under the cold, frosty conditions evident in the picture, rather a yellowed variegation. Other cultivars develop pink leaf tinges as a response to the cold.

Euonymus japonicus 'Bravo'. A vigorous, strong growing cultivar.

Euonymus japonicus EL DORADO ('Br2012'[PBR]). An exceptionally bright-yellow variegation.

Garden Design Considerations

This is an exceptionally useful group of evergreen shrubs, being reliable, tough and resilient in difficult planting situations. When planning their use in a design, care should be taken not only to consider leaf colours and size, but also the overall height and potential spread of the plants. Note – they do vary considerably.

The *Euonymus fortunei* cultivars are good for relatively small-scale ground-cover. Their ability to produce adventitious roots and to climb walls and fences of all orientations, including north-facing, can be very useful. This propensity can, however, also be a nuisance, requiring clearance by hand where it is not required or where the adventitious roots may cause damage, e.g. to the soft mortar in an old wall. They respond well to light clipping in late spring. This means they can be used for simple topiary: three-dimensional shaping, almost cloud pruning, on walls.

A recent development is to use these shrubs to form a clipped, smooth, undulating surface, emulating the sea. With a taller, upright-growing shrub planted amongst this low-growing ground-cover, the intended effect is analogous to a rock emerging from water.

All these variegated, evergreen shrubs can be exceptionally useful in an area intended to provide winter colour and interest. The pink leaf coloration is a distinct result of winter cold and can be an asset here. Their carpet of foliage can be used as a base to highlight the coloured stems of coppiced *Cornus alba* cultivars.

The larger cultivars of *Euonymus japonicus* can be employed as larger scale ground-cover. Those with a very formal upright habit, when planted closely, form effective low hedging. Their distinctive habit can also be employed when a contrast in plant habits is required. The newer semi-dwarf cultivars are excellent for container-growing and patio pots.

More recently, cultivars of *Euonymus fortunei* and *Euonymus microphyllus* are being considered as suitable alternatives to dwarf box hedge where the cultivation of box has become difficult because of pest and disease problems.

Euonymus japonicus 'Green Rocket'. An upright-growing cultivar, useful for low evergreen hedging or mixed with more rounded shrubs for a contrast in form.

Deciduous Species and Cultivars of *Euonymus*

Hardiness: USDA 6b/7a, RHS H6.

Garden Design Considerations

A group of shrubs whose principal attribute is the provision of autumn and early winter colour. The vast majority of trees and shrubs grown for autumn colour require acid soil and *Euonymus* is particularly useful because of its acceptance of alkaline soil.

Where sufficient space is available, *Euonymus alatus* is good for providing autumn leaf colour. The winged-stem characteristic provides a degree of winter interest when caught in direct sunlight. Often one side of the 'wings' or 'blades' is illuminated and the other in shade, giving a dramatic contrast. In this species, this is a variable feature, and it is worth looking through a batch of seedlings to find an individual where this is strongly developed. *Euonymus alatus* 'Blade Runner' is a clone of a plant where this 'winged stem' is exceptionally well developed.

For compact spaces, *Euonymus alatus* 'Compactus' is outstanding, providing a vivid splash of bright-red and orange leaf colour. During the growing season it fades into the background, but in autumn it comes into its own and fully justifies the limited space it occupies. A position in full sun is essential if maximum autumn leaf colour is to be obtained.

Euonymus europaeus is native to Britain and Europe, and is frequently found as a component of hedgerows, especially in chalk or soft limestone areas. When planning to plant mixed species' hedging, this should be included. Ideally, use plants grown from seed, rather than plants propagated vegetatively, in order to maintain natural genetic diversity. From an aesthetic point of view, they will increase autumn leaf and fruit colour. From a biological viewpoint, the plants will help the environment, acting as a valuable food source for dependent insects and birds.

In the garden, the cultivar *Euonymus europaeus* 'Red Cascade' is the best choice. It makes a large shrub but can be easily kept to a single stem and trained into a small tree. As such, it occupies less ground space. This selection produces a tremendous autumnal display. Autumn leaf colour is more intense and reliable than with the species. Fruit production is also much greater. As with others in this genus, it must be placed in a sunny position for maximum development of both leaf colour and fruit. The fruits, in contrast to the leaves, persist into early winter where they continue to provide colour and, indeed, interest, the orange seeds emerging from the purple-red lobes, finally hanging by a tip before falling to be dispersed.

Euonymus alatus. A relatively small example of this shrub, which produces vivid autumn leaf colour when grown in sun. The shrub can grow much larger than shown.

Euonymus alatus. A close-up shot showing both the coloured autumnal leaves and the winged stems, which become a much more obvious feature after leaf fall.

Euonymus alatus 'Compactus'. A cultivar that is much smaller and more compact than the straight species, but still produces very vivid autumn foliage colour and has winged stems. This is the cultivar to plant in small gardens.

Euonymus europaeus. The European species.

Euonymus europaeus 'Red cascade'. This cultivar produces more intense autumn leaf colour and more fruits than the species. For colour in the garden, 'Red Cascade' is best, but in the wider landscape, the species is better planted to conserve species' diversity.

A NOTE OF CAUTION

Unfortunately, in the USA and Canada, both *Euonymus alatus* and *Euonymus europaeus* have spread following their introduction and become serious weeds. Their planting is not, therefore, to be encouraged.

TABLE 3.12 DECIDUOUS SPECIES AND CULTIVARS OF *EUONYMUS*

Name	Habit	Height	Spread	Leaves	Flowers/Fruit	Remarks
E. alatus (winged spindle)	Large, deciduous shrub	1.5–2.5m (5–8ft)	1.5–2.5m (5–8ft)	Oval, turn brilliant orange-red in autumn	Small, green flowers. A few red fruits with orange seeds	Some branches have corky ridges, often referred to as 'wings'
E. alatus 'Blade Runner'	Large deciduous shrub	1.5–2.5m (5–8ft)	1.5–2.5m (5–8ft)	Oval, turn brilliant orange-red in autumn	Small, green flowers. A few red fruits with orange seeds	A selected form, where the branches have broad corky wings
E. alatus 'Compactus' AGM	Small deciduous shrub	0.5–1m (1½–3ft)	1–1.5m (3–5ft)	Oval, turn brilliant orange-red in autumn	Small, green flowers. A few red fruits with orange seeds	A compact selection
E. europaeus (spindle tree) (syn. *E. bulgaricus*)	Large shrub or small tree	2.5–4m (8–13ft)	2.5–4m (8–13ft)	Narrowly ovate, each 3–8cm (1–3in) long and 2–3cm (¾–1in) long. Turn yellow, orange and red in autumn	Panicles of small, yellow flowers, followed by many pinkish-red, 4-lobed fruits, each lobe containing one orange seed	Native to UK. Wood very hard and formerly used to make wooden spindles for spinning wool and butchers' skewers
E. europaeus 'Red Cascade' AGM	Large shrub or small tree	3m (10ft)	2m (6½ft)	Narrowly ovate, each 3–8cm (1–3in) long. Leaves turn orange and red in autumn	Panicles of small, yellow flowers, followed by masses of pinkish-red, 4-lobed fruits, each containing one orange seed	

GLEDITSIA

About ten species from Asia and North America, plus a single species from South America. As a genus they consist of deciduous trees or shrubs with pinnate or bipinnate leaves and insignificant green flowers, which may produce pods. Some are armed with formidable thorns.

Growing Requirements

Hardiness: USDA 6b/7a, RHS H 6.
Aspect: A site of any orientation is accepted.
Shelter: No specific shelter needs to be provided.
Sun: Full sun is required.
Soil: Soil of any texture is acceptable, provided it is well drained.
pH: A wide pH range is tolerated.

Gleditsia triacanthos (Honey Locust)

A fast-growing, deciduous tree.
Height: 20–30m (65–100ft).
Spread: 15m (49ft). Thorny with 3–10cm (1–4in) thorns growing out of the stems. The thorns may be single or branched, and may be clustered together to form a thorny mass. The form *inermis* is thornless.
Leaves: Pinnate or bipinnate.
Flowers: Green and insignificant.
Fruit: A pod containing several seeds.
The tree is popular for planting in areas of North America where it is difficult to grow other species. It is tolerant of urban conditions, including heat and drought. In addition, it is tolerant of compacted soil, alkaline soil and road salt.

In marked contrast, this species is regarded as an invasive weed species in the Mid-West of the USA and in Australia.

TABLE 3.13 CULTIVARS OF *GLEDITSIA TRIACANTHOS* F. *INERMIS*

Name	Habit	Height	Spread	Leaves	Remarks
G. triacanthos f. *inermis* 'Sunburst'	Bush-headed, but open	8–10m (26–30ft)	4–8m (15–26ft)	New foliage bright-yellow turns green later. Leaves produced later in the season are yellow. In summer have older green leaves and yellow leaves on younger growth	Late into leaf. Leaves often attacked by gall midge. Causes swelling of leaflets, but not defoliation. Branches less brittle than *Robinia pseudoacacia* 'Frisia'
G. triacanthos 'Rubylace'	Broadly conical. Upright spreading	8–10m (26–30ft)	4–8m (15–26ft)	New foliage ruby-red, later becoming bronzed green	As for 'Sunburst'. Can hard prune annually in winter to stimulate vigorous brightly coloured foliage from resultant new growth
G. triacanthos f. *inermis* 'Spectrum'	Bush-headed but open	8–10m (26–30ft)	4–8m (15–26ft)	New foliage is yellow and remains yellow during the growing season	Suggested to supersede *G. triacanthos* f. *inermis* 'Frisia'

Cultivars of *Gleditsia triacanthos* f. *inermis*

The cultivars listed in Table 3.13 are all deciduous trees with pinnate foliage, without thorns and whose flowers and fruit are insignificant.

Garden Design Considerations

Gleditsia triacanthos f. *inermis* 'Sunburst' may be considered an alternative to *Robina pseudoacacia* 'Frisia'. It does not suffer from dieback as 'Frisia' is reported to do, and its wood is less brittle, therefore, less liable to wind damage. In addition, the leaf colouring of *Robinia pseudoacacia* 'Frisia' fades in the leaves produced early in the season but is present in leaves produced later in the season. This results, by midsummer, with a tree having a curious two-tone effect with older green, plus younger yellow leaves on the shoot peripheries.

Robinia pseudoacacia 'Spectrum' has been introduced as its foliage does not fade to green and the two-tone effect is not present, but its wood is still brittle. In the end 'beauty is in the eyes of the beholder' and the choice open to the garden designer or gardener.

Gleditsia triacanthos f. *inermis* 'Rubylace' does offer a clearly different foliage colour, but growth has to be encouraged to experience the most strongly coloured foliage.

As with *Robinia*, *Gleditsia* offers the same advantage of only casting light, dappled shade, which allows valuable space for underplanting with a range of smaller subjects.

Gleditsia triacanthos f. *inermis* 'Sunburst'. Pictured here contrasting well against a dark background.

Gleditsia triacanthos f. *inermis* 'Sunburst'. Detail of the pinnate leaves showing their bright-yellow colour.

Gleditsia triacanthos 'Rubylace'. The purple-coloured foliage early in the growing season. Unfortunately, this colour changes to a rather dull green later in the year.

KOELREUTERIA PANICULATA (PRIDE OF INDIA, GOLDEN RAIN TREE)

This species, introduced from China in 1747, is a small, deciduous tree eventually making a height of 8–12m (26–40ft) with a width of 6–8m (20–40ft). The leaves, which emerge quite late in spring, are at first pink-tinted, then a light-yellow, finally changing to mid-green. They are pinnate with between nine and fifteen leaflets, each entire leaf being 15–20cm (6–18in) long and only casting light shade.

Flowering occurs in late summer and conspicuous, large panicles of numerous, small, yellow flowers are produced, hence the tree's common names. Each flower is star-shaped with a tiny, orange-red splash at the base of each petal. The flowers are followed by conical, three-valved capsules, looking like papery, pale-coloured, bladder-like fruits, each containing one black seed. The tree can be propagated readily from this seed, if required.

In early autumn the leaves take on a light-brown colour, sometimes with orange and red shades, before falling.

The tree's habit is distinctive: the trunk being slender and the head bushy and rounded in shape.

Growing Requirements

Hardiness: USDA 7b/8a. RHS H5.

Aspect and shelter: A site of any aspect is satisfactory, this being sheltered or exposed.

Sun: The main requirement is for full sun. The more sun the tree receives, the more flowers are produced. Growth is poor and weak even in part shade.

Soil: A soil of any texture is accepted but it must be well-drained, indeed soils that dry out in summer are well tolerated.

pH: A very wide range of soil pH is accepted.

Garden Design Considerations

This tree's main attribute is the late summer flowers, a time when very few other trees are in flower. When flowering well in full sun, the display can be very attractive.

As a small tree it makes an unusual and interesting specimen, contributing coloured spring foliage, late summer flowers, plus autumn leaf colour and unusual fruits, which always cause interest.

It is easy to grow and may also be incorporated into the back of a large, mixed border where it adds height, plus its other attributes, while its light canopy allows some underplanting.

There are a few cultivars available of which *Koelreuteria paniculata* 'Coral Sun' has more colourful spring foliage, first emerging bright-pink before turning green and later the same yellow flowers; in autumn, leaf coloration is more pronounced.

The utility of this species may increase with the advent of hotter, drier summers that are predicted, as this tree flourishes in these conditions.

Koelreuteria paniculata. A mature tree breaking into leaf in spring.

Koelreuteria paniculata. Here, not only the autumn colour, but the nature of the pinnate leaf can be clearly seen.

LABURNUM

A genus of only two species of deciduous trees, *Laburnum alpinum* and *Laburnum anagyroides*, from the mountains of South and Central Europe, plus western Asia. The leaves are composed of three leaflets. Yellow pealike flowers are produced in pendulous racemes in late spring and early summer.

The trees are short-lived. The life of *L. anagyroides* can be extended by removing the flower racemes once finished, where feasible. This relieves the tree of the considerable burden of seed production. All parts of the trees are highly toxic. The seeds, in particular, present a hazard to children

Growing Requirements

These trees are of very easy cultivation, being undemanding in their requirements.
Hardiness: USDA 6b/7a. RHS H6.
Aspect and shelter: The site can be of any orientation and there is no specific requirement for shelter.
Sun: Full sun is required.
Soil: The soil can be of any texture but must be well drained.
pH: A wide pH range from acidic to alkaline is accepted.

Laburnum × watereri 'Vossii' AGM

A hybrid that combines the longer racemes of *L. alpinum* with the denser flowers of *L. anagyroides*. A small, broad-headed tree. This cultivar is very free-flowering and produces only comparatively few seeds. It is by far the most widely planted of the laburnums.
Height: 4–8m (15–26ft).
Spread: 4–8m (15–26ft).
Leaves: Trifoliate and glossy.
Flowers and fruit: Flowers produced in early summer; pendulous racemes up to 60cm (24in) long of yellow, scented flowers. This is a sterile hybrid, and it produces only a light crop of poisonous seed.

Garden Design Considerations

Laburnums are easy to grow, small trees, producing vivid flower colour in late spring and early summer. The bright-yellow flowers contrast strongly with blue-flowering shrubs that flower at the same time, such as *Ceanothus* 'Skylark' AGM.

For profligacy of flowering, *Laburnum × watereri* 'Vossii' AGM is the best selection. The poisonous nature of the whole tree, especially the seeds, is a constraint where young children are likely to be present but, again, *L. × watereri* 'Vossii' is to be favoured due to its comparatively low seed production.

Where space is very limited, *Laburnum anagyroides* 'Yellow Rocket' may be preferred due to its upright habit and comparatively narrow head.

A young tree of *Laburnum × watereri* 'Vossii' in full flower.

TABLE 3.14 *LABURNUM ALPINUM* (SCOTTISH LABURNUM OR ALPINE LABURNUM) AND A CULTIVAR

Name	Habit	Height	Spread	Leaves	Flowers	Remarks
L. alpinum	A small, broad-headed tree	8m (26ft)	8m (26ft)	Trifoliate leaves, bright-green top surface, paler beneath	Racemes 15–40cm (6–16in) long of yellow, pealike, scented flowers produced in early summer	Seed produced in flattened pods. Seeds poisonous
L. alpinum 'Pendulum'	A small tree. Develops a dome-shaped head with stiffly pendulous branches	1.5–2.5m (5–8ft)	1.5–2.5m (5–8ft)	Trifoliate leaves, bright-green top surface, paler beneath	Racemes 15–40cm (6–16in) long of yellow, pealike, scented flowers produced in early summer	Seed produced in flattened pods. Seeds poisonous

TABLE 3.15 *LABURNUM ANAGYROIDES* (COMMON LABURNUM) AND A CULTIVAR

Name	Habit	Height	Spread	Leaves	Flowers	Remarks
L. anagyroides	A small, spreading tree	8m (26ft)	8m (26ft)	Small, trifoliate dull-green	Profuse, but only 10–30cm (4–12in) long racemes of yellow flowers in late spring	Numerous seed produced in rounded hairy pods. Seed is poisonous
L. anagyroides 'Yellow Rocket'	Narrow, conical head	6m (20ft)	4m (13ft)	Small trifoliate dull-green	Profuse, but only 10–30cm (4–12in) long racemes of yellow flowers in late spring	Good for where space is limited. Same comment for seed as for *L. anagyroides*

TABLE 3.16 *LABURNUM* × *WATERERI* 'VOSSII'

Name	Habit	Height	Spread	Leaves	Flowers	Remarks
L. × *watereri* 'Vossii' AGM	A small, broad-headed tree	4–8m (15–26ft)	4–8m (15–26ft)	Trifoliate and glossy	Produced in early summer. Pendulous racemes up to 60cm (24in) long of yellow, scented flowers	This is a sterile hybrid, and it produces only a light crop of poisonous seed

Picture showing the number of flowers produced by *Laburnum* × *watereri* 'Vossii'.

LIQUIDAMBAR

A genus of fifteen species occurring in the Mediterranean, Eastern Asia, North and South America. Superficially, they appear like *Acer* (maples) in that they have maple-like leaves, but these are arranged alternately, rather than in opposite pairs, as in *Acer*.

The insignificant flowers are arranged in globose heads. The female flowers develop into rounded, prickly, dark, woody capsules of some decorative merit.

Growing Requirements

Hardiness: USDA 6b/7a. RHS H6.
Aspect: The growing position can be of any orientation.
Shelter: There is no specific requirement for shelter.
Sun: Full sun or part shade.
Soil: The soil can be of sandy, clay or loamy texture, it must be well drained, but retain moisture in the summer.
pH: Soil pH should be acid or neutral, chalky thin soils that dry out in the summer should be avoided.

Liquidambar orientalis (Oriental Sweet Gum)

A slow-growing bushy tree or large shrub, giving good autumn leaf colour. Note – in warm climates this species can grow into a large tree.
Height: 4–8m (15–26ft).
Spread: 2.5–4m (8–13ft).

Garden Design Considerations

Where space permits, these trees make excellent stand-alone specimens. They grow into well-proportioned trees that look good all the year round and they are outstanding during autumn. Indeed, they look so good that there is a strong temptation to plant them where space does not permit.

Where space is limited, the slightly smaller *L. styraciflua* 'Palo Alto' would be a good selection, but where space is very limited, *L. styraciflua* 'Slender Silhouette' offers the least spread.

Liquidambar styraciflua 'Corky' offers the added attraction of reliably producing a markedly corky bark, which is valuable for winter interest.

Liquidambar styraciflua in early summer, showing how it will grow into a moderately large tree.

The corky bark developed by some trees of *Liquidambar styraciflua*. This is not developed by all trees, but where it does occur, it is another interesting feature, particularly in the winter.

The foliage of *Liquidambar styraciflua* 'Lane Roberts' showing its very attractive autumn leaf colour.

TABLE 3.17 *LIQUIDAMBAR STYRACIFLUA* (SWEET GUM) AND CULTIVARS

Name	Habit	Height*	Spread*	Leaves	Flowers	Remarks
L. styraciflua	Broad-headed	6m (20ft) 12m+ (40ft+)	8m (26ft)	5–7 lobes	Insignificant. Black spikey fruits	When grown from seed, the plants produced are variable. They may or may not produce good autumn leaf colour and a distinctive corky bark
L. styraciflua 'Corky'		4m (13ft) 8–12m (26–40ft)	4–8m (13–26ft)	Deep-red to burgundy autumn leaf colour		Produces a markedly corky bark
L. styraciflua 'Gum Ball'	Compact and round headed	1.5–2.5m (5–8ft)	1.5–2.5m (5–8ft)	Good autumn colour		Often grown as a standard pom-pom tree
L. styraciflua 'Lane Roberts' AGM	Broad-crowned	6m (20ft) 12m+ (40ft+)	8m (26ft)	Shining green in growing season. Blackish-red in autumn		
L. styraciflua 'Palo Alto' AGM	Neat, conical, fairly upright	6m (20ft) 12m (40ft)	8m (26ft)	Orange and scarlet autumn leaf colour		
L. styraciflua 'Slender Silhouette' AGM	Strongly fastigiate	6m (20ft) 12m (40ft)	2.5–4m (8–13ft)	Reddish-purple autumn leaf colour		
L. styraciflua 'Worplesdon' AGM	Broad crowned	6m (20ft) 12m (40ft)	8m (26ft)	Orange and yellow autumn leaf colour	Regularly bears black, spikey fruits	

* The upper figure is the estimated growth after ten years. The trees will continue to grow considerably larger after ten years. The lower figures are the estimated ultimate size.

Liquidambar styraciflua 'Lane Roberts' is a cultivar selected because it does reliably develop good autumn leaf colour. It will, however, grow into quite a large tree.

Liquidambar stryraciflua 'Palo Alto'. A middle-aged tree, the picture shows a more moderate-sized tree with an upright habit and good autumn leaf colour.

LUMA

This is now a genus of two evergreen shrubs or trees from Argentina and Chile. These two species have recently been taken out of the genus *Myrtus* and placed into the genus *Luma*.

Luma apiculata (Chilean Myrtle, Temu) (syn. *Myrtus luma*)

Height: An evergreen plant that can form a multi-stemmed tree up to 12m (40ft) tall.
Spread: In excess of 8m (26ft) wide under favourable conditions.
Leaves: Aromatic when crushed and dull green, elliptic and with a sharp distal point and 2.5cm (1in) long.
Flowers and fruit: Cup-shaped white flowers each 2cm (1in) across crowd the branches in late summer and early autumn, and in favourable conditions are followed by edible purple berries. The cinnamon-coloured bark peels off in patches to reveal a cream-coloured under-surface.

Growing Requirements

Hardiness: USDA 8b/9a. RHS H4.
Aspect: A site facing any orientation, except north, is acceptable.
Shelter: The site should be sheltered and warm.
Sun: Full sun or just partial shade.
Soil: The soil can be of any texture but must be well drained.
pH: A wide pH range is acceptable.

Garden Design Considerations

This is a genus that grows best in the warmer gardens of the UK. Speed of growth is slow in most gardens, but considerably faster in mild, sheltered gardens. In favoured gardens, a small multi-stemmed tree or large shrub will be formed. The variegated cultivars grow considerably more slowly but will, in time, make large shrubs.

Where conditions are ideal, as in southern Ireland, the species has become naturalised and fruits heavily.

One of the species' most attractive features is the light cinnamon-coloured bark, which flakes irregularly to reveal lighter cream-coloured patches. This is visible all the year round, but most noticeable in the winter.

These small trees and shrubs contribute evergreen structure, summer flowers and valuable winter-coloured bark interest to the garden.

They are an excellent choice for milder gardens in the south-west of England and climatically favoured coastal gardens where frost, prolonged winter cold and cold, drying winds are not frequently experienced. In such gardens, they can be used for hedging and areas that may be subject to a degree of summer drought.

In less favoured gardens, they are suited to sheltered wall-side planting, Mediterranean-style plantings, patio and container gardening. They are also appropriate for sheltered city and courtyard gardens.

TABLE 3.18 *LUMA APICULATA* AND CULTIVARS

Name	Habit	Height	Spread	Leaves	Flowers	Remarks
L. apiculata	Bushy	8–12m (26–39ft)	8m+ (26ft+)	Dull green, 2.5cm (1in) long	Produced in late summer and early autumn. Cup-shaped flowers, white, 2cm (1in) wide	Purple-black edible fruits are produced after flowering
L. apiculata 'Glanleam Gold'	Bushy	2.5m (8ft)	4m (13ft)	Creamy-yellow margin to leaf. Leaves tinged pink when young and in winter	Produced in late summer and early autumn. Cup-shaped flowers, white, 2cm (1in) wide	Less vigorous and slower growing than *L. apiculata*
L. apiculata 'Saint Hilary'	Bushy	2.5m (8ft)	4m (13ft)	Grey-green leaves with an irregular creamy edge	Produced in late summer and early autumn. Cup-shaped flowers, white 2cm (1in) wide	Less vigorous and slower growing than *L. apiculata*

Luma apiculata, developing into a small multi-stemmed tree.

The beautiful cinnamon-red-coloured stems of *Luma apiculata*.

Luma apiculata flowering in mid-summer.

Luma apiculata 'Glanleam Gold'. A slower growing, variegated cultivar.

MAGNOLIA

A large genus of deciduous and evergreen trees and shrubs found in the Himalayas, south and South-East Asia, China, Japan and, separately, North America and Brazil. Magnolias can be found in tropical, subtropical and temperate regions, but here only those from temperate areas, plus their hybrids, are included.

Magnolias are a primitive group of plants that emerged early in the evolution of flowering plants. Their flowers, developed for beetle pollination, have a spiral structure, and the inner and outer whorls of floral segments cannot be distinguished into sepals (the outer whorl) and petals (the inner whorl) as their appearance is identical. Hence, they are known collectively as the tepals.

As a group, they include some of the most magnificent flowering plants available to the gardener. They have been the subject of much plant attention by plant breeders and many cultivars have been produced. Recently, plants in the related genera of *Michelia* and *Manglietia* have been included into *Magnolia*. This information is included here to help those seeking to source these plants, since the old names are still frequently used.

Growing Requirements

Hardiness: The plants listed here range in their hardiness rating from USDA 8b/9a. RHS H4, to USDA 7b/8a RHS H6. Care needs to be taken in plant selection, since a plant overall may be rated USDA 7b/8a, RHS H5, but its flowers may be susceptible to late spring frost.
Aspect: A site facing south, west or east is acceptable, but some degree of shade is generally also necessary.
Shelter: Magnolias generally require some shelter, since their large leaves and flowers are susceptible to wind damage.
Soil: The soil they thrive in is often moderately heavy, moisture-retentive but well drained; others, such as *Magnolia stellata* and its hybrids, prefer a higher organic content.
pH: The great majority of magnolias require an acidic or neutral soil. A few will tolerate some lime, providing the soil does not dry out in summer.

Magnolias have thick, fleshy roots that are very liable to damage and subsequent decay during lifting, transport and storage prior to planting. Magnolias should not be supplied bare root, as the roots will inevitably dry out and transplanting is unlikely to be successful.

Plants can be successfully moved by root wrapping, also known as burlapping, the roots of the plant. The plant should be lifted carefully to retain a good volume of soil around the roots. This root ball is then carefully wrapped in material to retain the soil around the roots. This is a skillful job, requiring the nursery to part with not only the plant, but also some of its valuable nursery soil. The lifting task should be done whilst the plant is dormant and the soil is in a moist, but not waterlogged or very dry, condition. During transport and storage, the plants should be kept in a sheltered position, preferably under cover where they may be protected from drying winds, and the root balls kept just moist and not allowed to dry out. The plant must be protected from severe frost.

Plants may also be supplied as containerised, whereby the plants have been placed in pots and compost, but not given time to establish, as a method of storing the plants prior to sale. At the point of sale, it should be made clear that care will be needed by the purchaser to retain some of this compost and the entire root ball during planting.

Plants can be sold as container-grown plants. In this instance, the plants have been grown in their containers for some time and have become well established, but not root bound in their pots. This has the advantage that minimal damage and stress will be caused to the plant during the transplantation process and a very high degree of success should be achieved. Optimal times for transplantation are spring and autumn. When using container-grown plants, this can be done at any time, but great care needs to be taken to ensure that the plant roots do not become dry if planting is done in summer.

Mulching the plants generously in spring, before the soil moisture stored in the soil from winter rain has had opportunity to be lost by evaporation, is a good tip to encourage good new growth in old plants and especially young, establishing magnolias. Care should be taken to place the mulch around the root spread of the plants, but to avoid piling the mulch around the base of the stems, which may cause the stems to rot.

Magnolias Recommended for Today's Compact or Small Garden

Today *Magnolia* species are rarely grown in the average garden, their place having been taken by hybrids. These have been produced in a quest to produce smaller plants that begin to flower whilst still young and possess greater resistance to late spring frosts and damaging winds.

Perhaps most prominent amongst these breeders are Felix Jury and the next generation of Mark and Abbie Jury in New Zealand, who have produced many of the hybrid magnolias that are so widely enjoyed worldwide in our gardens. An indication of their breeding is often to be seen within the cultivar name with the letters 'jur' occurring.

A series of *Magnolia* hybrids bred by de Vos and Kosar at the Unites States National Arboretum in Washington DC were named in 1965 and subsequently known as the 'Eight Little Girls', of which *Magnolia* 'Susan' has proved to be a very garden-worthy and desirable plant.

Most magnolias are either shades of pink or white but, over recent years, there has been a drive to produce yellow-flowered hybrids using the yellow-flowered *Magnolia acuminata* as a parent. Some of these are described below. Coming with the genes for yellow flowers is the trait for flowering in late spring and early summer. This has the advantage of avoiding late spring frosts but may also cause the opening of the flowers to coincide with leaf emergence and the problem that the flowers may be obscured by the leaves.

Magnolia BLACK TULIP 'Jurmag'PBR

Hardiness USDA 7b/8a. RHS H5.
An upright, open-branched, deciduous plant making a tree 4–8m (15–20ft) tall and 4–8m (6–8ft) wide. Dark-purple flowers, the tepals of which are drawn together at their base and then expand outwards to form a distinctive neat, rounded shape. These flowers are freely produced from a young age from early to mid-spring. Unfortunately, the flowers have proved to be susceptible to late spring frost.

Magnolia × *brooklynensis* 'Yellow Bird'

Hardiness: USDA.6b/7a. RHS H6.
Upright-growing when young, it eventually grows into a pyramidal-shaped deciduous tree 10m (33ft) tall and 10m (33ft) wide at its base. Cup-shaped yellow flowers are produced from late spring to early summer, the depth of their colour depending on temperature. Cool temperatures result in creamy-yellow flowers, but greater warmth produces deeper yellow flowers. Flowering later than some magnolias the flowers may tend to become obscured by the developing foliage.

Magnolia BURGUNDY STAR ('Jurmag4')

Hardiness: USDA6b/7a RHS H6.
Initially upright growth produces, once mature, a small, narrowly columnar, deciduous tree 4m (13ft) tall and 1.2m (4ft) wide. An ideal size and shape for where space is restricted. Deep, dark-red, wine-coloured flowers, each approximately 10cm (4in) across and slightly scented are produced in early spring.

Magnolia BLACK TULIP 'Jurmag'PBR. A young plant well on the way to forming a small, upright tree.

A small tree of *Magnolia* × *brookleyensis* 'Yellow Bird'.

Magnolia BLACK TULIP 'Jurmag'PBR. The goblet-shaped, dark-coloured flowers are produced in early spring and may be liable to late frost damage.

The flowers of *Magnolia* × *brookleyensis* 'Yellow Bird' are produced in late spring as the leaves emerge. This means they are not usually liable to late frost damage.

The yellow and lime-green flowers of *Magnolia* × *brookleyensis* 'Yellow Bird' are held well clear of the developing leaves and can be clearly seen.

Magnolia 'Butterflies'

Hardiness: USDA 6b/7a. RHS H6.
A deciduous, upright tree, eventually 6m (20ft) tall and 4m (13ft) wide. In early spring, open, cup-shaped flowers are produced, each consisting of narrow, mainly upright, yellow tepals and red stamens.

Magnolia 'Elizabeth' AGM

Hardiness: USDA 6b/7a. RHS H6.
The first yellow-flowered hybrid *Magnolia*. A deciduous, upright-growing tree, eventually becoming approximately 8–12m (26–40ft) tall and 4–8m (13–26ft) wide. Yellow flowers, each 20cm (8in) across and scented, emerge in late spring. As with other yellow cultivars, the depth of yellow is affected by air temperature, high temperatures inducing a deeper colour.

Fairy Magnolias

Hardiness: USDA RHS H4.
A new group of magnolias bred by Mark Jury by crossing evergreen *Magnolia laevifolia* (syn. *Michelia yunnanensis*) with other small-growing, evergreen magnolias. In so doing, he produced this delightful group of slightly hardier, free-flowering, evergreen, large shrubs or small trees. These cultivars require a sheltered position, and some leaves may be lost in a cold winter, but the plant overall is hardy. They require the same soil conditions as other magnolias, but if plenty of organic matter is present and the soil does not dry out in summer, a slight degree of alkalinity may be tolerated.

Magnolia FAIRY MAGNOLIA BLUSH ('Murjur01'PBR)

Hardiness: USDA7a/8a. RHS H5.
Of compact and bushy growth, this cultivar will, if kept to one stem, form a small, evergreen tree about 2.4m (8ft) tall and 2.7m (9ft) wide. Many small, light-pink, scented flowers are produced in spring and intermittently through the growing season.

Magnolia FAIRY MAGNOLIA CREAM ('Murjur02'PBR)

Hardiness: USDA7a/8a. RHS H5.
Slightly more vigorous than the previous cultivar. If grown in the same manner, this will form an evergreen tree of 2.7m (9ft) tall and 2.7m (9ft) wide. Many small, cream-coloured, scented flowers are produced in spring and intermittently through the growing season.

A young plant of *Magnolia* 'Butterflies' that will grow into a multi-stemmed shrub.

Magnolia 'Butterflies'. The rather small, light-cream flowers that are stained lime-green at their base.

Magnolia 'Elizabeth', the first yellow-flowered hybrid magnolia to be bred. The flower colour is affected by temperature. The higher the temperature at flowering, the deeper the yellow colour.

Magnolia FAIRY MAGNOLIA BLUSH ('Murjur 01'PBR). This small-growing hybrid magnolia produces small, scented, pale-pink flowers.

Magnolia FAIRY MAGNOLIA CREAM ('Murjur 02'PBR). A young plant of this cultivar flowering in mid-spring and intermittently during summer.

Magnolia FAIRY MAGNOLIA CREAM ('Murjur 02'PBR). The picture shows the evergreen leaves and how the flowers are produced laterally along the main stem.

Magnolia FAIRY MAGNOLIA WHITE ('Murjur05'PBR)

Hardiness: USDA 7a/8a RHS H5.
This cultivar, if grown as a tree, will grow into an evergreen 2.4m (8ft) tall and 2.5m (8.2ft) wide. White, scented flowers are borne profusely in the spring and irregularly during the growing season.

Magnolia grandiflora 'Galissonniere'

Hardiness: USDA 7b/8a RHS H5.
A vigorous, evergreen *Magnolia* that, when grown as a free-standing plant, makes a cone-shaped tree 8m (26ft) tall and 5–6m (16–20ft) wide, growing well in full sun or part shade. During summer, the large, white, cup-shaped, scented flowers are produced intermittently. The large leaves are dark-green and shiny on the upper surface and have a layer of tiny, reddish-brown hairs known as indumentum on the underside, giving an attractive contrast, especially when seen under windy conditions or viewed from below.

This plant also makes a very attractive wall shrub, where the leaves and flowers can be viewed to best advantage. The vigorous growth is trainable and responds well to pruning. This cultivar is also often used as a large, pyramidal or cone-shaped specimen, positioned in a large Versailles-type container in a formal or courtyard garden, where the quality of the evergreen foliage and flowers can be inspected closely and admired.

Magnolia 'Genie'PBR

Hardiness: USDA.6b/7a. RHS H6.
A compact, upright-growing, deciduous magnolia growing 2m (6½ft) tall and 2.5m (8ft) wide. In late spring, dark-red, goblet-shaped flowers are produced. It grows well in acid or neutral soil, and in full sun or part shade.

Magnolia liliflora 'Nigra' AGM

Hardiness: USDA 6b/7a. RHS H6.
A deciduous magnolia making a compact but bushy tree, 2.5–4m (8–13ft) tall and 1.5–2.5m (5–8ft) wide. There is a long flowering period from spring into early summer with erect flowers whose upright, and then reflexed, tepals are deep-purple outside and lighter coloured on the inside. The plant is tolerant regarding soil type, accepting all well drained soils, except those that are chalky.

Magnolia FAIRY MAGNOLIA WHITE 'Murjur 05'PBR. The white-flowered fairy magnolia.

Magnolia grandiflora 'Galissonniere' will grow into a fairly large tree, but it can be grown in a large container where it will make an upright specimen or it can be grown as a wall shrub.

Magnolia grandiflora 'Galissonniere' with a green, shiny, upper surface and a distinctive reddish, hairy, undersurface. This feature makes this cultivar even more attractive as a wall shrub where its undersides may be clearly seen and appreciated.

Magnolia 'Genie'PBR. A compact, upright-growing magnolia, well-suited to a small garden.

The dark-purple flower of *Magnolia* 'Genie'PBR.

Magnolia × *loebneri* 'Leonard Messel' AGM

Hardiness: USDA 6b/7a. RHS H6.
A moderately vigorous, deciduous cultivar that will grow well on a site of any orientation and in full sun or part shade. Soil of any texture is accepted, providing that it is well drained, and acid to alkaline pH is tolerated, providing the soil does not dry out in summer. A spreading branched, small tree is produced reaching 4–8m (13–26ft) tall and 4–8m (13–26ft) wide. Small, dark-green leaves are produced after flowering.

The plant begins to flower when still very young, and flowering is profuse. Deep-pink buds open to a lighter pink in spring. The shade of pink is dependent on temperature, and the warmer the air temperature, the deeper pink the flowers. The flowers are fragrant, the amount of scent released again dependent on temperature – warmer conditions increasing fragrance. As the flower opens, the tepals reflex backwards, the flower at maturity having a distinctive star-shape. Fortunately, the flowers are resistant to light frost.

Due to its ease of cultivation, profuse flowering and moderate size, this cultivar is very popular with gardeners.

Magnolia × *loebneri* 'Merrill' AGM

Hardiness: USDA 6b/7a. RHS H6.
Initially slightly more vigorous selection than *Magnolia* 'Leonard Messel', but ultimately approximately the same overall size. White, scented flowers are produced instead of pink. Another very popular cultivar.

Magnolia 'Princess Margaret'

Hardiness: USDA 8b/9a RHS H4.
This magnolia flowers in early spring and will be prone to late spring frost damage where these are liable to occur. Growth is vigorous and fast, a tree of 7m (23ft) plus being rapidly produced. Very large flowers are produced, these being initially goblet-shaped and then opening further to a cup-and saucer-like flower. The tepals are purple on the outer surface and a paler shade on the inner surface.

If possible, as well as shelter from wind, overhead shelter, and dappled shade from the canopy of larger trees should be provided. Obviously, this is not often practical in smaller gardens, but in a few situations, protection can be provided by adjacent houses.

Magnolia × *loebneri* 'Leonard Messel' is a vigorous cultivar that flowers profusely from a young age and eventually forms an open-growing, small tree.

The soft-pink and white flower of *Magnolia* × *leobneri* 'Leonard Messel'.

Magnolia × *leobneri* 'Merrill' differs by having pure-white flowers.

Magnolia 'Princess Margaret'. Where space is available, this cultivar will produce its very large flowers whilst still young and grow on, eventually forming a moderate-sized tree.

Magnolia stellata (Star Magnolia)

Hardiness: USDA 7b/8a. RHS H6.
A slow-growing, deciduous magnolia that is very well suited to compact gardens. Ultimately making a shrub 1.5–2.5m (5–8ft) tall with a spread of 2.5–4m (8–13ft) after many years. The star-like flowers are white and 10cm (4in) across. It flowers in early spring, and these are susceptible to wind and frost damage. A sheltered site, not exposed to early morning sun, is required.

Magnolia stellata 'Royal Star' AGM

Hardiness: USDA 7b/8a. RHS H6.
This is similar to *Magnolia stellata*, but with slightly larger flowers, 12cm (5in) across and slightly more regularly shaped. It can be a little more vigorous, making a shrub 2.5–4m (8–13ft) high and with a width of 2.5–4m (8–13ft). The growing conditions required are the same as for *Magnolia stellata*.

Magnolia stellata grows only very slowly and takes many years to become a large shrub. It is very suitable for use where space is limited in a small garden.

Magnolia stellata, whose flowers are produced in early spring and are vulnerable to wind and frost damage.

Magnolia × *soulangeana*

Hardiness: USDA 6b/7a RHS H6.
Historically, the most widely planted deciduous magnolia. The name *Magnolia* × *soulangeana* covers a range of cultivars with varying vigour, flower colour and size. If purchased with just this name for information and not in flower, the gardener must wait to see exactly what he or she has bought! It may be white, a whole range of pinks or almost red, and may be upright or widely branching. The plant is usually seen as a large shrub or small tree with wide-spreading branches. Ultimate size is variable, dependent on the cultivar selected and site conditions, but generally in the order of 4–8m (13–26ft) tall and 4–8m (13–26ft) wide.

The flowers are produced in late spring before the leaves emerge. They are goblet-shaped, most often an attractive vinous-pink and up to 25cm (10in) across. The flowers may be damaged by late spring frosts or cold, drying wind at flowering time.

Magnolia stellata 'Royal Star' has slightly larger flowers with a more regular shape than *Magnolia stellata.*

Magnolia × *soulangeana*. This is a very variable group of fine hybrids. A plant from this group may well eventually grow into a large shrub or wide-spreading, small tree.

A wide range of soils are acceptable, with heavy soils that are well drained but do not dry out excessively in summer most favoured. Acid to neutral soils are needed; a little alkalinity may be tolerated but excess will result in poor, yellowed growth, due to lime-induced chlorosis.

Magnolia × *soulangeana* 'Rustica Rubra' AGM

Hardiness: USDA 6b/7a RHS H6.
An excellent selection from the magnolia cross above. A vigorous cultivar, which ultimately makes a tree within the range stated above. Large, cup-shaped flowers, 12cm

A typical flower of *Magnolia* × *soulangeana.*

The darker flower of *Magnolia* × *soulangeana* 'Rustica Rubra'. This cultivar usually grows with its own characteristic arching habit and can make a smaller shrub than other members of the group *Magnolia* × *soulangeana.*

(4½in) across and deep-pink with reddish tones are produced a little later than most *Magnolia* × *soulangeana* cultivars, making it a little less susceptible to frost damage.

The same soil conditions are required as for other *Magnolia* × *soulangeana* cultivars.

Magnolia 'Susan' AGM

This magnolia grows with a multi-stemmed, upright habit, making a large shrub or small tree with a height of 2.5–4m (8–13ft) and width of 2.5–4m (8–13ft). A deciduous plant with sickle-shaped buds that open before the leaves appear in late spring. The fragrant, upright flowers, each with six tepals, are reddish-purple on the outside and a lighter pink within.
Hardiness: USDA 6b/7a. RHS H6.

Garden Design Considerations

In spring, the precocious magnolias, those that flower before leafing, are visually dominant and, therefore, they do need very careful placement. Sporadic and summer-flowering magnolias are less dominant, and one can be perhaps more relaxed as to where they are positioned.

Most magnolias have white- or light-coloured flowers and benefit from a dark background. If this background also has a uniform medium to fine texture, contrasts of both colour and texture can be created. The darker colour and uniform texture of the background contrasts against, and highlights, the light-coloured and large magnolia flowers. Conifers chosen to be slightly taller than the magnolia can work well in this respect. If this is not planned for, the flowers can 'get lost' against the sky. This is much less of a problem with darker flowered magnolias.

Carefully positioned, a spring-flowering tree magnolia can make an excellent, if very seasonal, focal point. This emphasises the need to incorporate them into the overall garden planting scheme.

Summer-flowering magnolias, such as *Magnolia grandiflora*, set their flowers against their foliage and have an excellent appearance. Others, such as the large shrubs *Magnolia wilsonii and M. seiboldii*, have a much more subtle effect, their pendant white flowers in danger of being rather hidden by their foliage, but vividly revealed if positioned so that their flowers can be viewed from beneath.

The advent of fairy magnolias gives magnolias the possibility of another role in the garden, since these magnolias can not only be used individually as trees in small gardens, but where space and some shelter can be provided, they can also function as evergreen spring- and summer-flowering hedges or garden space-dividers.

When designing your overall scheme, consider points such as colour combinations and seasonality. If space permits, could several different magnolias be accommodated to extend their season and perhaps to include an evergreen magnolia to extend interest right into the dormant season? A magnolia used as a tree associates well not only with other shrubs, especially rhododendrons and camellias, to give tremendous spring colour, but also many summer-flowering shrubs to extend the season of interest. This is often a failing with gardens that focus solely on spring, leaving the other seasons rather devoid of interest.

When designing with magnolias included in the scheme, even more attention to detail is needed, since not only do the visual aspects mentioned above need attention, but also those of the magnolias' critical growing environment, especially the need for shelter and part shade for many, plus the need to avoid the problem of damage caused by late spring frost. Now we begin to see why the good garden designer earns his or her crust!

Magnolia 'Susan'. This cultivar is very popular with gardeners. It grows into a multi-stemmed, medium-sized shrub.

Magnolia 'Susan'. Flowers are produced relatively late in spring and will usually avoid frost damage. The flowers are typically slightly twisted to one side.

MALUS

Apples grown for dessert and culinary use, plus crab apples grown largely just for decorative effect, form a large genus spread throughout the Northern Hemisphere. Here, only decorative crab apple trees are considered, but some of these may be used to produce crab apple jelly.

Together with the genus *Prunus* (the cherries), *Malus* contributes the most spectacular display in our gardens in spring and early summer.

When in flower and in fruit, the two genera are easily distinguished. The apple flower has five styles and when each is pollinated and fertilised, this can lead to five seeds (pips) in the fruit. The cherry flower has just one style, which can lead to one seed (the stone) in the fruit.

Crab apples also give a colourful display in the autumn with numerous coloured fruit, which in many cases will persist well into winter. Autumn leaf colour is usually a restrained yellowing, but in a few cases, mentioned below, vivid autumn leaf colour is produced.

Growing Requirements

Crab apples are easy to grow successfully.
Hardiness: USDA 6a/7b RHS H6.
Aspect and shelter: A site with any orientation is acceptable and no extra shelter is required.
Sun: They grow best in full sun but will tolerate light shade.
Soil and pH: Providing the soil is well drained, soil of any texture and wide pH range is good for them to thrive in. To grow well, crab apples should be given an annual spring application of balanced fertiliser and, in the case of trees on very dwarfing rootstocks, will require extra feeding and care. Trees on very dwarfing rootstocks will produce healthy, very small trees but, perhaps counter-intuitively, they do require generous feeding regimes if adequate growth is to be produced.

Unfortunately, crab apples can be attacked by several diseases, those most serious being apple scab and apple powdery mildew. Some modern cultivars are resistant to these diseases and this is noted in the information given for each selection below.

The Range of Small and Very Small Trees Available

Commercially, all crab apple species and cultivars are produced on a range of rootstocks, as used for dessert and culinary apples. The selection of rootstock and subsequent production process will affect the size of plant offered for sale and the ultimate size of the tree once fully grown.

All the species and cultivars listed in Table 3.19 are moderate- to small-growing and it is assumed that they have been grafted on to semi-vigorous, semi-dwarf or, in the case of very small trees, top-worked on to a very dwarf rootstock.

Garden Design Considerations

Crab apples are a range of trees with a very wide range of applications in gardens. Their small size enables them to fit into informal and cottage gardens very well, combining two seasons of interest with flowers in spring or early summer followed by fruit and, in some instances, vivid leaf colour in autumn. In some cultivars the colourful fruits persist well into winter and in the case of *Malus* × *robusta* 'Red Sentinel' into the following early spring! By paths, roads and car-parking areas, fruits may be undesirable, the fallen fruit causing a problem that must be removed. In such a case, varieties can be selected that produce very little, if any, fruit.

In courtyards, formal patio settings and container-gardening, small or very small compact trees that provide strong seasonal interest may be required. By careful selection within *Malus*, the correct tree can be obtained.

Malus 'Adirondack' (syn. 'Admiration'). A young tree in fruit.

Immature fruit of *Malus* 'Adirondack' beginning to develop colour.

The flowers of *Malus* 'Adirondack', pink in bud opening to white.

A dwarf tree of *Malus* CORALBURST ('Coracole') in full flower.

The flowers of *Malus* CORALBURST ('Coracole'), pink in bud opening to a very soft pink.

The fruit of *Malus* 'Evereste'.

Malus × *floribunda*. A young tree displaying the appropriateness of its specific name.

The flowers of *Malus* × *floribunda*. These are followed by exceptionally small fruit, which cause little in the way of a slip hazard when falling on to paths.

Malus × *robusta* 'Red Sentinel'. A mature tree in full flower in spring.

Detail of the flowers of *Malus* × *robusta* 'Red Sentinel', which are large, pale-coloured and scented.

The same tree of *Malus* × *robusta* 'Red Sentinel' heavy with fruit in winter.

The crab apples of *Malus* × *robusta* 'Red Sentinel'. These remain hard and in good condition right through the winter and into early spring. Then they fall and provide a feast for birds.

Malus trilobata in autumn displays its brilliantly coloured autumn leaf colour. This unusual small tree deserves to be more widely planted.

The foliage of *Malus trilobata*. The leaves are a very atypical shape for a malus and the autumn leaf colour is well complemented by the plain-green fruit.

TABLE 3.19 SMALL AND VERY SMALL SPECIES AND CULTIVARS OF *MALUS*

Name	Height	Spread	Habit	Flowers, season and colour	Fruit	Autumn leaf colour	Remarks
Malus 'Adirondack' AGM (syn. *Malus* 'Admiration')	3–4m (10–13ft)	2–2.5m (6–8ft)	Dwarf and upright	Dark-red buds open white	Bronze-red. Persistent	Yellow	Fruits last until early winter
Malus Cinderella 'Cinzam'*	1m (3¼ft)	1m (3¼ft)		Red buds opening white	Small, yellow	Yellow	
Malus CORALBURST 'Coralcole'*	2.3–3m (7½–10ft)		Dense-growing, round-headed	Spring. White, pink on reverse	Small, orange-yellow	Yellow	
Malus 'Evereste' AGM (syn. *Malus* 'Perpetu')	4–8m (13–26ft)	4–8m (13–26ft)		Late spring. Red in bud opening white	Orange to orange-yellow		Good pollinator for dessert or culinary cultivars. Disease-resistant
Malus × *floribunda* AGM (Japanese crab apple)	4.8m (16ft)	8–12m (26–39ft)		Early spring. White and pink from red buds	Small, red-yellow		Prone to scab
Malus JELLY KING ('Mattfru') AGM	2.5–4m (8¼–13ft)	2.5–4m (8¼–13ft)		Masses of scented, white blossom	Large orange pink fruit. Persistent	Yellow	Makes good pink jelly
Malus × *moerlandsii* 'Liset'*	4–8m (13–26ft)	4–8m (13–26ft)	Compact, dense-growing, round-headed	Early summer	Very dark-red, persistent	Yellow	
Malus × *robusta* 'Red Sentinel' AGM*	4–8m (13–26ft)	4–8m (13–26ft)		White, scented	Cherry like, red and glossy 2.5cm (1in) across. Very persistent		Fruit gradually develops dark-red colour. Disease-resistant
Malus 'Royal Beauty'	1.5–2.5m (5ft–8ft)	1.5–2.5m (5ft–8ft)	Small, weeping tree	Mid-spring. Red-purple	Small, dark-red, ovoid	Yellow	Leaves copper-red when young, later green
Malus sargentii 'Tina'* (syn. *Malus toringo sargentii* 'Tina')	1.5m (5ft)	2.1m (6¾ft)		White	Small red		
Malus × *scheideckeri* 'Red Jade'* (syn. *Malus* 'Red Jade')	4m (13ft)	4m (13ft)	Small, weeping tree	White from pink buds	Small, red, shiny. Persistent		
Malus transitoria AGM (cut-leaf apple)	4–8m (13–26ft)	4–8m (13–26ft)	Small, graceful tree	White, profuse	Tiny, yellow	Yellow	Leaves cut into 3 lobes
*Malus triobata** (Lebanese wild apple)	12m (39ft)	4–8m (13–26ft)		Early summer. White, single, 4cm (1½in) across	Small, green but often with red coloration	Very good, red	Leaves maple-like. Good patio tree when on very dwarfing rootstock
Malus × *zumi* 'Golden Hornet'	8–12m (26–39ft)	8m (26ft)		White, flushed pink	Bright-yellow, 2.5 cm (1in) across	Yellow	Good pollinator for dessert and culinary. Fruit retained until early winter

* Size as stated but also available on semi-dwarfing or very dwarfing rootstocks as top-worked trees to form very small trees.

A mature tree of *Malus* × *zumi* 'Golden Hornet' in autumn.

The bright-yellow fruit of *Malus* × *zumi* 'Golden Hornet'. The fruit is preceded by profuse flowering in spring and this cultivar is a good pollinator for many *Malus domestica* (eating apple and culinary) cultivars.

In orchards of dessert or culinary apples, crab apples may be gainfully included. Where greater floral display and additional autumn colour is required, crab apples will provide this. Several crab apple cultivars can act as pollinators for the cropping trees, their profuse flowers providing extra pollen to improve the set of the dessert or culinary apples. Where the space of such a pollinator crab apple cannot be spared, scions of the crab apple can be grafted into the heads of the cropping trees.

Crab apples also have a role in helping to improve the biodiversity of our gardens. In spring their flowers provide a bounty for pollinating insects, including bees, and in the late autumn and winter their fruits provide a welcome feast for birds and other wildlife.

MESPILUS

A genus of only two species of which only *Mespilus germanica* (the medlar or Dutch medlar) is commonly grown. A small tree native of Europe and Asia and naturized, but not native in southern England.

A deciduous, large shrub or small tree with its own distinctive character. The bark develops parallel, vertical cracks. The flowers are large and solitary, being followed by uniquely shaped, brown fruits.

Growing Requirements

Hardiness: USDA 6b/7a. RHS H6.
Aspect: The tree will grow well on a site of any orientation.
Shelter: Some shelter is required since the leaves and fruit are liable to wind damage.
Sun: Full sun is required.
Soil: A soil of any texture but well drained is satisfactory.
pH: The tree has a wide pH tolerance but avoid a very limey soil.
Fruit: This is rounded with a broad, open eye and a persistent calyx. When ripe the fruit is russetted and brown. Pick the fruit after some frost has occurred. The fruit is tart and only 'Royal' is recommended fresh. Traditionally, after harvesting, the fruit is stored cool and left until 'bletted'. This is when the flesh is brown but before it has become rotten. The taste is not to everyone's liking but is said to be acquired and greatly liked by some. The fruit can also be made into jelly.

The tree is best grown as a standard or half-standard on the rootstocks Quince A or BA29 to provide greater stability.

Garden Design Considerations

This small tree is unique in its appearance at all seasons and always elicits comment. It manages to be attractive in flower and fruit, has good autumn leaf colour a distinctive pattern to its bark and provides a crop that is praised by some. Where a small talking point is required, it definitely fits the bill.

Mespilus germanica in flower in early summer.

Autumn leaf colour and fruit of *Mespilus* in early autumn.

The large, white flowers and foliage of *Mespilus germanica*.

A medlar, the fruit of *Mespilus germanica*. This must be left until it is bletted before it is considered fit to eat. This point is reached when the flesh of the fruit is brown, but not completely rotten.

TABLE 3.20 *MESPILUS GERMANICA* AND CULTIVARS

Name	Habit	Height	Spread	Leaves	Flowers/Fruit	Remarks
Mespilus germanica	Bushy	6m (20ft)	8m (24ft)	6–12cm (2–5in) long. Downy, especially on under surface. Produce good reddish-brown autumn leaf colour	Flowers early summer, solitary, white 3–4cm (1–1.5in) across, fruit 2.5cm (1in) diameter	Has spines 1–3cm (0.5–1in) long
M. germanica 'Flanders Giant'		4–6m (13–20ft)	6–8m (18–24ft)		Very large flowers and fruit	
M. germanica 'Nottingham' AGM	Semi-weeping, flat-topped tree	4–6m (13–20ft)	6–8m (18–24ft)	Larger than *M. germanica* and also produces good autumn leaf colour	Flowers white 5cm (2in) across, fruits larger than *M. germanica*	Best flavour. Less thorny than species
M. germanica 'Large Russian'	Spreading with pendulous branches				Fruits 7cm (2.5in) across	Large fruit, good flavour
M. germanica 'Royal'	More compact, upright to spreading	4m (13ft)				Fruit smaller than Nottingham, but best for eating fresh

MORUS

A genus of thirteen species from North and South America of which five are hardy, two are of horticultural importance plus one new cultivar of as yet uncertain provenance.

Growing Requirements

Hardiness: *Morus alba* USDA 7b/8a. RHS H5. *Morus nigra* USDA 6b/7a. RHS H6. *Morus* 'Waisei-kirishima-shikinari' USDA 7b/8a. RHS H5.

Aspect and shelter: A warm, sheltered site facing any orientation, except north, is suitable.

Sun: Full sun is preferred.

Soil: The soil can be of any texture providing that it is well drained.

pH: The plants have a wide pH tolerance.

Garden Design Considerations

Morus alba may be desirable because of its association with silk production in the East and is said to have been introduced to Britain in 1596 to help found silk production in the UK. The tree only grows well in the warmer parts of the UK and has never been used to raise silkworms successfully in England. Whilst very decorative, it soon becomes too large for small gardens or where space is restricted.

Morus alba 'Pendula' is ideal as a small tree. The branches fall almost perpendicularly from the head, which is formed by top-working, forming a cascade of shining, green leaves. The height is determined by the height at which the pendulous cultivar is grafted on to the upright stem of *Morus alba*. This pendulous tree contrasts strongly in habit against any other tree or shrub and its height and spread can be tailored to the space available.

Morus nigra has great character as a tree, assuming a distinct gnarled appearance, with branches falling to the ground and self-laying. The tree, however, becomes too expansive and large for where space is restricted.

A small, weeping tree of *Morus alba* 'Pendula'.

The almost perpendicular branches of *Morus alba* 'Pendular'.

Fruit ripening on a white mulberry, *Morus alba*.

TABLE 3.21 *MORUS ALBA, MORUS NIGRA* AND *MORUS* 'WAISEI-KIRISHIMA-SHIKINARI'

Name	Habit	Height	Spread	Leaves	Fruit	Remarks
M. alba (white mulberry)		9–12m (30–40ft)		Heart-shaped and very variable in shape and size	A cluster of drupes, enclosed by persistent, succulent sepals. Fruit at first white, then pink and sometimes becoming black. Edible, but insipid	Too large and spreading for a small garden or compact space
M. alba 'Pendula' (weeping white mulberry)	Narrow head and stiffly pendulous branches	2.5–4m (8–15ft)	4–8m (15–26ft)	Heart-shaped, turning yellow in autumn	Same as *M. alba*	Height and spread depends on the height at which the tree was initially top-worked
Morus nigra (black mulberry)	Wide-spreading, snarled, self-layering. Long-lived and architectural	8m (26ft)	10m (33ft)	Heart-shaped, turning yellow in autumn	As in *M. alba*, but black, sweet-tasting and edible	Too large and spreading for a small garden or compact space
Morus 'Waisei-kirishima-shikinari'	Small, widely branched shrub or small tree	2–4m (6.5–15ft)	1.5–3m (5–10ft)	Heart-shaped	Small, but sweet. Produced from early summer to early autumn	Potentially ideal for pot culture and the small garden

Morus 'Waisei-kirishima-shikinari' (also known as *Morus* Charlotte Russe, *Morus* 'Matsunaga' and *Morus* Mojo Berry). This plant was launched at the Chelsea Flower Show in 2017 and became Chelsea Plant of the Year 2017. The plant is claimed to be the result of development work in Japan from *Morus rotundiloba*. The plant is naturally small-growing and appears to be ideal in a pot or container-grown fruiting bush or small tree, and as such is perfect for the modern small or patio garden. The plant produces small, but sweet fruit from a very young age over the growing season. So far results would appear to be very mixed in the UK; it may be a very good new introduction, but at the time of writing, 2022, more time is necessary to fully evaluate this plant

PITTOSPORUM

A large genus of evergreen shrubs and small trees native to New Zealand, Australia, East and South-East Asia and Africa. Many are unfortunately not hardy in temperate climates. Horticulturally, the most important species come from New Zealand. Only those that are suitable for gardens of limited size and have proved reasonably hardy are considered here.

Growing Requirements

Aspect: Any orientation of site is acceptable except a cold north-facing situation.
Shelter: Shelter from cold, desiccating winds is an important requirement to note. They will grow well in the presence of wind but cannot stand low temperatures and are well adapted to coastal planting, which may be windy but not generally subject to very low temperatures in winter because of the modulating effect of the sea.
Sun: Full sun or part shade is suitable, but not a deeply shaded position.
Soil: The soil can be of any texture, including sandy or chalky.
pH: As a group they grow well in soils of widely varying pH.

Pittosporum tenuifolium (Tawhiwhi)

Hardiness: USDA 8b/9a. RHS H4.
This is the hardiest and most widely planted species. An evergreen shrub or small tree, with a densely branched structure, the young wood being almost black. When allowed to fully develop into a small tree or large shrub, it can reach a height of 4–8m (13–26ft) with a spread of 2.5–4m (8–13ft). The light-green leaves are 2.5–5.7cm (1–2¼in) long, oblong and have an entire, but wavy, edge. The combination of light-green leaves and undulating edges against the almost black of the peripheral stems gives this plant a unique appearance.

The flowers, produced in summer, are not a prominent feature, being small with chocolate brown, nearly black petals and almost lost within the foliage. They do, however, emit a pleasant scent during the evening.

Pittosporum tenuifolium 'Golf Ball'[PBR]

Hardiness: USDA 8b/9a. RHS H4.
A cultivar with bright-green foliage and compact growth. Height 0.5–1m (1½–3ft). Spread 0.5–1m (1½–3ft). As its name suggests, it lends itself very well to being clipped to form very neat, symmetrical ball shapes and is useful as a low hedge in seaside gardens.

There are several variegated cultivars of which some of the most widely grown are discussed here.

Pittosporum tenuifolium 'Irene Patterson' AGM

Hardiness: USDA 8b/9a RHS H4.
Slow-growing, eventually attaining 3m (10ft) with a width of about 75cm (2½ft). The leaves emerge creamy-white, later becoming green with white marbling. If exposed to cold in winter, the leaf edges becoming tinged pink. Overall, very attractive.

Pittosporum tenuifolium 'Tom Thumb' AGM

Hardiness: USDA 8b/9a RHS H4.
A dwarf cultivar, which can surprise if seen for the first time in late spring. The young leaves emerge bright-green, which then contrast very strongly with mature, deep-purple leaves. As the season progresses, the young leaves gradually change to the dark purple.

Pittosporum tenuifolium 'Silver Queen'

Hardiness: USDA 8b/9a RHS H4.

Growth similar to the species but less vigorous. Height 2.5–4m (8–13ft), spread 1.5–2.5m (5–8ft). Leaves greyish-green, margined creamy-white. Flowers deep-purple coloured.

Pittosporum tenuifolium 'Variegatum' AGM

Hardiness: USDA8b/9a. RHS H4.
Height: 4m (13ft); spread 2m (6ft).

Like the species, leaves undulating, but edged with creamy-white.

Pittosporum tobira AGM

Hardiness: USDA 9b/10a RHS H3.
A slow-growing, but eventually large shrub. Height 2–10m (6–40ft). Spread 1.5–3m (3–10ft). Obovate, bright-green leaves are held in whorls up the stems. In summer, creamy-white, orange blossom scented flowers are produced. After opening, the flowers become yellow before falling.

Pittosporum tobira 'Nanum'

Hardiness: USDA 9b/10a RHS H3.
A low-growing, compact cultivar of the above species forming a very neat and attractive mound of foliage. Height 0.5–1m (1½–3ft). Spread 0.5–1m (1½–3ft). The bright-green leaves cover the surface of the domed plant. Flowers the same as the species are produced in the summer.

Garden Design Considerations

A group of evergreens that is very prominent and useful for furnishing the garden in winter and giving all year-round structure, where necessary. In summer, the taller cultivars can act as a uniform backdrop against which to set the colourful flowers of herbaceous and smaller summer-flowering shrubs.

Pittosporum tenuifolium or its larger cultivars, when clipped to a regular upright shape, can be used to function as an accent plant. Their use as evergreen hedging in mild, seaside gardens should be considered.

One of the taller cultivars could be used as a specimen near to a flight of steeply rising steps. When walking up the steps, the dark flowers could be more clearly seen against the light-coloured foliage and the nearby scent from the flowers enjoyed.

Pittosporum can be used in a mixed border for leaf colour and textural contrasts. *Pittosporum tenuifolium* 'Tom Thumb' can be very striking when used as an isolated specimen, the seasonal leaf colour changes being very arresting in spring. When deliberately used in plant associations, the prevailing deep-purple leaf colour lends itself to very strong contrasts.

Topiary and clipping can be very effective, especially when used in sheltered or seaside gardens. The cultivar *P. tenuifolium* 'Golf Ball'[PBR] is exceptional when used in this manner. Winter cold may cause some dieback but pruning back to where still alive in late spring will usually result in rapid regrowth.

Pittosporum tobira is useful as an evergreen flowering hedge in seaside gardens provided it is protected from the worst winter cold winds and sea-salt spray. *Pittosporum tobira* 'Nanum' is very popular as a container plant for a sunny patio, but winter protection in the form of a move to a cold greenhouse or conservatory will be needed in all but the mildest locations.

All those mentioned above, coupled with some less hardy species in very favoured gardens, are successful and luxuriate in seaside gardens along with many other representatives from the New Zealand flora, e.g. any *Phormium*, New Zealand Flax.

TABLE 3.22 *PITTOSPORUM TENUIFOLIUM* AND *PITTOSPORUM TOBIRA*, PLUS CULTIVARS

Name	Habit	Height	Spread	Leaves	Remarks
P. tenuifolium	Upright and densely branched	4–8m (13–26ft)	2.5–4m (8–13ft)	Light-green, oblong, 2.5–3cm (1–1¼in) long, wavy edge	Flowers small, dark, scented in evening
P. tenuifolium 'Golden Ball'	Compact	1–1.5m (3–4ft)	1–1.5m (3–4ft)	Golden-yellow young foliage, ageing to yellow-green	Possible substitute for *Buxus* (box) balls
P. tenuifolium 'Golf Ball'[PBR]	Very compact	0.5–1m (1½–3ft)	0.5–1m (1½–3ft)	New leaves golden-yellow, ageing to yellow-green	Possible substitute for *Buxus* (box) balls
P. tenuifolium 'Irene Paterson' AGM	Upright, slow-growing	3m (10ft)	75cm (2½ft)	New leaves emerge white, later becoming green with white marbling	In cold weather leaf edges become tinged pink
P. tenuifolium 'Silver Queen' AGM	Like straight species, but slower growing	2.5–4m (8–13ft)	1.5–2.5m (5–8ft)	Greyish-green, margined creamy-white	
P. tenuifolium 'Tom Thumb' AGM	Compact dwarf	0.5–1m (1½–3ft)	0.5–1m (1½–3ft)	Young leaves emerge bright-green, contrasting against mature deep-purple leaves. Later young leaves age to deep-purple	Great for foliage colour contrast in small plant associations
P. tenuifolium 'Variegatum'	Upright and fast-growing	4m (13ft)	2m (6ft)	Leaves edged creamy-white	
P. tobira AGM Hardiness: USDA 9a/10a RHS H3		2–10m (6–40ft)	1.5–3m (3–10ft)	Leaves obovate, bright-green, in whorls	White orange blossom scented flowers, in summer. Flowers become yellow before falling
P. tobira 'Nanum' Hardiness: USDA 9a/10a RHS H3	Low growing, compact	0.5–1m (1½–3ft)	0.5–1m (1½–3ft)	Leaves obovate, bright-green, in whorls	Flowers the same as the species above

A clipped, ball-shape of *Pittosporum tenuifolium* 'Golf Ball'PBR.

Foliage of *Pittosporum tenuifolium* 'Golf Ball'PBR a few weeks after having been clipped in spring.

Foliage of *Pittosporum tenuifolium* 'Irene Patterson'.

A neat, complete, clipped ball of *Pittosporum tenuifolium* 'Tom Thumb'.

Foliage of *Pittosporum tenuifolium* 'Tom Thumb' in summer. In spring, the new growth is very light-coloured and contrasts strongly against the old, dark-purple foliage.

A pleasing contrast of leaf colours and textures between *Pittosporum* 'Tom Thumb' and *Skimmia japonica*.

A clipped shape of *Pittosporum tenuifolium* 'Variegatum'.

Detail of the foliage of *Pittosporum tenuifolium* 'Variegatum'.

A container-grown specimen of *Pittosporum tobira* 'Nanum'.

Details of the foliage of *Pittosporum tobira* 'Nanum'.

The flowers of *Pittosporum tobira* 'Nanum' when newly open are white and then change to white. They have a pleasant fragrance of orange blossom.

PRUNUS

A very large genus of mainly deciduous trees and shrubs, plus a few evergreen species, that naturally occur over the north temperate region. There are many species and a very large number of hybrids of both decorative and fruit plants that could be mentioned. This includes those native to the UK, which are *Prunus avium* (gean or mazzard), *Prunus domestica* (plum, bullace, greengage, etc.), *Prunus padus* (bird cherry) and *Prunus spinosa* (blackthorn or sloe).

Here only deciduous decorative cherry trees well suited to modern compact and small gardens will be considered.

Growing Requirements

Decorative cherries are relatively undemanding as to their requirements for good growth.
Hardiness: USDA 7b/8a or 6b/7a.
RHS H5 or H6.
Aspect and shelter: The planting site can face any direction and whilst a sheltered site will be acceptable, no shelter is required – exposed sites being equally acceptable. Note, however, that *Prunus mume* and its cultivars do require some shelter, principally because of their very early flowering.
Sun: An open, sunny position is required by all cherries. Part shade is tolerated by a few, but full shade is not suitable. Flowering will be more profuse and autumn leaf colour better developed in full sun.
Soil: The soil can be of any texture providing that it is well drained.
pH: A very wide pH range is tolerated, including a high soil pH.

The roots of many cherries often appear on the soil surface. This is entirely natural but can cause problems when the tree is growing in a lawn. In this situation, a circular area equal to the width of the canopy should be left clear of grass. This will avoid spoiling the appearance of the lawn and hitting the roots when mowing.

The Range of Shrubby, Very Small and Small Trees Produced, Plus Flowering Times

Here it should be noted that how the tree is produced on the nursery and, later, how the tree is grown and pruned by the cultivator may greatly affect the eventual size and shape of the tree. When ordering a cherry, it is very important to inspect the tree prior to purchase or at least check how the plant has been propagated and grown on, up to the point of sale. Otherwise, the tree you receive may be very different to what you expect!

Some, such as *Prunus nipponica* var. *kurilensis* 'Brilliant', when young are shrub-like in form, producing many stems from at or near the base, but each stem will eventually grow to an upright main branch that may be several metres tall.

Some, especially small and pendular cultivars, may be top-worked, forming a small tree with a straight stem and branches arising from the top. The height of the stem may be as little as 45cm or 1m (17½/39in) but can be more. The eventual height of these small trees is largely determined by the height at which they are top-worked. Many are pendular and their height only extends a little above the top-worked position. The width of these pendulous trees is again determined by the degree to which the branches weep downwards.

Cherry trees are also produced as conventional feathered and standard trees, which will grow on to form more normally shaped trees.

Cherries, in general, do not respond well to pruning and should only be pruned during dry weather in the summer to avoid entry of silver leaf and bacterial canker diseases. Some, as mentioned in Table 3.23, can be pruned in summer and this can be used to effectively reduce their size.

'Tailored' cherry trees are produced by a limited number of nurserymen for extra decorative effect. First, *Prunus serrula* is budded or grafted on to a rootstock and grown upwards to produce a stem. Once this stem is produced, another decorative flowering cherry cultivar is budded on to the top of the *Prunus serrula* stem. A tree formed in this manner will then have several seasons of display. In the winter, following leaf fall, the highly decorative shiny trunk of *Prunus serrula* is revealed. In the spring, the cultivar forming the head of the tree produces the

Prunus 'Accolade' in full flower. An excellent, broad-headed, early spring-flowering cherry.

Prunus 'Accolade' produces pendant clusters of semi-double, pink flowers.

Prunus 'Amanogawa' is an upright-growing cherry that matures into a small, columnar tree.

mass of flowers that cherries are renowned for. Finally, in the autumn, the season is be capped by very good autumn leaf colour. Such a tree is not cheap to produce, needing much skill and care by the nurseryman, but will provide an excellent specimen tree.

Cherries are notorious for their individually short flowering season. In Japan, the Land of Cherry Blossom, the answer is to watch the nightly weather forecasts, which include a mention of where in the country the display will be best. This position gradually moves northwards as spring comes to different parts of the country. In other countries where this is not an option, a longer season of flower can be obtained by selecting several different trees from early, mid and late spring-flowering species or cultivars.

Prunus 'Amanogawa' has semi-double, shell-pink flowers in late spring.

Prunus 'Fugenzo' (syn. *P.* 'Shirofugen') typically grows into a small,flat-topped tree, as shown here. Flowering late, the pale-pink flowers are produced on long pedicles.

Prunus 'Fugenzo' (syn. *P.* 'Shirofugen') flowers. Here the contrast between the pale-pink flowers and the copper-red young leaves can be seen. This contrast makes the whole tree stand out when it is in flower.

Prunus 'Fugenzo' (syn. *P.* 'Shirofugen') follows the spring display with short-lived but vivid autumn leaf colour.

Prunus 'Fugenzo' (syn. *P.* 'Shirofugen'). Here the autumn leaf colour, which is reliably produced every autumn, is seen in detail.

Prunus incisa 'Kojo-no-mai'. This cultivar is slow-growing and eventually develops into a shrub or a small tree.

Prunus incisa 'Kojo-no-mai'. In early spring, this cultivar is covered in pale-pink, single flowers.

Prunus incisa 'Kojo-no-mai'. In autumn, leaf fall is regularly preceded by the development of red autumn foliage colour.

Prunus incisa 'Kojo-no-mai'. A branch displaying autumn leaf colour. Once the leaves fall, the plant's zig-zag pattern of stem growth is revealed. This is another attraction in the winter garden.

TABLE 3.23 *PRUNUS* SPECIES AND CULTIVARS

Name	Habit	Height	Spread	Flower colour and season	Leaf Colour	Autumn leaf colour	Remarks
Prunus 'Accolade' AGM	Open and broad-headed	4–8m (13–26ft)	4–8m (13–26ft)	Mid-spring. Pale-pink	Green	Orange-red	
Prunus 'Amanogawa' AGM	Narrowly fastigiate	4–8m (13–26ft)	2.5–4m (8–13ft)	Late spring. Semi-double, scented, shell pink	Green	Orange-red	
Prunus FRILLY FROCK ('Fpmspl')	Very small or small weeping tree	2m*or 1m (6 or 3ft)	3m or 0.5m (10ft–20in)	Early spring. White	Dark-green with light-green margins	Orange-red	
Prunus 'Hally Jolivette'	Small, round-headed tree	5m (16ft)	2–3m (6½–10ft)	Mid-spring. Soft pink, semi-double	Green	Orange-red	Hardiness only USDA 7b/8a. RHS H5
Prunus incisa 'Kojo-no-mai' AGM	Small, bushy shrub or small tree on short stem	1.25m* (4ft)	2m (6½ft)	Early spring. Pale-pink	Green	Red	Zig-zag branch growth noticeable in winter
Prunus incisa 'Paean'	Small, multi-stemmed tree	3m (10ft)	2–2.5m (6½–16ft)	Early spring. Dark-pink			
Prunus incisa 'Pendula'	Small, weeping tree on a short stem	1.25m* (4ft)	2m (6½ft)	Early spring, white			
Prunus 'Kursar' AGM	Small, single-stemmed tree	4–8m (13–26ft)	4–6m (13–20ft)	Spring, single, deep-pink	Young foliage red/bronze, turning green	Orange-red	
Prunus 'Matsummae-fuki' (syn. 'CHOCOLATE ICE')	Umbrella-shaped, spreading tree	4–8m (13–26ft)	5–6m (16–20ft)	Spring, single, blush-white	Deep coppery brown young leaves, turning green	Red	Flowers contrast against young foliage
Prunus mume 'Beni-chidori' AGM (syn. *Prunus mume* 'Beni-shidare')	Small, irregularly shaped tree	1.5–2.5m* (5–8ft)	1.5–2.5m (5–8ft)	Early spring. Single, almond scented, rich madder pink	Green	Often orange-red	Cultivar of the Japanese apricot, may produce edible, bitter yellow fruits
Prunus mume 'Omoi-no-mama'	Small, irregularly shaped tree	1.5–2.5m* (5–8ft)	1.5–2.5m (5–8ft)	Early spring. Semi-double white with occasionally part pink or totally pink flowers	Green	Often orange-red	This and the tree above only USDA 7b/8a. RHS H5. Some shelter is also advised
Prunus nipponica var. *kurilensis* 'Brilliant'	Dense, upright shrub or bushy tree	2.5–4m (8–13ft)	1.5–2.5m (5–8ft)	Spring, single, white	Rusty red in spring, turning green later	Orange-red	

Prunus incisa 'Paean' grows into a small, multi-stemmed tree.

Prunus incisa 'Paean' flowers in early spring with dark-pink flowers.

Prunus incisa 'Pendula' produces early white flowers on steeply pendulous stems.

Name	Habit	Height	Spread	Flower colour and season	Leaf Colour	Autumn leaf colour	Remarks
Prunus 'Pandora' AGM**	Initially vase-shaped, broadening with age	8–12m (26–39ft)	4–8m (15–26ft)	Spring, single, pink	Bronze-red in spring turning green later	Orange-red	
Prunus serrula (Tibetan cherry) (syn. *Prunus tibetica, Prunus serrula* var. *tibetica*)	Bushy, round head.	8–10m (26–34ft)	8m (26ft)	Late spring, often lost in foliage. Small, white	Green, willow-shaped	Yellow	Main feature is coppery brown, polished bark. Also available as multi-stemmed trees, intended to highlight the bark
Prunus serrula 'Branklyn' AGM		8–10m (26–34ft)	8m (26ft)		Green willow shaped		A selected form with dark mahogany shiny bark
Prunus serrula 'Dorothy Clive' AGM		8–10m (26–34ft)	8m (26ft)		Green, willow-shaped		A selected form with lighter copper-coloured, shiny bark
Prunus 'Shogetsu' AGM (syn. Blushing Bride, *Prunus serrulata* 'Longipes')	A small tree, but with a wide-spreading canopy	4–8m (13–26ft)	4–8m (13–26ft)	Late spring. Double, pale-pink fading to white, hanging on long stalked clusters. Each flower 5cm (2in) across	Green	Orange-red	A superb tree where horizontal space is available
Prunus × *subhirtella* 'Autumnalis'** (winter-flowering cherry)	Bushy, irregular canopy	4–8m (13–26ft)	2.5–4m (8–13ft)	Flowers open in bursts during mild weather between late autumn and early spring. White, semi-double	Green	Yellow	Needs a dark background to contrast with and show off flowers
Prunus × *subhirtella* 'Autumnalis Rosea'**	Bushy, irregular canopy	4–8m (13–26ft)	2.5–4m (8–13ft)	As above, but flowers pale-pink	Green	Yellow	As above
Prunus 'The Bride' AGM (syn. *Prunus incisa* 'The Bride')	Small, round-headed tree	1.2m* (4ft)	2m (6½ft)	Early spring. White, single flowers with prominent red anthers	Green	Orange-red	

* Size very dependent on the height at which the tree is top-worked.
** Also available as the top component of 'tailored' trees.

Prunus 'Kursar'. A young tree in flower. It will develop only into a small tree.

Prunus 'Kursar'. The flowers are well displayed against the emerging foliage.

Prunus 'Kursar'. A good display of autumn foliage colour.

Prunus 'Matsumae-fuki' (syn. 'CHOCOLATE ICE'). Detail of the brilliant autumn leaf colour.

Prunus nipponica var. *kurilensis* 'Brilliant'. This often makes a dense, upright-growing shrub; the white flowers are followed by vivid autumn leaf colour.

Prunus 'Kursar'. Orange and red autumn foliage colour is reliably produced.

Prunus 'Matsumae-fuki' (syn. 'CHOCOLATE ICE'). The large, white flowers are well displayed against the dark, emerging foliage.

Prunus mume 'Beni-chidori'. Very early flowering and only grows into a small tree.

Prunus mume 'Omoi-no-mama'. A white-flowered cultivar. Some shelter is recommended to protect the very early flowers.

Prunus 'Matsumae-fuki' (syn. 'CHOCOLATE ICE'). This will grow into a beautiful round-headed tree.

Prunus 'Matsumae-fuki' (syn. 'CHOCOLATE ICE'). The spring display is followed by an equally spectacular autumn splash of leaf colour.

Prunus mume 'Beni-chidori'. Cultivars of this cherry have been long cultivated by the Chinese and Japanese. They are greatly appreciated as they flower at the end of what, for many, was a long and hard winter.

Prunus 'Pandora'. This normally grows to make a broad, vase-shaped tree.

Prunus 'Pandora' can be described like so many cherries as a 'dual season' plant producing attractive flowers in spring and good leaf colour in autumn.

Prunus serrula. The Tibetan cherry is grown for its shiny, mahogany-coloured bark, which is very effective in the winter. The tree in the picture is just a representative of the species and has not been specially selected.

Prunus × *subhirtella* 'Autumnalis Rosea'. The flowers are produced over a long season from late autumn, through the winter and into early spring. Frost damages the flowers that are out, but these are replaced by more flowers in the next mild spell.

Prunus serrula 'Dorothy Clive'. A selection with a distinctive, lighter coloured bark.

Prunus × *subhirtella* 'Autumnalis Rosea'. The winter-flowering cherry grows into a small- to medium-sized tree. It needs careful placement against a dark background to show off its small, light-coloured flowers.

Prunus 'The Bride'. This develops into a small, round-headed tree.

Prunus 'The Bride'. A mass of early flowers are produced.

ROBINIA (FALSE ACACIA)

A genus of four species only found in North America and North Mexico. Their characteristics in common are pinnate leaves, spiny stems and pendant racemes of pea-like flowers.

Growing Requirements

Hardiness: USDA 6b/7a. RHS H6.
Aspect: A site of any orientation can be selected.
Shelter: The growth is brittle and shelter from strong winds is required.
Sun: Full sun is required.
Soil: Soil of any texture is acceptable, but it must be well drained.
pH: A wide soil pH range, from acidic to alkaline is accepted.

The genus has nitrogen-fixing bacteria present in root nodules and this enables it to grow in poor nitrogen-deficient soils. This will help establishment in difficult urban settings where soils may be poor, but this also allows it to spread and colonize where soils are naturally low in nitrogen. This has led to *Robinia pseudoacacia* becoming an invasive weed in some situations.

The trees as a genus are drought-resistant and this may enable them to be grown where natural rainfall is low.

Robinia pseudoacacia 'Frisia'. Just how large this tree becomes depends very much on local conditions. Here it has grown into a fairly large tree, but it is often much smaller.

Robinia Species and Cultivars

All have pinnate foliage and are deciduous.

Garden Design Considerations

Robinia pseudoacacia 'Frisia' is a very popular tree for small or medium-sized gardens, contributing vivid leaf colour and casting only light, dappled shade. This light shade enables a wide range of plants to be grown beneath it, an important consideration where space is at a premium. A word of caution – its growth is very brittle and subject to breakage by wind. Select a sheltered site. It may be advisable to initially shorten the new growth that a young tree makes. This reduces the chance of wind damage and encourages the tree to develop a strong initial framework before growing a larger head. Problems with tree dieback have been reported. No specific cause for this has as yet been determined, but good initial site preparation, plus subsequent care and maintenance should help to prevent this problem. This tree is deservedly popular, as it does have colourful, finely divided foliage and a light airy habit. Its vivid, yellow leaves contrast well against purple-leaved subjects, such *Cotinus coggygria* 'Royal Purple', *Cotinus* 'Grace' or *Cercis canadensis* 'Forest Pansy'. Before planting, it should be noted that this tree may eventually become quite large, as shown in Table 3.24. The individual size of a tree is difficult to predict because of the great variability of site situations and growing conditions.

Robinia pseudoacacia 'Lace Lady' is a much slower growing, smaller tree, whose height and spread depend largely on the height at which it is grafted on to the stem of a rootstock. When grafted only a little above ground level, it can make an attractive and very small tree, suitable for use as a specimen in a container. The contorted growth and twisted leaves give the plant a very distinctive habit and structure, which can be employed to contrast against plants with more conventional growth.

Robinia hispida can be trained and pruned to eventually produce a very attractive and unusual small tree, with fine, dark-green,

Robinia hispida. A very small tree with very brittle branches, but very beautiful flowers.

Robinia pseudoacacia 'Lace Lady'. The curiously contorted pinnate foliage of this tree.

Robinia pseudoacacia 'Frisia'. The pinnate, light-yellow foliage that is so popular with gardeners and only casts light shade.

Robinia pseudoacacia 'Lace Lady' only slowly grows into a curiously shaped, small tree. Many consider it is well worth the wait!

TABLE 3.24 *ROBINIA* SPECIES AND CULTIVARS

Name	Habit	Height	Spread	Foliage	Flowers	Remarks
R. × *ambiqua* 'Decaisneana'	Upright	10–11m (30–36ft)	5–10m (16–32ft)	Green, downy on underside	Long racemes of pink flowers	
R. hispida (rose acacia)	Suckering shrub or small tree	1.5–2.5m (5–8ft)	1.5–2.5m (5–8ft)	Green. Stems covered with bristly glandular hairs	Large, deep-pink flowers. Early summer	Slow-growing, can be a suckering shrub or trained into a small tree. Also makes a good wall shrub
R. × *marginata* 'Pink Cascade'	Bush-headed tree	5–6m (16–20ft)	5m (16ft)	Dark-green	Racemes of pink flowers. Early summer	Fast-growing
R. pseudoacacia (common acacia, false acacia, black locust)	Upright	12m+ (40ft+)	8m+ (26ft+)	Green	Racemes of white, fragrant flowers. Early summer. Flowers more abundant in climates with warm summers	Too large for small or compact gardens. May form a fast-growing suckering tree. Bark deeply furrowed. Widely naturalised. May become an invasive species in warm, dry climates
R. pseudoacacia 'Frisia'	Upright	12m (40ft)	4–8m (13–26ft)	Bright-yellow in spring, then paler, golden in autumn	A few white racemes of white flowers	There have been problems with dieback with this cultivar
R. pseudoacacia 'Lace Lady' (syn. 'Twisty Baby')	Contorted branches	1m + (39in+)	0.5m+ (20in+)	Dark-green. Pinnules twisted		Often top-worked on to *R. pseudoacacia*. Eventual height depends largely at what height the tree is grafted. Usually makes a small or very small tree. Can be suitable for large containers

pinnate foliage and short racemes of large pink flowers. It does, however, need a very sheltered site to grow successfully as a tree. Against a warm wall it can be trained to produce a very successful wall shrub. When trained and pruned to grow up and over an archway, the plant makes a very desirable garden feature.

Used with care in their selection, trees from the genus *Robinia* can make a very successful addition to a small or compact garden.

SORBUS

This is a large genus of deciduous trees and shrubs, spread widely throughout the north temperate regions. Many are of horticultural importance, being cultivated mainly for their autumn leaf and fruit colour. Flowers are produced in late spring and early summer but are usually small, white and not of major decorative importance.

Most authorities currently agree that the genus should be split botanically into three sections, the first two sections containing trees whose overall appearance is distinctive as a result of their differing leaves:

- **Section Aria**, where the leaves are simple; in some species these may be toothed or lobed, but never divided into leaflets.
- **Section Aucuparia**, where the leaves are pinnate with several or many leaflets.
- **Section Micromeles**, a smaller, horticulturally less important group, where the fruits have deciduous calyces in contrast to section Aria.

The ecology of *Sorbus* has a distinct bearing on their successful cultivation. In the wild they seldom form a pure stand, but occur mainly on rugged ground, such as cliffs or screes, and are tolerant of cold, exposed situations. In heaths and woodlands, they occur only as scattered trees, mixed with many other species of trees, one of these other species being dominant. In these shaded situations, seedlings of *Sorbus* may survive as suppressed, dwarfed plants. If a gap in the tree canopy should occur, due to the collapse of another larger tree, these previously suppressed trees may then begin to grow rapidly and form a clump, but subside again once regrowth or regeneration of the larger species takes place. From this we may correctly conclude that when cultivated, *Sorbus* is generally tolerant of cold, but not of competition or shade.

Sorbus naturally grows on very poor, infertile soils, which may also be poorly drained. As a response to these challenges, trees often have a superficial root system, thus avoiding the waterlogged conditions, but which is also efficient in obtaining enough nutrients for the tree. Such areas often have a high rainfall to which several species are adapted. These adaptations may then result in some of these species being susceptible to drought when in cultivation.

Speaking generally, we find that *Sorbus*, to grow well, needs open, sunny situations, are tolerant of poor soils over a wide pH range but do poorly when subject to competition and need adequate moisture.

The size to which the trees may grow is often very variable in the wild, particularly under extreme conditions, but in cultivation the trees may grow to be larger than in the wild. Trees from within the Aucuparia section may grow to a smaller size and be shorter lived when grown on calcareous (limey) soils. This may be partly due to the high pH, but also because of the excessive drainage and consequent drought in these lighter soils.

In the wild, *Sorbus* often grows as multi-stemmed trees. Each stem may have a short life, but the life of the tree is extended by other younger stems taking over. When in cultivation, *Sorbus* is most often grown as single-stemmed trees, which may result in it being rather short-lived when compared to other genera of trees; fifty years being considered a good age for a cultivated *Sorbus* tree.

Growing Requirements

In the garden and landscape, *Sorbus* proves easy to cultivate, requiring relatively undemanding conditions.

Hardiness: USDA 7b/8a or 6b/7a.
RHS H5 or H6.

Shelter: Almost any degree of exposure can be accepted, and sheltered or fully exposed sites are all acceptable.

Sun: *Sorbus* strongly prefers to grow in full sun. Light-dappled shade will be accepted but results in slower growth and it will not grow well in shade.

Soil: The soil should be well drained, sand, loam or clay. *Sorbus aucuparia* (mountain ash or rowan) and several, but not all, of its cultivars grow less well and are shorter lived on soils derived from chalk or soft limestone, e.g. Cotswolds' brash, not liking their tendency to dry out in summer.

pH: Soils of a low pH or neutral are acceptable, but again *Sorbus aucuparia* and several of its cultivars do not thrive in a high pH.

Sorbus aria (whitebeam) and its cultivars are exceptions, thriving both on the soils liked by *Sorbus aucuparia* and thin, well-drained soils with a high pH.

In the case of *Sorbus cashmiriana*, there have been reports of this species not growing well and being short-lived when grown on a rootstock. When other species are budded or grafted on to a rootstock, the fast-growing, single-stemmed tree produced is very atypical of the plant when on its own roots. When on their own roots, *Sorbus* species often grow more slowly and may produce a multi-stemmed or bush-like plant. The multi-stemmed trees look as they would in nature and convey the natural habit of the species. If required, the laterals producing the multi-stems can be pruned off whilst the tree is young and a 'conventional' standard or half-standard produced.

Apomictic *Sorbus* species (producing seed asexually and hence coming 'true' from seed), which includes all those tabulated under 'Species', except *S. pseudovilmorinii* and *S. koehneana*, come readily from seed. Ample seed is produced naturally by these trees. Extraction of the seed from the fruit

Sorbus aucuparia. A young tree in fruit in late summer.

Sorbus aucuparia. The pinnate foliage is clearly seen, as are the early stages of infection by a leaf gall in the leaflets. The flowers are strongly scented, liked by some, but not by all.

Sorbus aucuparia. The orange fruit are soon eaten by birds for whom it is a valuable source of food.

TABLE 3.25 *SORBUS AUCUPARIA* AND CULTIVARS

Name	Common name	Height	Spread	Foliage	Autumn leaf colour	Fruit	pH Tolerance *	Remarks
S. aucuparia	Mountain ash, rowan	10–12m (40–47ft). Very variable according to conditions	4–8m (13–26ft)	Pinnate	Yellow may be orange	Red	Acid to neutral	Native. Good for wildlife. By far the most widely planted species
S. aucuparia 'Asplenifolia'		10–12m (40–47ft)	4–8m (13–26ft)	Pinnate, each leaflet deeply cut	Yellow	Red	Acid to neutral	
S. aucuparia CARDINAL ROYAL ('Michred')		6–10m (20–33ft)	3–4m (10–13ft)	Pinnate	Orange-red	Red	Acid to neutral	Upright habit

* *See* note on Table 3.27.

Sorbus aria 'Lutescens'. This cultivar is frequently planted as a street tree, but perhaps it is rather too large for a small garden.

Sorbus aria 'Lutescens'. Early in the growing season, the hairy surface of the leaf gives the foliage a light, silvery appearance. As the season advances, the hairs are removed by wind and rain, and the leaves then have a mid-green appearance.

Sorbus 'Joseph Rock'. The distinctive yellow fruit of this cultivar. As the season advances, the fruit colour may deepen to a darker yellow, as the foliage develops red and yellow autumnal colour.

A middle-aged tree of *Sorbus* 'Chinese Lace', showing the tree's typical habit.

Sorbus AUTUMN SPIRE ('Fanrock')AGM. A heavy crop of light-orange fruit on display.

The foliage of *Sorbus ulleungensis* 'Olympic Flame' lives up to the cultivar's name and changes from green to bright shades of yellow, oranges and red as the season moves on from summer into autumn.

Sorbus 'Chinese Lace'. Here the pinnate foliage, with each leaflet having a serrated edge, is shown as is the good crop of orange fruit.

Sorbus AUTUMN SPIRE ('Fanrock')AGM. In this young tree, the narrowly columnar habit is developing and a crop of fruit has been produced.

and germinating the seed are relatively simple and a method is clearly described in McAllister (2005). The trees or shrubs produced are typical of the species and much to be preferred to budded or grafted trees. The trees may be multi-stemmed, as often in nature, or restricted by formative pruning to single-stemmed trees. The growth rate of the seed-raised plants is normal and fine trees or shrubs with a normal long life are the result.

Seed of these species is readily available, and this is the route to be preferred for gardeners prepared to wait for their tree. Unfortunately, commercial growers are reluctant to adopt propagation by seed – it requires a change and complication to their production process. The seed-raised plants are slower to grow to a saleable size, meaning that the cost of production and hence selling price must be higher. Perhaps if nurserymen could promote these species as the smaller, more delicate plants they really are, all would benefit.

Garden Design Considerations

Sorbus can be usefully employed in a wide range of garden and landscape situations. A broad choice of trees has been deliberately selected in the tables above, leaving the final choice to the would-be planter who can further select to find the ideal tree or trees for a precise location.

The plants listed include moderately sized but compact trees, small trees, compact small trees, very small trees and those that straddle the boundary between trees and shrubs. In addition, many of these can be formed into single- or multi-stemmed trees.

The larger trees can make effective screens or provide a degree of shelter for a garden. They find a ready application for planting in exposed or very windy and seaside situations.

Within a garden they can be used to create not excessive, but dappled, shade so loved by many understory plants.

The fact that they enjoy acid soil conditions can be employed with advantage when considering the planting of other ericaceous plants such as rhododendrons. *Sorbus* provide light shade for the underplanting and extend the largely spring and summer flowers of these plants by producing a vivid autumn fruit and foliage colour.

Sorbus with narrow or very upright habit make excellent specimen or accent plants for an autumn fruit and foliage colour display, but those plants with pink or white fruit need a strong contrasting background to highlight their more subtle display. These upright small trees are ideal for today's compact or small garden.

The very small plants in the *Sorbus* selection are ideally suited to the rock garden or raised bed where their tiny charms may be fully appreciated.

Native species of *Sorbus*, and those with a loose, informal habit, are well suited to naturalistic plantings and excellent for encouraging wildlife. The nectar-rich flowers encouraging bees and other insects in spring and early summer, these being followed by the bountiful fruit produced much to the delight of many birds.

TABLE 3.26 *SORBUS* CULTIVARS

Name	Height	Spread	Foliage	Autumn foliage colour	Fruit	pH tolerance*	Remarks
S. aria 'Lutescens'	8–12m (26–39ft)	4–8m (13–26ft)	Silvered in spring becoming grey-green	Yellow	Orange-red	Wide range tolerated, including high pH	From a distance magnolia-like in spring. Cultivar of native species. Good for wildlife
Sorbus AUTUMN SPIRE ('Flanrock') AGM	4m (13ft)	1m (3ft)	Pinnate	Orange-red	Orange-red	Wide pH range tolerated including high pH	Upright habit
Sorbus 'Chinese Lace'	6m (20ft)	5m (16ft)	Pinnate, pinna deeply cut giving lace like appearance	Purple-red	Orange-red	Avoid high pH	
Sorbus 'Glendoick Spire'	10m (33ft)	7m (23ft)	Pinnate	Orange-red	Bright-pink		Upright habit
Sorbus 'Joseph Rock'	8–12m (26–40ft)	4–8m (13–26ft)	Pinnate	Orange, red purple	Yellow	Avoid high pH	Upright habit
S.ulleungensis 'Olympic Flame' AGM (syn. *S. commixta* 'Olympic Flame', *S. dodong* 'Olympic Flame')	4–8m (13–26ft)	2–4m (6.5–13ft)	Pinnate	Orange-red	Red	Wide pH range tolerated	Upright habit

* *See* note on Table 3.27.

TABLE 3.27 *SORBUS* SPECIES

Name	Height	Spread	Foliage	Autumn foliage colour	Fruit	pH tolerance*	Remarks
S. bissetii	4–5m (13–17ft)	2–4m (6½–13ft)	Pinnate, fern-like	Red/purple	Pale-pink becoming white	Wide range tolerated	
S. cashmiriana AGM**	4m (13ft)	3m (10ft)	Pinnate		White and large	Avoid high pH	Avoid soils that dry out excessively in summer. Pale-pink flowers
Sorbus frutescens AGM**	2–2.5m (6.5–8ft)	2–2.5m (6½–8ft)	Pinnate	Orange/red	White		
S. × kewensis (syn. *Sorbus pohuashanensis*)	6m (20ft)	4m (13ft)	Pinnate	Yellow	Red and very profuse	Avoid alkaline soil	Spectacular when in fruit
*S. koehneana*** (syn. *S. fruticosa* 'Koehneana', *S. pogonopetala*)	4–8m (13–26ft)	2.5–4m (8–13ft)	Finely pinnate	Plum purple	White with red stalks		A variable sexual diploid. Plants from later collections are much to be preferred
S. pseudohupehensis (syn. *S. pseudohupehensis* 'Pink Pagoda', syn. *S. hupehensis* 'Rosea')	4–8m (13–26ft)	4–8m (13–26ft)	Blue-green	May turn orange-red	Pink slowly turning to white	Avoid high pH	Needs a contrasting background to show off the fruit
S. pseudovilmorinii	4–7.5m (13–25ft)	3m (10ft)	Finely pinnate	Bronze red	Crimson turning to white		Arching branch habit
S. reducta AGM	1–1.5m (3–5ft)	1–1.5 (3–5ft)	Pinnate	Shades of red	Pale-pink to white	Wide pH range tolerated	A dense-growing suckering shrub
*S. sargentiana***	10m (33ft)	10m (33ft)	Coarsely pinnate	Red	Large bunches of small red fruit		Large sticky buds in winter
*S. vilmorinii***	2.5–4m (8–13ft)	2.5–4m (8–13ft)	Finely pinnate	Some years orange	Pink turning to white	Avoid high pH	Short-lived on shallow chalky soils. Sensitive to summer drought

* There is a definite correlation between high soil pH, soils that dry out excessively in summer and poor growth of many *Sorbus* species and cultivars. Many *Sorbus* species originate in areas of high rainfall and are well adapted to this. Consequently, they do not thrive in soils that dry out in summer. Many calcareous soils are also very well or excessively well-drained and dry out in summer. To an extent, both these limitations may be reduced by the addition of copious amounts of organic matter into the soil at planting and after planting, and keeping an area around the base of the tree clear of all vegetation. Mulching and irrigation during establishment will also help.

** In normal commercial tree production, most *Sorbus* cultivars and species are propagated by budding on to a *Sorbus aucuparia* rootstock. This is satisfactory for *Sorbus aucuparia* cultivars and other *Sorbus* cultivars. It provides the nurseryman with a method of producing a uniform crop of single-stemmed uniform plants wanted by the tree-growing public. It is much less satisfactory for many *Sorbus* species.

Sorbus cashmiriana. A small tree bedecked with fruit. This tree is best propagated from seed. This is in contrast to grafting or budding on to another *Sorbus* species, which is used as a rootstock. Trees produced by the latter process appear to fail fairly quickly.

Sorbus cashmiriana. The large, white, spherical fruit are a joy to see in early autumn.

Sorbus koehneana. When grown from seed this forms a small, shrub-like tree.

Sorbus koehneana. Occasionally, autumn leaf colour is well developed, but this will last only a few days before falling.

Sorbus koehneana. The picture shows the fine, pinnate foliage and small, white fruit.

Sorbus pseudohupehensis. This picture is intended to show how a good contrasting background is necessary to display a tree such as this. Here, the tree, laden with colourful fruit, is lost against an indifferent background.

Sorbus pseudohupehensis. Here, masses of pink, changing to white fruit, present on the tree shown previously, but not clearly seen.

Sorbus reducta. A true dwarf, suckering, shrub-like tree.

Sorbus reducta. Fruit are produced in just the same manner as for larger trees.

Sorbus sargentiana. A small, young tree ablaze with autumn colour. Note how this small tree is already developing this species' characteristic broad head.

Sorbus sargentiana. In late summer, the orange fruit show clearly against the still-green leaves.

Sorbus sargentiana. In late autumn, the orange fruit, although still present, tend to become 'lost' amongst the bright-orange foliage.

Sorbus vilmorinii. A small, dainty tree. Here, in early autumn, the fruit are pink.

Sorbus vilmorinii. Later in autumn, the fruit change colour to white and autumn leaf colour begins to develop.

STYPHNOLOBIUM

A genus of trees native to North and Central America and China. The leaves are alternate and pinnate with nine to 21 leaflets and up to 30cm (1ft) long. The flowers are pea-like and produced in late summer and autumn.

Growing Requirements

Hardiness: USDA7b/8a. RHS H5.
Aspect: A site of any orientation is satisfactory.
Shelter: No special provision of shelter is required.
Sun: Full sun.
Soil: A soil of any texture is satisfactory providing that it is well drained.
pH: A soil of alkaline to acid pH is accepted.

Garden Design Considerations

Styphnolobium japonicum is too large for consideration in a small garden or compact space; however, *Styphnolobium japonicum* 'Pendulum' is much smaller. The pendular form is very well suited for the small garden. It naturally makes a garden arbor. Its habit is very distinct, stiff and formal. As such it makes an excellent isolated specimen in a lawn or perhaps planted adjacent to another tree or shrub with a layered or tabulate habit, e.g. *Cornus controversa* 'Variegata' or *Viburnum plicatum* f. *plicatum* 'Mariesii'. It is frequently planted next to water where its perpendicular branches contrast with the horizontal plane of water.

Styphnolobium japonicum 'Pendulum'. This small tree next to water, which is where it is often placed.

TABLE 3.28 *STYPHNOLOBIUM JAPONICUM* (SYN. *SOPHORA JAPONICA*) AND *STYPHNOLOBIUM JAPONICUM* 'PENDULUM' (SYN. *SOPHORA JAPONICA* 'PENDULA')

Name	Habit	Height	Spread	Foliage	Flowers	Remarks
S. japonicum (Japanese pagoda tree)	A large tree with a rounded head	10–20m (30–60ft)	10–20m (30–60ft)	Deciduous. Pinnate	White, pea-like, scented in late summer and autumn	Despite its name originally coming from China. Flowers only produced on mature trees
S. japonicum 'Pendulum'	Small, weeping tree. Branches stiffly pendulous	4.5–7.5m (15–25ft)	4.5–7.5m (15–25ft)	Deciduous. Pinnate	White, pea-like	Flowers seldom produced. Often grafted on to a stem of *S. japonicum*. Height of tree depends on height it is grafted at

STYRAX (STORAX, SNOWBELL)

A genus of about 130 species mainly from warm, temperate and tropical regions of the northern hemisphere. As a genus it has alternate leaves, can be deciduous or evergreen and has pendant white flowers.

Styrax japonicus (Japanese Snowbell)

A small- to medium-sized deciduous tree. A spreading tree with a slow to medium growth rate, ultimately with a height of 8–12m (26–40ft) and a spread of 4–8m (12–26ft). Leaves ovate, finely pointed, 10cm (4in) long, deciduous. Flowers produced in late spring or early summer are bell-shaped, pendant, hanging beneath the branches, white, 1.5cm (0.75in) across with yellow stamens and slightly fragrant. Fruits are drupes, greyish-green and resemble small olives.

Growing Requirements

Hardiness: USDA 7b/8a. RHS H5.
Aspect: A site facing any orientation, but north is not favoured.
Shelter and sun: Full sun or part shade and some shelter, especially from cold winds and early morning sun following late spring frosts is required.
Soil: The soil can be loam, clay or sandy, but must be well drained in winter, humus-rich and moisture-retentive in summer.
pH: Acid or neutral is accepted, but alkaline soils should be avoided.

Garden Design Considerations

A small group of choice slow-growing, late spring or early summer flowering trees that are well suited to a small woodland garden or where the pendant flowers can be viewed from below, such as the top of a retaining wall with a path beneath. The tree should be planted far enough away from the wall to avoid any later damage to the roots.

They would also make a feature in a sheltered town or courtyard garden, where the tree's elegant growth habit and summer flowers could be fully appreciated.

TABLE 3.29 *STYRAX JAPONICUS, STYRAX JAPONICUS* CULTIVARS AND *STYRAX OBASSIA*

Name	Habit	Height	Spread	Leaves	Flowers	Remarks
S. japonicus	Spreading, fan-like branches	8–12m (26–40ft)	4–8m (12– 26ft)	Ovate, finely pointed, 10cm (4in) long. Yellow and red autumn leaf colour	Bell-shaped, white with yellow stamens, pendant. Slightly fragrant	Greyish-green fruits, looking like small olives
S. japonicus (Benibana Group) 'Pink Chimes'	Spreading, fan-like branches	8–12m (26–40ft)	4–8m (12– 26ft)		As above, but a delicate pink	
S. japonicus (Benibana Group) 'Fargesii' AGM	Spreading, fan-like branches	8–12m (26–40ft)	4–8m (12– 26ft)		White flowers	More robust-growing than *S. japonicus*
S. japonicus 'Purple Dress' AGM	Spreading, fan-like branches	8–12m (26–40ft)	4–8m (12–26ft)	Leaves plum-purple	Pale-pink to white flowers	
S. japonicus 'Sohuskan' AGM (syn. 'Emerald Pagoda')	Spreading, fan-like branches	8–12m (26–40ft)	4–8m (12–26ft)	Slightly larger leaves	White flowers	A vigorous selection
S. obassia	Rounded head, small tree	8–12m (26–40ft)	4–8m (12–26ft)	Rounded, broadly ovate leaves, 20 cm (8in) long. Yellow and red autumn leaf colour	Bell-shaped, fragrant, white, 2.5cm (1in) across in racemes 20cm (8in) long	

Styrax japonicus (Benibana Group) 'Pink Chimes'. The flowers of this small tree are best viewed from the side or underneath.

Styrax japonicus (Benibana Group) 'Pink Chimes'. Close up the detail and beauty of the flowers are clearly seen.

CHAPTER 4

DIRECTORY OF SUGGESTED SHRUBS

ABELIA

A genus of about twenty flowering shrubs from South-East Asia. The leaves are opposite or in whorls of three, each leaf being 1.5–8cm (¾–3in) long. The corolla is bell-shaped or tubular. The calyx is composed of two to five sepals and may remain on the plant after the corolla has fallen. At this stage the remaining calyx may still look attractive, prolonging the period of floral display. Where there is sufficient sun, flowering is profuse from early summer into autumn.

Growing Requirements

Hardiness: USDA7b/8a. RHS H5.
Aspect: A south-facing site is required for best flowering.
Shelter: A sheltered site is required. A sheltered, warm, sunny situation is necessary to ensure that the plant overwinters satisfactorily and flowers well during the growing season.
Sun: Full sun.
Soil: A soil of any texture is satisfactory, but it must be well-drained.
pH: A wide range of pH from acid to alkaline conditions is accepted.

Garden Design Considerations

Providing a sheltered, warm, sunny position can be provided, then profuse flowering is assured. If the site is warm in winter, semi-evergreen cultivars will probably retain their leaves, but if the weather is cold, the leaves will drop.

Abelia mosanensis 'Korean spring' begins the flowering season in spring to be followed by the cultivars of *Abelia* × *grandiflora* in summer. These then provide flower colour and a limited amount of fragrance into the autumn when some cultivars will produce autumn leaf colour. *Abelia chinensis* appears to have been almost superseded by *Abelia* × *grandiflora*, the Chinese species requiring an especially favoured site to do well. In China it flowers profusely, as evidenced by its performance in Beijing Botanical Gardens. In a Royal Horticultural Trial, the then new cultivar *Abelia chinensis* 'China Rose' briefly appeared. This was clearly superior to the straight species and was promptly awarded the Award of Garden Merit, AGM, but then it apparently disappeared from cultivation. Perhaps this is an opportunity for an enterprising nurseryman to reintroduce this cultivar?

These relatively small, free-flowering shrubs have a wide range of possible uses in today's smaller gardens, being suitable for inclusion in cottage-style, informal gardens, flower beds generally and, in particular, warm wall-side borders. Here they will produce a long season of flower and, in many cases, foliage colour.

Caryopteris × *clandonensis*. A beautiful small shrub giving, in late summer, deep-blue flowers.

TABLE 4.1 *ABELIA* SPECIES AND CULTIVARS

Name	Habit	Height	Spread	Foliage	Flowers	Remarks
A. chinensis	A small shrub	1–1.5m (3–5ft)	1–1.5m (3–5ft)	Oval, dark-green	White, fragrant	Needs a sunny, warm position to flower well
A. 'Edward Goucher' AGM	Arching, compact, bushy shrub	1.5–2.5m (3–8ft)	1.5–2.5m (3–8ft)	Semi-evergreen, glossy, bronzed when young	Lilac-pink. Profuse flowering, summer into autumn	
A. × *grandiflora* (glossy abelia)	A medium-sized shrub with arching stems	2.5–4m (3–13ft)	2.5–4m (3–13ft)	Semi-evergreen. Leaves deep-green, oval and glossy. Summer into autumn	Pale-pink and fragrant. Summer into autumn.	A large plant from which smaller and, in some cases, variegated plants have been selected, named and propagated
A. × *grandiflora* CONFETTI ('Conti'PBR)	Low-growing	0.5–1m (20–40in)	1–1.5m (3–5ft)	Semi-evergreen. Leaves green and margined creamy-white	White tinged pink. Summer into autumn	
A. × *grandiflora* 'Francis Mason'	Rounded, densely branched	1–1.5m (3–5ft)	1–1.5m (3–5ft)	Semi-evergreen. Leaves lance-shaped, yellow with a central green splash	Pale-pink. Summer into autumn, fragrant	
A. × *grandiflora* 'Hopley's'PBR AGM	Arching stems. Dome-shaped	1–1.5m (3–5ft)	1–1.5m (3–5ft)	Green with an irregular yellow margin	White tinged with lilac. Summer into autumn	Has a stable variegation
A. × *grandiflora* 'Kaleidoscope'PBR	Dome-shaped	0.5–1m (20–40in)	1–1.5m (3–5ft)	Semi-evergreen. Leaves irregularly variegated yellow and green, becoming more orangey towards autumn	White, tinted pink, fragrant. Summer into autumn	Summer into autumn
A. × *grandiflora* 'Lake Maggiore' AGM	Upright	1.5–2.5m (5–8ft)	1–1.5m (3–5ft)	Semi-evergreen. Dark-green, oval and pointed	White tinged pink, slightly fragrant. Summer into autumn	
A. mosanensis 'Korean Spring' AGM (Korean abelia)	Arching stems producing a rounded shrub	1.5–2.5m (5–8ft)	1.5–2.5m (5–8ft)	Ovate glossy, green. Deciduous. Orange-red autumn leaf colour	White flushed pink. Fragrant. Flowering from spring into early summer	

ANISODONTEA

A genus of twenty shrubs and woody perennials from South Africa, which are evergreen, with alternate leaves. The flower consists of five petals forming a cup-shaped, mallow-like bloom.

Growing Requirements

Hardiness: *Anisodontea capensis* USDA 10b. RHS H2.
Anisodontea 'El Rayo' USDA 9b/10a. RHS H3.
Anisodontea 'Large Magenta' USDA 10b. RHS H2.
Aspect: A planting site facing south or west is required.
Shelter: A sheltered, warm site is required.
Sun: Full sun is required.
Soil: This can be of any texture but must be well drained.
pH: This genus will grow well in a wide range of pH from acid to alkaline conditions.

Garden Design Considerations

Both *Anisodontea capensis* and *Anisodontea* 'Large Magenta' are best treated as summer bedding plants to supply flower colour in the garden during summer. They should be planted out after the last spring frost has passed and lifted before the first frost of autumn. After lifting, the plants may be potted up and kept in frost-free conditions over winter or regarded as disposable. If grown in a very sheltered, warm situation outside, and given cold protection, they may survive a mild winter.

Anisodontea 'El Rayo' has proved hardier. If given a sheltered, warm garden position it will survive the winter outside. Following pruning in spring, it will grow on to flower continuously throughout the summer and well into the autumn. It is a valuable small shrub for summer flower colour.

All three plants are very well suited to patio and container plant-growing; *A. capensis* and *Anisodontea* 'Large Magenta' on a summer season only basis, but *A.* 'El Rayo' as part of a permanent planting.

Anisodontea 'El Rayo'. Pictured in a late-summer garden, where it contributes much needed flower colour.

TABLE 4.2 *ANISODONTEA CAPENSIS* AND OTHER *ANISODONTEA*

Name	Habit	Height	Spread	Foliage	Flowers	Remarks
A. capensis AGM (African mallow)	Small, erect, sub-shrub	60–90cm (24–36in)	60–90cm (24–36in)	Mid-green, triangular, 2.5cm (1in) long	Pink to purple, 2.5cm (1in) across	Flowers from summer to autumn. May survive winter outside, better treated as a summer bedding plant
A. 'El Rayo'	Upright sub-shrub	0.5–2m (1ft 8in–7ft)	0.5–2m (1ft 8in– 7ft)	Mid-green, lobed, 5cm (2in) long	Pink, 3.5cm (1½in) across	Flowers from summer until late autumn. Prune back to where still alive in spring
A. 'Large Magenta'	Compact sub-shrub	75cm (30in)	75cm (30in)	Lobed, mid-green	Rose-magenta with darker veins	Flowers from summer to autumn. May survive outside, but better treated as a summer bedding plant

Anisodontea 'El Rayo'. Flowers are produced throughout the summer and into the autumn.

BERBERIS THUNBERGII

A deciduous, small shrub introduced from Japan in about 1864. The shrub is of close compact habit with stiff branches. Green, crowded leaves, long and rounded or spathulate in shape are produced in tufts along the branches. The foliage reliably turns fiery red in the autumn. Small, yellow, not showy, flowers are produced in spring, these being followed by bright-red fruits in autumn.

Thorns about 1cm (½in) long are produced along the branches, usually single but occasionally three-pronged.

Berberis thunbergii f. *atropurpurea* differs from the type species in that it has reddish-purple foliage in spring and summer. The cultivars with reddened foliage mentioned below are attributable to this form. Hence, *Berberis thunbergii* f. *atropurpurea* 'Atropurpurea Nana' is the correct full name for this plant.

Growing Requirements

Hardiness: USDA 6a-1. RHS H. 7.
Shelter: No special degree of shelter is necessary, the species being tolerant of exposed sites.
Sun: The plant will grow well in full sun or part shade. For cultivars with yellow foliage, best colour will be developed in full sun, as shade may result in some greening of the foliage.
Soil: Plants of very easy cultivation. The main requirement is for a well-drained soil, since waterlogging during winter will not be tolerated.
pH: Soil pH is not critical.
Pruning: When grown as a free-growing shrub, little annual pruning will be required. Occasionally, dead, diseased or damaged growth may be removed and in older plants removal of about one-third of the branches at ground level in summer will result in new, vigorous growth and rejuvenation of the plant. When grown as a formal hedge, annual trimming in spring after flowering will suffice to keep a tidy shape.

Cultivars Available

Numerous cultivars have been produced, some for more compact growth habit, some for vivid foliage colour, both in summer and in winter. Some cultivars combine both the favorable characteristics of compactness and vivid foliage colour.

Garden Design Considerations

The cultivars listed in Table 4.3 comprise a very useful group of plants with a wide range of garden design applications.

Those with a more rounded shape are useful as colourful, low-maintenance ground-cover. Their tolerance to a range of growing conditions makes them useful for covering banks and slopes, where excessive drainage may be a problem. Their tolerance of exposure makes them eminently suited to windy sites and seaside planting.

The dense growth habit of these plants makes them useful for hedging, both informal and formal. Formal hedges are clipped to produce a relatively narrow, crisp, well-defined shape. The plants in informal hedges are left to grow naturally and a wider, less regular shape overall is produced; but if those cultivars with a narrow, upright habit are planted closely together, then a still relatively narrow hedge with a slightly irregular height will be produced. These plants' spiny nature is useful when a hedge is intended as a barrier or when employed as a block planting, greatly inhibiting unwanted ingress.

Buxus sempervirens and its dwarf form *Buxus sempervirens* 'Suffuticosa' have been loved by gardeners as the ideal plant for hedging, especially dwarf hedging as employed in the parterres of formal gardens. The plant was ideal in that it responded well to being clipped to form compact, dense, evergreen hedges. Unfortunately, box is now frequently attacked by box blight caused by the fungi *Cylindrocladium buxicola* and less seriously by *Volutella* sp. and eaten by the caterpillars of the box moth, *Cydalima perspectalis*. Both forms of attack have proved difficult to prevent or to control. The hunt to find an alternative to box is on! The *Berberis* cultivars mentioned in Table 4.3, whilst not evergreen, have many of the other desired characteristics and may make very well-suited alternatives. As such, many of these *Berberis* plants have featured in a trial run by the RHS at their Wisley garden to find plants suitable for low hedging.

Those *Berberis* cultivars with an upright habit and bright foliage colour are also valuable as accent plants in border design. When considering foliage colour to create strong colour contrasts, the plants in Table 4.3 are an obvious choice.

Berberis × *carminea*

This group of hybrids resulted from the cross: *B. aggregata* × *B. wilsoniae*. Of these, *Berberis* × *carminea* 'Pirate King' is a particularly good selection.

The plant forms a dense-growing, spiny, deciduous shrub with a height of 60–90cm (2–3ft) and spread of 90–120cm (3–4ft). Light-green foliage sets off yellow flowers in spring, these being followed by masses of fiery orange-red berries in autumn.

Growing Requirements

Hardiness: USDA 6b-7a. RHS H6. An exceptionally tough plant.
Shelter: Exposed sites are well tolerated.
Sun: Sun or part shade is acceptable, but more fruits will be produced under sunny conditions.
Soil: Good soil drainage is required.
Pruning: Very little is required. After ten or more years, plants may be rejuvenated by cutting out one-third of the oldest stems down to ground level after flowering. This may, however, be difficult because of the very spiny nature of the plants!

Garden Design Considerations

The plant has a wide range of garden applications, as detailed for *Berberis thunbergii* cultivars.

TABLE 4.3 CULTIVARS AVAILABLE OF *BERBERIS THUNBERGII*

Cultivar	Shape	Height	Spread	Flower and berry colour	Foliage
B. thunbergii f. *atropurpurea* 'Atropurpurea Nana'	Dense, rounded bush	0.5–1m (20–40in)	0.5–1m (20–40cm)	Small, yellow flowers followed by red berries	Purplish-red, intensifying in autumn
B. thunbergii 'Golden Torch'	Narrowly upright-growing	1.0–1.15m (20–40in)	0.5–1m (20–40in)	Small, white flowers followed by red berries	Bright-yellow in spring, turning to yellowish green in summer, in autumn flushed with pink and orange
B. thunbergii f. *atropurpurea* 'Harlequin'	Dense bush	0.5–1m (20–40in)	0.5–1m (20–40in)	Small, yellow flowers followed by dark-red berries	Deep-purple, variously splashed with cream and pink
B. thunbergii f. *atropurpurea* 'Helmond Pillar'	Upright-growing	1.0–1.5m (20–40in)	0.5–1m (20–40in)	Small yellow flowers followed by red berries	Brownish red, becoming brighter red in autumn
B. thunbergii 'Orange Rocket'PBR	Upright-growing	1–1.5m (20–40in)	0.5–1m (20–40in)	Small, yellow flowers followed by red berries	Rusty orange in spring. Green in summer. Orange-red in autumn
B. thunbergii 'Orange Sunrise'PBR	Strongly upright	1–1.5m (20–40in)	0.1–0.5m (4–20in)	Small, yellow flowers followed by red berries	Coral-pink in spring, dusty-red with a narrow gold ring in summer, red in autumn
B. thunbergii f. *atropurpurea* 'Redtorch'PBR	Upright-growing	0.7–1m (20–40in)	0.5–1m (20–40in)	Small, yellow flowers followed by red berries	Small, red leaves, becoming green towards the base of the plant
B. thunbergii f. *atropurpurea* 'Rose Glow'	Dense bush	1–1.5m (20–40in)	1–1.5m (20–40in)	Small, yellow flowers followed by red berries	Young leaves purple, irregularly marked with white and pink, later purple all over. Reddens in autumn

Berberis thunbergii 'Aurea'. A small shrub, very good for colour contrasts in early summer.

Berberis thunbergii 'Aurea'. The leaves are very bright-yellow when they first appear, may scorch in strong sunlight but become pale-green by late summer.

Berberis thunbergii 'Golden Torch'. An upright-growing cultivar whose coloured leaves are scorch-resistant.

Berberis thunbergii 'Orange Rocket'[PBR] is a compact cultivar that is promising for use as a low, clipped hedge.

Berberis thunbergii 'Orange Sunrise'[PBR]. Here seen providing island highlights in a berberis and grasses' border.

Berberis thunbergii f. *atropurpurea* 'Helmond Pillar' a cultivar with a narrow, upright habit and purple foliage forming an interesting combination of colours and shapes with an upright, yellow-leaved conifer and two rounded, spikey, grey clumps of *Festuca glauca*.

Berberis thunbergii 'Orange Sunrise'[PBR]. Another coloured leaf cultivar with an upright habit.

Berberis thunbergii f. *atropurpurea* 'Rose Glow' AGM. Shows an unusual marbling of a purple leaf.

Berberis × *carminea* 'Pirate King' giving a late-autumn splash of berry colour.

Berberis thunbergii 'Redtorch'[PBR]. A very upright cultivar with small, red leaves.

Berberis thunbergii f. *atropurpurea* 'Rose Glow' forming a pleasant colour combination with the blue flowers of *Ceratostigma willmottianum*.

Berberis × *carminea* 'Pirate King'. The glorious orange fruits glowing in the autumn sunshine.

BRACHYGLOTTIS

Native to New Zealand and Tasmania, this is a genus of evergreen trees, shrubs and herbaceous perennials. Plants in this genus were, until recently, placed in the genus *Senecio* and many gardeners will still be familiar with these shrubs, but under this old generic name. All produce yellow or white daisy-like flowers, but in a few the usually prominent ray florets may be lacking.

In common with many plants originating in New Zealand they cannot tolerate low temperatures but are resistant to salt-laden wind.

Growing Requirements

Hardiness: All USDA 8b/9a. RHS H4, except *Brachyglottis* (Dunedin Group) 'Drysdale', which is USDA 9b/10a. RHS H3.
Aspect: The site can face any direct except north.
Shelter: Shelter is not specifically required and exposure to salt-laden winds tolerated well but exposure to cold, drying winds should be avoided.
Sun: Full sun is best. Part shade may be acceptable, but growth may become leggy.
Soil: The soil may be of any texture but must be well drained. They are drought-tolerant.
pH: Soil of any pH will be accepted.

Garden Design Considerations

Brachyglottis (Dunedin Group) 'Sunshine' is a very widely planted, evergreen shrub and performs excellently in well-drained sunny positions, including dry, sunny banks. Evergreen ground-cover is its major use, where it contributes foliage colour and evergreen interest, especially in winter, and bright cheery flowers in summer. Some do not like the yellow flowers, in which case the plants can be lightly clipped prior to flowering. Unfortunately, this use as ground cover is frequently seen in unsuitable shaded or poor-drainage conditions resulting in leggy, gappy, unattractive plantings. *Brachyglottis* (Dunedin Group) 'Drysdale' has the advantage of being slightly more compact but is less hardy.

Brachyglottis monroi is a sadly under-used shrub. It provides the same benefits as *B.* 'Sunshine', plus arguably more attractive leaves. Grey on top and felted below, it has the added attraction of wavy edges and is undeniably attractive to our tactile sense. *Brachyglottis monroi* is also more compact than the Dunedin Group and well suited for use where space is limited, such as patio and container-growing.

Brachyglottis WALBERTON'S SILVER DORMOUSE ('Walbrach'[PBR]) is a fairly recently introduced cultivar. It combines all the attractiveness and hardiness of *B.* 'Sunshine' but is more compact. Truly, a tailor-made shrub for small sunny gardens.

Brachyglottis is very well-adapted to coastal gardens, where salt-laden windy conditions can be expected, but not cold, drying winds. These adaptations, plus drought-tolerance, make *Brachyglottis* also useful in Mediterranean and 'Mediterranean-style' gardens, where hot, dry, sunny, well-drained conditions are the norm.

TABLE 4.4 *BRACHYGLOTTIS* SPECIES AND CULTIVARS

Name	Habit	Height	Spread	Foliage	Flowers	Remarks
B. (Dunedin Group) 'Drysdale'	Bushy	05–1m (20–40in)	0.5–1m (20–40in)	Oval, slightly scalloped edge. Green top surface. White, felted bottom surface	Bright-yellow daisies in summer	Slightly more compact than *S.* 'Sunshine' but less hardy
B. (Dunedin Group) 'Sunshine' AGM	Bushy, spreading	1–1.5m (3–5ft)	1–1.5m (3–5ft)	Ovate, 8cm (3in) long. Greyish top surface. White, felted bottom surface	Bright-yellow daisies in loose clusters in summer	Rather lax habit. Responds well to light clipping just after flowering. This removes the old flower heads and results in more compact, bushy growth
B. monroi (Monro's ragwort)	Small, rounded bush	0.5–1m (20–40in)	1–1.5m (3–5ft)	Oblong, 5cm (2in) long. Dark-green top surface. White, felted bottom surface. Leaf wavy edged	Bright-yellow daisies, each 2cm (1in) across in clusters	
B. WALBERTON'S SILVER DORMOUSE ('Walbrach'[PBR]) AGM	Compact bush	0.5–1m (20–40in)	1–1.5m (20–40In)	Oval. Silvery-grey top surface. White, felted bottom surface	Bright-yellow daisies	

Brachyglottis (Dunedin Group) 'Sunshine'. A plant in full flower in mid-summer. Some gardeners remove the flowers before they open, preferring to keep the plant's overall grey-leaved effect.

Brachyglottis (Dunedin Group) 'Sunshine'. Showing the bright-yellow daisy-like flowers and the grey-green felted leaves beneath.

Brachyglottis monroi. A smaller shrub than *B.* 'Sunshine' with similar flowers but distinct, smaller leaves with wavy edges.

BUDDLEJA (SYN. BUDDLEIA)

This is a genus found in North America, South America, Africa and Asia consisting of about ninety species. All, with the sole exception of *Buddleja alternifolia*, have opposite leaves and can be deciduous or evergreen. The tubular, fragrant flowers are grouped densely into terminal or axillary racemes or panicles, or, as in the case of *Buddleja globose*, into globular heads.

Growing Requirements

These recommendations are specifically for *Buddleja davidii* and its cultivars; most other species are the same in their requirements. There are a few exceptions, and these are indicated in Tables 4.5–4.9.
Hardiness: Most are USDA, 6b/7a. RHS H6, but exceptions are shown in Tables 4.5–4.9.
Aspect: A site of any orientation is acceptable.
Shelter: No specific degree of shelter is required, and the plants will grow well in windy, coastal gardens.
Sun: Full sun or part shade.
Soil: The soil may be of any texture but must be well drained.
pH: The soil pH can be acid, neutral or alkaline and high soil fertility is not required.

Pruning: Pruning is best done every spring, pruning the plant back to a basic, permanent framework. This keeps the plant in check, limits overall size and will help prevent wind damage, which may occur if the plants are allowed to grow too freely over several years. Spent flower heads are best removed once finished to encourage more flowering.

Buddleja davidii (Butterfly Bush)

This is a potentially large deciduous shrub.
Height: 2.5–4m (8–13ft).
Spread: 4–8m (13–26ft).
Leaves: During the growing season, long, arching shoots are produced with opposite, lance-shaped green or grey-green leaves up to 25cm (10in) long.

Buddleja davidii 'Harlequin'. The inflorescence is slightly smaller than other cultivars, but the bright leaf variegation more than makes up for this.

Buddleja davidii 'Empire Blue'.

Buddleja davidii 'Pink Delight'. A big, bold inflorescence.

Buddleja davidii 'Pink Delight'. An excellent cultivar producing a great many flowers.

TABLE 4.5 *BUDDLEJA DAVIDII* AND LARGER CULTIVARS

Name	Habit	Height	Spread	Leaves	Flowers	Remarks
B. davidii	Open and arching	2.5–4m (8–13ft)	4–8m (13–26ft)	Lance-shaped, grey-green	Shades of purple in summer and autumn	
B. davidii 'Black Knight' AGM	Upright and arching	2.5–4m (8–13ft)	2.5–4m (8–13ft)	Lance-shaped	Fragrant, deep-purple in elongated panicles	
Buzz Series *B. davidii* BUZZ CANDY PINK 'Tobudsopin'	Upright, bushy, dwarf	1.2m (4ft)	1.2m (4ft)	Lance-shaped small	This cultivar has pink flowers throughout summer and autumn	Sterile, no seed produced
B. davidii 'Dartmoor' AGM	Upright and arching	2.5–4m (8–13ft)	2.5–4m (8–13ft)	Lance-shaped, white beneath	Fragrant, light-purple in branched panicles	
B. davidii 'Empire Blue'	Upright and arching	2.5–4m (8–13ft)	2.5–4m (8–13ft)	Lance-shaped	Fragrant, violet-blue, elongated panicles	
B. davidii 'Harlequin'	Upright and arching	2.5–4m (8–13ft)	2.5–4m (8–13ft)	Leaves variegated green and creamy-white	Fragrant, dense red-purple panicles	
B. davidii NANHO BLUE ('Mongo')	Upright, bushy, semi-dwarf	1.5–2m (5–6½ft)	1–1.5m (3–5ft)	Willow-like, silvery-green	Panicle 25–30cm (9–12in) long. Fragrant, lavender-blue with an orange eye	
B. 'Pink Delight' AGM	Upright and arching	2.5–4m (8–13ft)	2.5–4m (8–13ft)	Lance-shaped	Panicles of pink flowers	

Flowers: In summer and early autumn, dense sprays of small, fragrant flowers of shades of purple are produced. The flowers are very attractive to butterflies, but the rest of the plant is not a food source for the larvae of these insects.
Seeds: The species seeds itself about very freely and can be invasive. In Oregon and Washington States the plant is classified as a noxious weed.

Plant breeders have been very active in trying to produce cultivars that are much better suited to cultivation in compact and small gardens. The cultivars are smaller than the species and many are sterile, ensuring that they cannot spread by seed and become weedlike.

Tables 4.5–4.9 contain details about these smaller cultivars, plus some of the larger, but still garden-worthy cultivars.

The Buzz Series

This series are all *Buddleja davidii* cultivars and have been bred to provide a dwarf, compact range of plants that are ideal for the smaller garden or where space is restricted.
Height: 1–1.5m (3–5ft).
Spread: 1–1.5m (3–5ft).
Leaves: Foliage grey-green above and woolly-green below. They are available in a range of colours, those currently listed are:

- Buzz Candy Pink 'Tobudsopin': flowers pink
- Buzz Hot Raspberry: flowers deep-pink with an orange eye
- Buzz Indigo: flowers dark-blue
- Buzz Ivory 'Tobudivory'PBR: flowers ivory-white
- Buzz Lilac
- Buzz Magenta Improved 'Tobudmagen'
- Buzz Magenta 'Tobudpipur'PBR: flowers purple-red
- Buzz Sky Blue 'Tobudskybi'PBR: flowers pale violet-blue
- Buzz Velvet 'Tobudvelve'
- Buzz Violet 'Tobudviole': flowers violet

The Nanho Group

A group of *Buddleja davidii* hybrids bred to be smaller and more compact than typical *Buddleja davidii* hybrids. Well suited to modern smaller gardens.

Buddleja davidii Buzz Series Buzz Magenta ('Tobudipur'PBR).

Buddleja davidii NANHO BLUE ('Mongo'). A good well-established cultivar.

Buddleja davidii NANHO BLUE ('Mongo').

English Butterfly Series

A series of hybrids cultivars of *Buddleja davidii* all bred to be more compact, but equally floriferous to *Buddleja davidii*. Well suited to the smaller garden.

Buddleja Dreaming Series

A series of complex hybrids bred to be very compact and trailing making them suitable for pots and containers. They have also been bred to be sterile or of low fertility, ensuring that spread by seed should not be a problem.

The series was developed by Peter Podaras at Hines Nursery, California, USA. At present they are only offered by very few nurseries in the UK. Due to their complex breeding, their hardiness at present is questionable and full sun is recommended for best results.

Buddleja alternifolia AGM

This is a large, deciduous shrub found originally in north-west China and introduced in 1915. Here it is recommended to be grown as a small tree, a single stem being selected from a young plant and trained vertically until the desired height is attained. When grown in this manner, the ultimate height is determined by the cultivator and the spread is 2.5–3m (8–10ft).

Once at the height required, this leader is then tipped and the resultant side branches allowed to form the head of the small tree. These branches will grow into graceful, arching stems, with long, narrow, alternate, dark-green leaves. The cultivar *Buddleja alternifolia* 'Argentea' is even more desirable – its leaves being covered with small, silky hairs giving the leaves a silvery sheen.

In early summer, a mass of small, pale-lilac, scented flowers cover the previous year's growth, giving a most attractive display, which is prolonged after flowering by the silvery leaves.

Rapid new growth is produced after flowering. If required to limit the spread of the head of the tree, some or all of this can be pruned later in the summer.

When grown in this manner, the main stem must be securely staked and some shelter is recommended to prevent excessive stress on the single, upright stem.

Both *B. lindleyana* and *B. lindleyana* 'Little Treasure' could also be cultivated in this manner to produce small or, in the case of *B. lindleyana* 'Little treasure', very small trees.

TABLE 4.6 *BUDDLEJA DAVIDII* CULTIVARS OF THE NANHO GROUP

Name	Habit	Height	Spread	Leaves	Flowers
B. davidii NANHO BLUE ('Mongo')	Compact, bushy, arching	1.5–2m (5–6½ft)	1–1.5m (3–5ft)	Willow like, silvery-green	Panicle 25–30cm (9–12in) long. Fragrant, lavender-blue with an orange eye
B. davidii NANHO PURPLE 'Monum' AGM	Compact, bushy, arching	1–1.5m (3–5ft)	1–1.5m (3–5ft)	Willow like, silvery-green	Panicle 15–20cm (6–8in) long. Fragrant, purple with an orange throat
B. davidii NANHO WHITE 'Monite' AGM	Compact, bushy, arching	1–1.5m (3–5ft)	1–1.5m (3–5ft)	Willow like, pale-green	Panicle 15–20cm (6–8in) long. Fragrant, white with an orange throat

TABLE 4.7 *BUDDLEJA DAVIDII* CULTIVARS OF THE ENGLISH BUTTERFLY SERIES

Name	Habit	Height	Spread	Leaves	Flowers
B. davidii CAMBERWELL BEAUTY ('Camkeep') AGM	Compact, upright, then arching	1.5m (5ft)	1.5m (5ft)	Willow-shaped, grey-green	Fragrant, purple
B. davidii MARBLED WHITE 'Markeep'	Compact, upright, then arching	1.5m (5ft)	1.5m (5ft)	Willow-shaped, grey-green	Fragrant, white
B. davidii PEACOCK 'Paekeep'[PBR]	Compact, upright, then arching	1–1.5m (3–5ft)	1–1.5m (3–5ft)	Willow-shaped, grey-green	Fragrant, pinkish-purple
B. davidii PURPLE EMPEROR ('Pyrkeep')	Compact, upright, then arching	1.5–2.5m (5–8ft)	1–1.5m (3–5ft)	Willow-shaped, grey-green	Fragrant, bright-purple

TABLE 4.8 *BUDDLEJA* CULTIVARS OF THE *BUDDLEJA* DREAMING SERIES

Name	Height	Spread
Buddleja DREAMING LAVENDER ('Hindebud[1]'[PBR])	0.5–1m (1½–3¼ft)	10–50cm (4–20in)
Buddleja DEAMING PURPLE ('Hindbud[4]'[PBR])	0.5–1m (1½–3¼ft)	10–50cm (4–20in)
Buddleja DREAMING WHITE ('Hindbud[3]'[PBR])	0.5–1m (1½–3¼ft)	10–50cm (4–20in)

Garden Design Considerations

Buddleja davidii cultivars have long been favourites amongst gardeners largely because of their attractiveness to butterflies, hence their common name of butterfly bush. The larger, older cultivars are excellent plants for the larger garden, but their arching and spreading nature is a problem for small gardens or where space is limited.

In recent years, plant breeders and selectors have put efforts into producing the many, much more compact cultivars now available. These are designed to be appropriate for today's smaller garden and still provide colour and scent over a long summer and autumn season. Several sterile cultivars have also been produced and this should help overcome the problem of *Buddleja* self-seeding, spreading and becoming a threat to the native flora.

Whilst they do produce a very good nectar source for pollinating insects, the plants do not provide a food source for their larvae. This should be remembered when planning a garden that favours pollinating and other beneficial insects.

As well as developing smaller, compact plants, some very dwarf *Buddleja* may find a use as seasonal plants in pots, planters, raised beds and bedding displays. This is the intended application for the new Dreaming Series.

Finally, it is suggested that *Buddleja alternifolia* and *Buddleja lindleyana*, plus their cultivars, can be grown both as shrubs but also trained to form very attractive and unusual small or very small trees. The arching, flowing, fragrant, flowering branches of a tree-like *Buddleja alternifolia* swaying gently in a balmy summer breeze is indeed very apt in an informal country garden.

Buddleja alternifolia. This species can be grown as a shrub or as a small tree.

TABLE 4.9 FURTHER *BUDDLEJA* SPECIES AND CULTIVARS

Name	Habit	Height	Spread	Leaves	Flowers	Remarks
B. alternifolia AGM	Open and arching	2.5–4m (8–13ft)	2.5–4m (8–13ft)	Deciduous. Narrow, grey-green	Lilac and sweetly scented in early summer	
B. alternifolia 'Argentea'	Open and arching	2.5–4m (8–13ft)	2.5–4m (8–13ft)	Silvery appearance due to the presence of tiny hairs adpressed to the surface of the leaf	Lilac and sweetly scented in early summer	Leaves a definite plus factor
B. lindleyana	Open and arching	1.5–2.5m (5–8ft)	1.5–2.5m (5–8ft)	Deciduous. Leaves slender and 20cm (10in) long	Produced on slende, curved racemes. Each flower violet-coloured and 2cm (¾in) long	USDA 8b/9a. RHS H4.
B. lindleyana 'Little Treasure'	Open and arching dwarf bush	1m (3ft)	0.75m (30in)	Deciduous. Leaves slender and 20cm (10in) long	Produced on slender curved racemes. Each flower violet-coloured and 2cm (¾in) long	

Buddleja lindleyana. A graceful inflorescence, but rather small flowers.

BUXUS

This is a genus of about thirty species found in the Old World, West Indies and Central America. The genus comprises shrubs and small trees all of which are evergreen, have opposite leaves and small apetalous flowers.

Growing Requirements

Hardiness: USDA 6b/7a. RHS H6.
Aspect: A site of any aspect is suitable, as is full sun to part shade.
Shelter: No particular degree of shelter is required.
Sun: Growth may be scorched by strong sun or strong wind when the soil is dry, but growth will soon recover.
Soil: Box will grow well in soil of chalk, clay, loam or sandy texture. The soil should retain adequate moisture in the summer, be high in organic matter and well drained.
pH: A very wide range of soil pH from acid to alkaline conditions is satisfactory.

Garden Design Considerations

For centuries, box along with yew have been the standard hedging plants used for formal gardening and topiary. Box is used in a wide range of garden settings, including patio and container gardens, small city and courtyard gardens, edging, hedging and screening. Yew (*Taxus baccata*) was, and still is, widely used when a dark-green, formal hedge of tall to intermediate height is required. For shorter and dwarf hedging, as frequently found in parterres, box (*Buxus sempervirens*) and its dwarf cultivar, *Buxus sempervirens* 'Suffruticosa', was the turn-to species, being unrivalled in its ability to form dense, regular, small and dwarf hedges.

For topiary, where all manner of regular and irregular shapes can be produced, box and yew are the best species. Yew is favoured for larger topiary, but for smaller and more intricate shapes, box is the best species.

To produce a box hedge is a relatively simple and easy process. Once a suitable space has been selected and prepared, planting is done at a spacing of 30–40cm (14–16in) apart. Clipping of subsequent growth will produce the desired hedge. Many small plants are required to initiate the formation of such a hedge and plants are available in several forms. Individual, container-grown plants will give the quickest result but are the most expensive to purchase. In autumn and early spring, bare root, root balled and rooted cuttings should be available to purchase. These are all cheaper than container-grown plants but may take longer to form the hedge desired.

Pest and Disease Problems with Box

Box Blight

Comparatively recently, box has been assailed by several pests and diseases. Box blight, caused by the fungus *Cylindrocladium buxicola*, and to a less serious extent by the fungus *Volutella* sp., has in many cases devastated established box plantings. The fungus attacks both the leaves and young stems causing brown leaves, brown streaks and dieback on young stems and eventually large, dead, bare patches.

Control is not easy. Initially, great care should be taken to ensure that only healthy plants are planted. Good cultural practice will help keep the plants healthy. If the disease appears, control may only be achieved by employing a variety of fungicides. Frequent applications may be necessary, and control may not be achieved.

Box Tree Caterpillar

The moth responsible is *Cydalima perspectalis*, which appears to have been unfortunately introduced into Britain around 2007 and has since become widespread in London and much of the surrounding area. It is now present in Wales, Scotland and Ireland. Damage is first seen as webbing over the plant growth, plus the leaves being eaten and the bark being stripped of young stems by the caterpillars. Damage is rapid and serious.

Control can be achieved by chemical and biological methods.

Alternatives to Box

These problems with the growing of box have led to a search for an alternative plant to produce these hedges so loved by gardeners worldwide. The Royal Horticultural Society at its garden in Wisley, Surrey, England, is at the time of writing (November 2021) conducting a trial on many plants that may prove a suitable alternative. No plant will produce a hedge exactly similar to that of box, but some may prove a good alternative in some situations.

Listed below are some of the plants included in this trial plus others that may be suitable. Information about these plants will be found under their listing in this book.

- *Berberis darwinii* 'Compacta'
- *Berberis thunbergii* cvs.
- *Euonymus fortunei* and *Euonymus microphyllus* cvs.
- *Corokia* × *virgata* 'Frosted Chocolate'
- *Ilex crenata*
- *Phillyrea angustifolium* f. *rosmarinifolia*
- *Pittosporum tenuifolium* 'Golf Ball'PBR

TABLE 4.10 *BUXUS SEMPERVIRENS* AND CULTIVARS

Name	Habit	Height	Spread	Leaves	Remarks
B. sempervirens (common box)	Dense bushy	4–8m (13–26ft)	2.5–4m (8–13ft)	Dark-green	A variable species. For topiary, a form with small leaves and dense growth should be selected
B. sempervirens 'Elegantissima' AGM	Dense bushy	1–1.5m (3–5ft)	1–1.5m (3–5ft)	Small, undulating, dark-green with irregular cream-coloured margins	Very attractive, making a dense bush. Contrasts well against plain-green species
B. sempervirens 'Suffruticosa' (dwarf box or edging box)	Dwarf, compact	0.5–1m (1ft8in–3ft)	1–1.5m (3–5ft)	Dark-green, 3cm (1in) long	Slow-growing compact form

Box parterre at Bourton House, Gloucestershire, England.

Box topiary garden at Bourton House, Gloucestershire, England. (Unfortunately, since being photographed, this topiary was badly attacked by box blight and has had to be removed and replaced by another garden feature.)

Box topiary of Hens and Chicks at Bourton House, Gloucestershire, England.

CAMELLIA

A genus of about 120 species of evergreen shrubs coming from south and South-East Asia. Leaves are alternate and are usually toothed. Flowers are often large and may be solitary or clustered.

Camellia japonica × *Camellia azalea* Cultivars

This is potentially a new range of camellias derived by crossing hardy species, such as *Camellia japonica*, with the subtropical *Camellia azalea* (syn. *C. changii*). So far only one cultivar is available, *Camellia* 1001 Summer Nights Jasmine ('Yulan01'). This cultivar was short-listed in the RHS Chelsea Flower Show Plant of the Year 2021. As yet it is little tested under UK conditions, but it could represent the beginning of a new range of summer-flowering camellias. The growing conditions above must be regarded as the best estimate of these plants' requirements.

Growing Requirements

Hardiness: Yet to be determined.
Aspect: Avoid a north- or east-facing site.
Shelter: A sheltered site is required.
Sun: Full sun or part shade.
Soil: This can be clay, loam or sandy textured. The soil must be well drained but should retain adequate moisture in the summer.
pH: Soil pH should be acid or neutral. Avoid an alkaline pH; grow in pots or containers in ericaceous compost.

TABLE 4.11 *CAMELLIA* 1001 SUMMER NIGHTS JASMINE ('YULAN01')

Habit	Height	Spread	Foliage	Flower	Flowering period	Remarks
Compact, bushy	2.5m (8ft)	2.5m (8ft)	Glossy and dark-green	Single/semi-double pink, 8–10cm (3–4in) across	Summer and autumn	Unique flowering period

Camellia azalea photographed in Kunming Botanical Gardens, PDR China. It was crossed with *Camellia japonica* to produce Camellia 1001 SUMMER NIGHTS JASMINE ('Yulan01').

TABLE 4.12 CULTIVARS OF *CAMELLIA JAPONICA*

Name	Habit	Height	Spread	Foliage	Flower	Remarks
'Adolphe Audusson' AGM	Fairly upright	4–8m (13–26ft)	4–8m (13–26ft)	Dark-green	Red, semi-double 10–12cm (4–4½in) across with yellow central stamens	Very floriferous
'Alba Simplex'	Upright, bushy	4–8m (13–26ft)	4–8m (13–26ft)	Dark-green	White, single with central yellow stamens	Avoid early morning sun
'Brushfield's Yellow'	Bushy	2.5–4m (8–13ft)	4m (13ft)	Dark-green	White outer petals with pale-yellow central ruff	Avoid early morning sun
'Dr Burnside'	Dense-growing	2.5–4m (8–13ft)	2.5–4m (8–13ft)	Dark-green	Dark-red, semi-double	
'Dr King'	Compact	2m (6ft)	2m (6ft)	Dark-green	Red single with yellow central stamens	
'Kramer's Supreme' AGM	Bushy	2.4–3m (7–10ft)	2.4–3m (7–10ft)	Dark-green	Red, double 12cm (4½in) across	
'Margaret Davis' AGM	Upright	2.5–4m (8–13ft)	2.5–4m (8–13ft)	Dark-green	White with picotee pink edge	Best in full sun for full flower colour development
'Nuccio's Gem' AGM	Bushy	4m (13ft)	4m (13ft)	Dark-green	White, double	Avoid early morning sun
'Nuccio's 'Pearl' AGM	Compact, upright	2.5–4m (8–13ft)	2.5–4m (8–13ft)	Dark-green	Formal double, white flushed pink	Avoid early morning sun
'Silver Anniversary' AGM	Compact	2.5m (8ft)	2.5m (8ft)	Dark-green	White, peony form	Avoid early morning sun

Camellia japonica 'Adolphe Audusson' AGM. Produces large, red, semi-double flowers.

Camellia japonica 'Alba Simplex'. Plant in a position where the blooms will not get damaged by frost followed by early morning sun.

Camellia japonica 'Nuccio's Gem' AGM. Produces large, formal, double-white flowers.

Camellia japonica 'Adolphe Audusson' AGM. This makes a magnificent specimen plant.

Camellia japonica 'Juno'.

Camellia japonica and *Camellia × williamsii* Cultivars

Growing Requirements

Hardiness: *Camellia japonica* USDA 7b/8a. RHS H5. *Camellia × williamsii* USDA 7a/8b. RHS H5.
Aspect: A site facing north, west or east, but not south, is acceptable.
Shelter: A sheltered site is required. Avoid a site that is liable to late spring frosts or cold winds.
Sun: Full shade or part shade is required. Avoid a site that is open to early morning sun. This is likely to cause flower damage following night frost at flowering time.
Soil: The soil must be well drained but retain adequate moisture in the summer. A clay, loam or sandy soil is satisfactory.
pH: The soil must be acid or neutral. Avoid an alkaline soil. Camellias are a little more tolerant of a little lime in the soil than Rhododendrons, but it is much better to avoid this problem. In tubs and containers, grow in an

4.13 CULTIVARS OF *CAMELLIA × WILLIAMSII*

Name	Habit	Height	Spread	Foliage	Flower	Flowering period	Remarks
'Anticipation' AGM	Upright-growing	2.5–4m (8–13ft)	1.5–2.5m (5–8ft)	Dark-green	Deep-pink, peony form, 12cm (4½in) across	Late winter	
'Bow Bells'	Dense-growing	4–8m (13–26ft)	2.5–4m (8–13ft)	Dark-green	Small, single, pink	Mid-winter to mid-spring	Many flowers produced
'Brigadoon'	Compact	2.5–4m (8–13ft)	1.5–2.5m (5–8ft)	Dark-green	Large, 15cm (5¾in) across, semi-double pink	Spring	
'Debbie' AGM	Compact, dense-growing	4m (13ft)	4m (13ft)	Dark-green	Large, pink, peony form	Late spring	
'Donation' AGM	Strong growing	4–8m (13–26ft)	2.5–4m (8–13ft)	Dark-green	Light-pink, semi-double, 12cm (5in) across	Late winter/ early spring	Very floriferous
'E.T.R. Carlyon' AGM	Upright, bushy	2.5–4m (8–13ft)	1.5–2.5m (5–8ft)	Dark-green	White, semi-double, 10cm (4in) across	Spring	

Camellia × williamsii 'Anticipation' AGM.

Camellia × williamsii 'Bow Bells'. Produces many flowers.

Camellia × *williamsii* 'Celebration'. What flowering ability this cultivar has!

Camellia × *williamsii* 'Donation' AGM. Ultimately a large shrub but produces masses of flowers from a young age.

Camellia × *williamsii* 'Jury's Yellow' AGM. Take care to avoid early morning sun damage to frosted flowers.

Camellia × *williamsii* 'Saint Ewe' AGM. This cultivar produces many flowers.

Camellia × *williamsii* 'Brigadoon' AGM. Produces rose-pink, semi-double flowers.

ericaceous compost.

Camellia sasanqua Cultivars

Growing Requirements

Hardiness: *Camellia sasanqua* USDA 8b/9a. RHSH4.
Aspect: A site with a south- or west-facing orientation is preferred.
Shelter: A sheltered site is required. Avoid a site exposed to early morning sun, cold winds or late frosts.
Sun: Full sun or part shade is required. Full shade will result in little flowering.
Soil: This can be clay, loam or sandy textured. The soil must be well drained but should retain adequate moisture in the summer.
pH: Soil pH should be acid or neutral. Avoid an alkaline pH. Grow in pots or containers in ericaceous compost, if necessary.

The cultivars listed in Table 4.14 have been bred to be a little hardier than *Camellia sasanqua* itself.

Camellia transnokoensis AGM

Growing Requirements

Hardiness: *Camellia transnokoensis* USDA 8b/9a. RHS H4.
Aspect: A north- or west-facing site is preferred.
Shelter: A sheltered site is necessary.
Sun: Part shade is required.
Soil: The soil can be clay, loam or sand, but it must be well drained and retain adequate moisture in the summer.
pH: Soil pH should be acid or neutral. Avoid an alkaline pH. Grow in pots or containers in ericaceous compost.

Camellia × *vernalis* 'Yuletide'

Growing Requirements

Hardiness: *Camellia* × *vernalis* USDA 8b/9a. RHS H4.
Aspect: Avoid an east-facing site.
Shelter: Shelter is required. Avoid exposure to cold easterly or northerly winds.
Sun: Part sun. Avoid full shade or full sun.
Soil: The soil can be clay, loam or sand, but it must be well drained and retain adequate moisture in the summer.
pH: Soil pH should be acid or neutral. Avoid an alkaline pH. Grow in pots or containers in ericaceous compost, if necessary.

Camellia sasanqua cultivar flowering in early winter at the Jardin des plants, Paris, France.

Camellia sasanqua 'Crimson King' AGM.

Camellia sasanqua 'Maiden's Blush'.

Camellia sasanqua 'Shiro-wabisuke'.

Camellia Cultivars with Complex or Unknown Parentage

Growing Requirements

Hardiness: USDA 7b/8a or 7a/8b. RHS H5 or H4.

All other growing requirements, as for *Camellia japonica* and *Camellia* × *williamsii* cultivars.

Garden Design Considerations

There are several different groups of camellias and an enormous number of cultivars. This gives the garden planner a great opportunity to select the ideal plant or plants for his or her particular situation. They are all evergreen shrubs, but there is an enormous range in flowering time, flower colour and form available. Constraints are the need for an acid, or at least neutral, soil and for most some shade.

In the woodland or semi-shaded garden, be it large or small, they excel. Using the range of cultivars available, flower colour can be had from autumn, through the winter and on to late spring. If new cultivars, such as *Camellia* 1001 SUMMER NIGHTS JASMINE ('Yulan01'), are included, then flowering can continue into summer and early autumn!

Camellia transnokoensis AGM. From coloured buds, many small flowers are produced, giving a most delicate effect. Ideal for a large container in a sheltered position.

TABLE 4.14 *CAMELLIA SASANQUA* CULTIVARS

Name	Habit	Height	Spread	Foliage	Flower	Flowering period	Remarks
'Cleopatra'	Compact, upright	2.5m (8ft)	1.8m (6ft)	Dark-green and glossy	Semi-double, rose-pink with a yellow centre	Autumn and winter	
'Dwarf Shishi'	Compact, low-growing	0.5–1m (20–40in)	0.5–1m (20–40in)	Dark-green and glossy	Double, dark-pink	Mid to late winter	
'Hugh Evans'	Arching	1.2–1.5m (4–5ft)	1.5–2.5m (5–8ft)	Dark-green and glossy	Single, rose-pink	Late autumn to late winter	
'Jean May' AGM	Compact	1.8–3m (6–10ft)	1.5–2.5m (5–8ft)	Dark-green and glossy	Semi-double, soft-pink, slightly scented	Winter to early spring	Slightly fragrant
'Narumigata' AGM	Upright, vigorous	2.5–4m (8–13ft)	1.5–2.5m (5–8ft)	Dark-green and glossy	Single, white, pink-tinged	Autumn and winter	Slightly fragrant
'Paradise Helen'	Upright and narrow	1.5–1.75m (5–6ft)	1.5–2.5m (5–8ft)	Dark-green and glossy	White, double	Autumn and winter	
'Sekiyo'	Compact	1.6m (5ft)	1–1.5m (3–5ft)	Dark-green and glossy	Semi-double, pink. Flower 12cm (5in) across	Autumn and winter	
'Survivor'	Upright and bushy	1.5–2.5m (5–8ft)	1–1.5m (3–5ft)	Dark-green and glossy	Single white with a yellow centre	Late winter to early spring	Slightly fragrant

TABLE 4.15 *CAMELLIA TRANSNOKOENSIS* AGM

Name	Habit	Height	Spread	Foliage	Flower	Flowering period	Remarks
C. transnokoensis	Upright and elegant	2.5–4m (8–13ft)	2.4–5m (8–13ft)	Small-leaved, dark-green	Single, white, six-petals, scented, 3–4cm (1–1.5in) across	Winter and early spring	Flowers emerge from pink and white buds. USDA 8b/9a. RHS H4

TABLE 4.16 *CAMELLIA* × *VERNALIS* 'YULETIDE'

Name	Habit	Height	Spread	Foliage	Flowers	Flowering period	Remarks
C. × *vernalis* 'Yuletide'	Bushy	3m (10ft)	2.5m (8ft)	Dark-green	Small, single, red with a yellow centre	Early to late winter	USDA 8b/9a. RHS H4

TABLE 4.17 *CAMELLIA* CULTIVARS WITH COMPLEX OR UNKNOWN PARENTAGE

Name	Habit	Height	Spread	Foliage	Flowers	Flowering period	Remarks
'Black Lace' AGM	Compact, upright	1.5–2.5m (5–8ft)	2.5m (8ft)	Dark-green	Dark-red with a sheen, semi-double, 11cm (4¼in) across	Early to late spring	No stamens USDA7b/8a. RHS H5
'Cornish Snow' AGM	Open growing	1.5–2.5m (5–8ft)	1–1.5m (3–5ft)	Narrow, ovate, dark-green	Single, small, white flowers 4–5cm (1½–2in) across	Long season of flowering, starting in early winter. Flowers will withstand light frost	USDA 8b/9a. RHS H4. Masses of flowers produced
'Royalty' AGM	Small and lax growing	0.5–1m (20–40in)	0.5–1m (20–40in)	Dark-green	Very large, bright-red, semi-double	Mid to late spring	USDA7b/8a. RHS H5

When not in flower, they provide a dark-green backdrop upon which other smaller flowering plants may be contrasted.

Camellias are suited to formal and informal situations, being equally at home in formal city or courtyard gardens and informal or cottage gardens. In the formal situation, they can be clipped to a more regular shape, but flowering may be reduced. In the informal setting, the natural, often very attractive, habit and form may be allowed to develop. In mild, sheltered gardens, where cold winds present no problems, they can be employed as effective hedging of either a free-flowering, informal delight or a more restricted and restrained formal hedge. The latter can be seen in some of the famous traditional Japanese gardens of Kyoto.

The need of camellias for some shelter from cold winds and, in many cases, avoidance of early morning sun, does preclude their planting in some situations. In cold gardens, this can be overcome by growing them as wall shrubs and, by careful selection of cultivars plants, they may be grown successfully on north-, south- and west-facing walls.

Where space is limited, the native soil unsuitable or an isolated specimen plant is required, camellias are very suitable for patio, large pot and container-growing. Growth and flowering will be excellent, provided that long-lasting ericaceous compost is used and watering with lime-free water, plus later feeding is regularly attended to. Keeping the plant roots constantly moist during late summer and autumn when flower buds are forming is essential.

If it is necessary to reduce the size of plants, pruning immediately after flowering is successful. Moderate pruning will result in limited subsequent regrowth in the growing season following pruning and flowering the following year. Should plants become very overgrown or poorly shaped, very hard pruning, almost down to the ground, if necessary, can be done after flowering. Providing watering, feeding and mulching are attended to following pruning, then vigorous regrowth will occur. This new growth can then be pruned and trained to create an essentially new young shrub.

Winter-flowering, slightly tender plants, such as *Camellia* × *vernalis* 'Yuletide', *Camellia sasanqua* cultivars and *Camellia transnokoensis*, are exceptionally well-suited to container-growing. The mobility of the container allows them to be given the favoured site they need and deserve.

Camellia 'Cornish Snow'.

Camellia 'Cornish Snow' is also rather tender, but such a desirable plant that it is deserving of special treatment on a sheltered wall or well-placed container.

CARYOPTERIS

A genus of about seven species consisting of deciduous shrubs, sub-shrubs and herbaceous plants. They have opposite, aromatic leaves.

In horticulture, apart from *Caryopteris incana*, all the other species have been replaced by the hybrid *Caryopteris* × *clandonensis* (*C. incana* × *C. mongholica*) an accidental cross first made by Arthur Simmonds at Clandon in Surrey, UK, and the selection *Caryopteris* × *clandonensis* 'Arthur Simmonds' AGM named in commemoration. From this cross, numerous selections have been made, differing mainly in their shade of violet-blue flowers, although variegated forms have also appeared and the most popular of these are listed in Table 4.18.

Caryopteris × *clandonensis* 'Arthur Simmonds'

A deciduous, small, bushy shrub.
Height: 1–1.5m (39–59in).
Spread: 1–1.5m (39–59in).
Leaves: 2–5cm (1–2in) long in opposite pairs, dull-green upper surface and greyish beneath.
Flowers: Delicate cymes of violet-blue flowers produced in late summer and early autumn.
Pruning: After flowering, this and all cultivars die back during the winter and require pruning in early spring to where the stems are still alive. This point will be obvious by the appearance of small, grey buds from the still-living growth. From these pruned branches, rapid new growth will be made followed by flowers later in the season.

Once established, the plant is drought-tolerant and has been used in 'xeriscaping' in the USA – a form of landscaping or gardening that eliminates, or at least reduces, the need for irrigation.

Growing Requirements

Hardiness: USDA 8b/9a. RHS H4.
Aspect and shelter: A sheltered, south- or west-facing site is required for best results.
Sun: Full sun.
Soil: Soil of any texture is acceptable, but it must be well drained – poor drainage is not tolerated. The plant prefers a light, well-drained soil and grows particularly well on thin, chalky soils and light, sandy soils.
pH: Wide pH range is accepted.

TABLE 4.18 CULTIVARS OF *CARYOPTERIS* × *CLANDONENSIS* (BLUEBEARD)

Name	Habit	Height	Spread	Leaves	Flowers	Remarks
'Arthur Simmonds'	Bushy	0.5–1m (20–39in)	1–1.5m (39–59in)	Dull-green upper surface, greyish beneath	Violet-blue	The first selection to be named
'Dark Knight'	Upright	1.2m (47in)	1m (39in)	Silvery-green upper surface	Deep-purple blue	
'First Choice' AGM	Compact	0.5–1m (20–39in)	1–1.5m (39–59in)	Grey-green	Deep-blue, large	
GRAND BLEU ('Inoveris'PBR)	Compact, bushy	0.5–1m (20–39in)	1–1.5m (39–59in)	Dark-green	Intense deep-blue	
'Heavenly Blue'	More compact than 'Arthur Simmons'	0.5–1m (20–39in)	1–1.5m (39–59cm)	Grey-green	Deeper blue than 'Arthur Simmonds'	
HINT OF GOLD ('Lissaura'PBR) AGM	Upright, bushy	0.9m (35in)	0.9m (59in)	Gold to yellow-green	Cobalt-blue	
'Kew Blue'		0.5–1m (20–39in)	1–1.5m (39–59in)	Grey-green	Slightly deeper blue than 'Arthur Simmonds'	
STEPHI ('Lissteph'PBR)		0.5–1m (20–39in)	1–1.5m (39–59in)	Grey-green	Pale-pink	
STERLING SILVER 'Lissilv'PBR		0.5–1m (20–39in)	1–1.5m (39–59in)	Silvery-blue	Deep violet-blue	
'Summer Sorbet'PBR AGM	Bushy	0.5–1m (20–39in)	0.5–1.5m (20–59in)	Variegated with a yellow margin	Light-blue	
'Worcester Gold' AGM	Bushy	0.5–0.9m (20–35in)	1m (39in)	Golden-yellow foliage	Violet-blue	Full sun required for best foliage colour and flowering

Caryopteris × *clandonensis* 'Dark Knight'. A sun-loving shrub that flowers from late summer to autumn.

Caryopteris × *clandonensis* 'Dark Knight'. Showing the deep violet-blue of this cultivar's flowers.

Caryopteris × *clandonensis* 'Worcester Gold'. To obtain strong yellow foliage to contrast with the flowers, a sunny position is essential.

Caryopteris incana

A small, deciduous shrub.
Height: 1–1.5m (39–59in).
Spread: 1–1.5m (39–59in).
Leaves: Leaves in opposite pairs covered in a greyish pubescence.
Flowers: Dense, rounded cymes of violet-blue flowers produced from the upper leaf axils in late summer and early autumn.

Growing Requirements

Hardiness. USDA 8b/9a. RHS H4.
All other requirements the same as for *Caryopteris* × *clandonensis* and its cultivars.

Garden Design Considerations

A group of small, deciduous shrubs whose main garden attribute is the provision of late summer and autumn flower colour. Those with yellow or otherwise variegated foliage provide good colour contrasts between their flower and foliage, but need full sun to develop and display this effectively.

If full sun and good drainage can be provided, these are very suitable for inclusion in cottage or informal garden borders, sunny and sheltered coastal gardens and Mediterranean-style gardens.

Their resistance to drought, once established, makes them very useful in situations where this is likely to be experienced, e.g. sunny banks and slopes and on soils with poor moisture-retention, such as thin, chalky and sandy soils.

The small and compact nature of *Caryopteris* makes it very suitable for inclusion in plantings for sunny patios, raised beds and planters.

TABLE 4.19 CULTIVARS OF *CARYOPTERIS INCANA*

Name	Habit	Height	Spread	Leaves	Flowers
'Blue Cascade'	Weeping branches	1–1.5m (39–59in)	1–1.5m (39–59in)	Greyish-green	Violet-blue
'Jason'[PBR] (syn. SUNSHINE BLUE)	Upright and bushy	1–1.5m (39–59in)	1–1.5m (39–59in)	Bright-yellow becoming green-yellow later in season	Pale mid-blue

Caryopteris incana 'Blue Cascade'. Requiring the same growing conditions and flowering at the same time as *Caryopteris* × *clandonensis* but having a weeping habit.

CERATOSTIGMA

A genus containing herbaceous plants and small shrubs that originate from warm temperate and tropical regions, which accounts for the fact that it requires a warm and sunny position to grow well.

Growing Requirements

Hardiness: USDA 8b/9a. RHS H4.
Aspect and shelter: All require an east-, west- or south-facing, sheltered position to grow and flower well.
Sun: Full sun.
Soil: The soil may be of any type but must be well drained and retain some moisture in the summer.
pH: A wide pH range is tolerated.
Pruning: Pruning will be required in early to mid-spring. *Ceratostigma plumbaginiodes* is herbaceous and should be cut down to approximately 2.5cm (1in) above ground level, where small, red, new shoots will be apparent. *Ceratostigma willmottianum* and its cultivars, plus *Ceratostigma griffithii*, will die back during the winter. The degree to which they die will depend on the severity of the winter, but pruning should include pruning back to where shoots are still alive, plus a little extra shortening to ensure subsequent bushy regrowth.

Garden Design Considerations

Ceratostigma plants are great in providing late summer and autumn colour in the garden. Their small size makes them well suited for inclusion in a shrubbery, mixed border or herbaceous planting. The short stature of *Ceratostigma plumbaginiodes* makes it eminently suitable for the front of any border, linking well between a lawn and taller plants behind. A particularly pleasing combination of flower colour in the autumn can be achieved between the blue flowers of this plant and the the pink chalices of *Colchicum speciosum*, the large corms of which can be planted to come up between the short stems of the herbaceous perennial.

The deep-blue flowers of this genus are a particularly uncommon and a valuable colour in the garden. The contrast between this blue and the yellow leaves of some cultivars is potentially very attractive and it is to be hoped that new cultivars with this yellow foliage, but with more vigour, may soon be released. Those currently available may find a use in raised planters and containers, where they may not be swamped by stronger growing, adjacent plants, and more individual attention can be given to each plant.

Their flower colour is sufficiently dark to stand out in strong sunshine, but this requirement for both sun and shelter does preclude their use in shaded or exposed situations.

Their requirement for annual pruning means that they are not well-suited to very low-maintenance plantings.

TABLE 4.20 SPECIES AND CULTIVARS OF *CERATOSTIGMA*

Name	Growth form	Height	Spread	Leaves	Flowers	Remarks
C. griffithii (Burmese plumbago)	Semi-evergreen shrub	0.5–1m (1½–3ft)	0.5–1cm (1½–3ft)	Simple with a rounded tip, bristly with edge tinged red by the hairs	Blue. Late summer and autumn	
C. plumbaginiodes AGM (syn. *Plumbago larpentiae* and *C. larpentiae*)	Herbaceous perennial	0.1–0.5m (¼–1½ft)	0.1–0.5m (¼–1½ft)	Oval. Turning red-purple in autumn	Blue. Late summer and autumn	Included here as it can fulfil the function of a small shrub
C. willmottianum AGM (hardy plumbago and Chinese plumbago)	Deciduous shrub	0.5–1m (1½–3ft)	0.5–1m (1½–3ft)	Lanceolate. May colour well in autumn	Blue. Late summer and autumn	
C. willmottianum 'Forest Blue' AGM	Deciduous shrub	0.5–1m (1½–3ft)	0.5–1m (1½–3ft)	Lanceolate. May colour well in autumn	Blue. Late summer and autumn	More compact in growth and more floriferous than the straight species
C. willmottianum DESERT SKIES ('Palmgold'PBR)	Deciduous shrub	0.5–1m (1½–3ft)	0.5–1m (1½–3ft)	Lanceolate. Yellow	Blue. Late summer and autumn	Strong colour contrast between flowers and foliage. Found to be lacking in vigour
C. willmottianum SAPPHIRE RING ('Lissbrill'PBR)	Deciduous shrub	0.5–1m (1½–3ft)	0.5–1m (1½–3ft)	Lanceolate. Yellow	Blue. Late summer and autumn	Strong colour contrast between foliage and flowers. Too new to be able to fully appraise

Ceratostigma griffithii produces clear blue flowers followed by autumn leaf colour that often persists into the winter.

Ceratostigma plumbaginiodes flowering and flowing over the edge of a border in late summer.

Ceratostigma plumbaginiodes in autumn leaf colour.

Ceratostigma willmottianum is the taller plant in background and *Ceratostigma plumbaginiodes* the shorter plant in the foreground.

The mass of flowers produced by *Ceratostigma willmottianum* in late summer and into the autumn.

CERCIS

A genus coming from China, North America and the eastern Mediterranean that is easily distinguished by rounded leaves and pea-like flowers. The hardy members are all deciduous shrubs or small trees and the three species considered here are *Cercis canadensis*, *Cercis chinensis* and *Cercis silquastrum*.

Flowering occurs in late spring before the leaves emerge and is unusual as it exhibits cauliflory, meaning flowers are produced normally from buds, but also directly out of the bark of branches and the trunk. In recent years, *Cercis canadensis* has been the subject of much breeding and selective work; several new cultivars have been released that display new leaf coloration and/or altered tree habit.

These three species and their cultivars can be grown as shrubs, their size being controlled by pruning after flowering in summer. Young plants can also be restricted to one stem, which is then trained upwards to form a small tree.

Growing Requirements

Hardiness USDA 7a/8a. RHS H5.
Aspect: A site of any orientation is acceptable and may be exposed or sheltered.
Shelter and sun: Plants will grow in part shade or full sun, but full sun is recommended to encourage flowering and to fully harden the wood prior to winter. All three species are susceptible to the fungal pathogen coral spot (*Nectria cinnarbarina*), but wood thoroughly ripened before winter is less likely to suffer winter damage, which can predispose it to attack.
Soil: The soil can be of any texture but must be well drained. *Cercis* does have some resistance to drought and a soil that tends to dry out in summer may be tolerated.
pH: A wide range of soil pH is acceptable.

Cercis siliquastrum (Judas Tree)

A large, deciduous shrub or small tree. The name Judas tree comes from the legend that Judas Iscariot is said to have hanged himself from this tree. Another idea is from the alternative common name of Judea's tree, Judea is the mountainous part of the region of Palestine and is a location where this species is commonly found.
Height: 8–12m (26–39ft).
Spread: 8m+ (26ft+).
Leaves: Alternate, rounded, tip obtuse, base cordate, pale-green, providing light shade. Autumn leaf colour is yellow.
Flowers: Flowering occurs in late spring to early summer, the pea-like flowers being produced profusely before the leaves expand. Flower colour is a shade of pink, slightly variable when grown from seed.
Seeds: Seed pods, technically each a siliqua, the name for a long, three-walled pod, are produced; these are 12cm (5in) long, purple and persist until the following spring.
Varieties: *Cercis siliquastrum* 'Bodnant' AGM: a selection producing profuse deep-purple flowers.

Cercis siliquastrum 'White Swan': a selection producing white flowers.

Cercis chinensis

A large, deciduous shrub or small tree. Unfortunately, this species is rather susceptible to cold and requires a warm site.
Height: 6m (20ft).
Leaves: The leaves are very similar to those of *Cercis siliquastrum*.
Flowers: Profuse, bright-pink flowers are produced in early summer. The author has seen this species in Beijing Botanical Gardens, and it did produce an almost incredible number of flowers but note should be taken that there it does benefit from a hot, dry, late summer.
Varieties: *Cercis chinensis* 'Avondale' AGM. This is a more compact cultivar, height 3m (10ft). In early summer, deep-purple buds produce a profusion of pink flowers.

Cercis canadensis (North American Redbud)

A deciduous shrub or small tree.
Height: 5m (16ft).
Leaves: Alternate, rounded, 7–12cm (2¾–4¾in) long and wide, thinner and lighter green with an acuminate tip rather than obtuse as with *C. siliquastrum*.
Flowers: Flowering occurs in early summer before the leaves emerge. Unfortunately, these pink pea-like flowers are not freely produced in Great Britain.

Garden Design Considerations

A genus of plants greatly appreciated by modern garden designers and gardeners.

Cercis siliquastrum (the Judas tree) is well known and widely used to provide a mass of late spring flowers to continue the explosion of garden colour from spring into summer. As a single, large shrub, it can make a feature in a lawn or deep border. As a tree, it has a light canopy allowing dappled shade below and, as well as delivering flower colour, can also function as a shade tree for a sunny patio.

The slightly smaller *Cercis chinensis* has the same range of possible uses but needs a warmer, sunnier position.

The range of *Cercis canadensis* cultivars is still expanding and offers new and exciting opportunities for the gardener. Tree size varies with cultivar from small to very small and habit from open and wide-ranging to steeply pendulous. Unfortunately, the species and its cultivars do not flower as

Cercis siliquastrum having grown into a very colourful, wide-branching, small tree.

Cercis siliquastrum. Showing the characteristic pea-like flowers.

Cercis chinensis 'Avondale'. Grown here against a warm wall, the picture shows the very profuse flowering of this cultivar when given a favoured site.

Cercis canadensis 'Alley Cat'. An arresting, small, slow-growing, semi-pendulous, small tree, well placed against a dark background.

Cercis canadensis 'Alley Kat'. Here, the unusual variegation consisting of white, light-green and dark-green patches on the leaf can be clearly seen.

Cercis canadensis 'Forest Pansy'. Here, kept as an isolated specimen shrub by hard annual pruning in spring. This results in vigorous regrowth with large leaves and strong leaf colour.

Cercis canadensis 'Forest Pansy'. Leaves photographed in late summer showing full, dark, late-growing season pigmentation.

Early autumn leaf colour of *Cercis canadensis* 'Forest Pansy'.

Before leaf fall, the autumn leaf colour of *Cercis canadensis* 'Forest Pansy' may become lighter and the leaves semi-transparent.

Cercis canadensis 'Ruby Falls^{PBR} AGM'. The pendulous nature of this cultivar is apparent, and the dark foliage contrasts against the light-coloured, ground-cover planting.

Cercis canadensis 'Ruby Falls^{PBR} AGM'. Here, the foliage of this cultivar can be seen to be slightly smaller and more rounded that that of *C. canadensis* 'Forest Pansy'.

Cercis ETERNAL FLAME ('Nc2016-2'). The orange-red and orange leaves of this exciting new cultivar. An excellent candidate for some novel leaf-colour combinations with other coloured leaved shrubs.

TABLE 4.21 CULTIVARS OF *CERCIS CANADENSIS*

Name	Habit	Height	Width	Leaves	Remarks
C. canadensis 'Alley Cat'	Semi-weeping, slow-growing	5m (16ft)	8m (26ft)	Leaves irregularly shaped and undulating. Heavily variegated with patches of white, light- and dark-green	Full sun required. Small clusters of deep-pink flowers
C. Canadensis 'Forest Pansy' AGM	Open and ranging	4–8m (13–26ft)	4–8m (13–26ft)	Young leaves dark reddish-brown and shiny. Adult leaves deep-purple, becoming greener with age or shade	Branches thin with near black bark. Cluster of light-pink flowers. Autumn leaves lighten in colour, becoming orange-yellow and translucent
C. canadensis LAVENDER TWIST ('Covey')	Weeping, slow-growing with slightly contorted growth	1.5–2.5m (5–8ft)	1–1.5m (3–4ft)	Green, with good yellow autumn colour	Large clusters of lavender flowers
C. canadensis 'Ruby Falls'PBR AGM	Pendulous	1.25–2.5m (4–8ft)	1.25–2.5m (4–8ft)	Approximately 10cm (4in) long, slightly smaller but otherwise similar to 'Forest Pansy'	Lavender-pink flowers
C. canadensis ETERNAL FLAME ('Nc2016-2')	Open branching	3–4m (10–13ft)		Bright reddish-brown when young, changing to pale yellowish-red as they mature	Leaf coloration very striking. RHS Chelsea Flower Show Plant of the Year 2021*

* Developed as part of a long-term breeding programme at North Carolina State University in 2010.

freely as they do in North America, but the range of habit and foliage colour fully compensates for this. These small trees or multi-stemmed shrubs are excellent for patio and large container-growing, allowing close inspection and appreciation of the flowers and foliage. If necessary, these trees and shrubs can be reduced in size and kept in check by summer pruning after flowering. Such pruning should not result in excessive subsequent regrowth that growing season and will allow pruning cuts to heal before winter. In large beds or borders, they can be used as a series of upright features with a subsidiary underplanting of uniform height. The colours of their foliage are very vibrant and cry out to be used in colour associations with other shrubs with strongly coloured foliage. Used collectively, if space permitted, a virtual kaleidoscope of colours could be created that changed from spring to summer and into autumn.

A particular planting took the writer's interest and is recorded here so that others may experiment with it. Plants of *Cercis canadensis* 'Forest Pansy' were trained as informal cordons up and over a pergola. The dark-purple leaves made an attractive feature and provided welcome dappled shade in summer. In the autumn, the leaves cleared and bright shafts of direct sunlight shone through these transparent leaves giving the same wonderful effect as light shining through stained glass – magical and memorable.

CHAENOMELES

This genus consists of three species of deciduous spiny shrubs native to Eastern Asia. The leaves are alternately arranged. The flowers opening in late winter or early spring are 3–4.5cm (1–1¾in) in diameter and have five petals. The fruit is a pome, which, when ripe in the autumn, has a characteristic aroma. Common names that are frequently used are ornamental quince, flowering quince, Japanese quince and japonica.

Growing Requirements

Hardiness: USDA7b/8a. RHS 6.
Aspect: A planting position facing any direction is satisfactory.
Shelter: There is no specific requirement for shelter.
Sun: Full sun or part shade.
Soil: The soil can be of any texture but must be well drained.
pH: Acid, neutral and alkaline soils are accepted.

Garden Design Considerations

A cottage gardener's traditional favourite providing flowers very early in the year. Ornamental quince flower at the same time as spring bulbs and produce the bright colour combinations so welcome at the end of a long winter. They perform well at the base of walls, even north-facing walls, their tough nature and compact habit making them well suited to this position. The spiny and dense growth habit means that they can be used to form a low hedge; this, whilst looking attractive, is also an effective barrier to people.

Chaenomeles speciosa 'Nivalis'. The white flowers of this cultivar look very attractive on a dark-coloured wall or fence.

TABLE 4.22 *CHAENOMELES* SPECIES AND CULTIVARS

Name	Habit	Height	Spread	Foliage	Flower	Fruit
C. japonica (Japanese quince)	Spiny, dense bush	1.5–2.5m (5–8ft)	2.5–4m (8–13ft)	Glossy, dark-green, oval	Red, 4cm (1½in) diameter	Yellow/green, aromatic
C. japonica 'Orange Beauty'	Spiny, dense bush	1–1.5m (3–5ft)	1–1.5m (3–5ft)	Glossy, dark-green oval	Reddish-orange	Yellow, aromatic
C. japonica MADAME BUTTERFLY ('Whitice')	Dense bush	1–1.5m (3–5ft)	1–1.5m (3–5ft)	Glossy, dark-green oval	Coral-pink, with an irregular white edge	Golden-yellow aromatic
C. speciosa 'Nivalis'	Spiny, spreading bush	2.5m (8ft)	5m (15ft)	Glossy, dark-green oval	White	Green/yellow aromatic
C. speciosa 'Scarlet Storm'PBR	Dense-growing and non-thorny	1.2m (4ft)	2m (6½ft)	Glossy, dark-green	Red, semi-double	
C. × *superba* 'Crimson and Gold' AGM	Dense-spreading shrub	1–1.5m (3–5ft)	1.5–2.5m (5–8ft)	Glossy, dark-green, obovate	Crimson, 3.5cm diameter. Centre of golden anthers	Yellow, aromatic
C. × *superba* 'Pink Lady' AGM	Spreading shrub	2m (6½ft)	1.5m (5ft)	Glossy, dark-green, oval,	Early flowering. Rose-pink	Green, aromatic
C. × *superba* 'Rowallane' AGM	Small spreading shrub	0.1–1m (6–12in)	1–1.5m (3–5ft)	Glossy, dark-green, obovate	Crimson 4cm diameter	Yellow, aromatic

Chaenomeles × *superba* 'Crimson and Gold'. Here, well used as early spring-flowering ground-cover.

Chaenomeles × *superba* 'Crimson and Gold'. A winning combination of scarlet petals and a golden centre to each flower.

Chaenomeles × *superba* 'Crimson and Gold'. The fruit look attractive and can also be picked and brought inside where their aroma is more fully appreciated.

Chaenomeles × *superba* 'Rowallane'. Here, skillfully trained and pruned to form a wall shrub at Killerton, a National Trust property in Devon, England.

Chaenomeles × *superba* 'Rowallane'.

Chaenomeles × *superba* 'Pink Lady'. This cultivar is always very welcome as the first ornamental quince to flower in late winter.

CHOISYA

A genus of small- to medium-sized, evergreen, aromatic shrubs native to Mexico and south-western USA. Given their origin, these shrubs are surprisingly hardy. The flowers are sweetly scented leading to the common name of Mexican orange blossom for *Choisya ternata* and the leaves emit a strong smell when crushed, which some like and others find too strong.

Recent years have seen a flurry of plant breeding within this genus, producing a range of small, compact, evergreen shrubs that are ideal for today's garden, where space is so often at a premium.

Growing Requirements

These are easy to grow shrubs.
Hardiness: All those mentioned are USDA 8b/9a. RHS H4.
Aspect and shelter: Any orientation is acceptable, but some shelter must be provided, and exposed or cold windy sites avoided. Mild winds are tolerated, and they make good coastal plants where low temperatures are not experienced. Cold winds, and especially cold, desiccating winds, can cause shoot death in winter. Less severe, but still chilling winds in winter can cause leaf scorch and leaf loss. Yellow-leaved forms may be damaged or suffer leaf scorch from late spring frosts, especially if this is followed by early morning sunshine. Polluted atmospheres and dry conditions are quite well-tolerated.
Sun: They thrive in full sun and will tolerate light shade. Yellow-leaved forms are less tolerant of any shade and may turn pale-green, even in light shade.
Soil and pH: The soil must be well drained but, providing this requirement is met, a wide range of soil type and soil pH is perfectly acceptable.
Pruning: Any dead growth and winter damage are best removed immediately after flowering in late spring or early summer. Where plants have become too open or leggy, growth may be reduced by up to half at the same time and rapid new growth will be made.

Species and Cultivars of *Choisya*

In general, cultivars of the hybrid group *Choisya × dewitteana* have more finely cut foliage and are smaller in height and width than *Choisya ternata* and its cultivars. The flowers of all plants in Table 4.23 are sweetly scented.

Choisya ternata. A mainstay evergreen shrub for gardeners that is also generous with its flowers.

Choisya ternata SUNDANCE ('Lich') AGM. Provides bright-yellow leaf colour all year round.

Choisya × dewitteana 'Aztec Pearl'. A great combination of fine leaf texture with profuse flowering.

Choisya × dewitteana WHITE DAZZLER ('Londaz'[PBR]). Very compact and very free-flowering, great for pots.

Garden Design Considerations

Choisya ternata (the Mexican orange blossom) has been a mainstay shrub of gardeners and landscapers for a long time. It is very useful in obtaining a balance between evergreen and deciduous plants in a design and giving structure to, plus furnishing, the garden in winter. The plant is also generous with its flowers: its main flush is in late spring, but flowers will also appear in the autumn and at other odd times.

Choisya ternata SUNDANCE ('Lich') AGM' represented a break in evergreen shrub foliage colour, its yellow foliage being a wake-up call to the gardening world. It is still very popular, providing bright foliage colour all year round. It is slightly more tender that *Choisya ternata* and should be kept away from cold, drying winds.

From *C.* SUNDANCE there have been developments leading to finer leaf texture, increase in flowering and reduction in overall size. The smaller, more compact cultivars are ideal for large pots and patio gardens, promising foliage colour, finer leaf texture and increased flowering. The choice is yours!

CISTUS

The rock roses. This is a genus of about twenty species native to the Mediterranean region and the Middle East. It consists of evergreen shrubs with opposite leaves. The flowers are single, rose-like with five tissue-like petals and numerous central stamens. Each flower only lasts a day at most, but they are quickly replaced in a succession giving a continuous display in early summer.

The species hybridise readily and have given rise to many hybrids. In Table 4.24, an effort has been made to be selective and only those considered to be most garden-worthy are included.

Growing Requirements

Hardiness. USDA9b/10a. RHS H4.
Aspect: The planting site should face south or west.
Shelter: Shelter from cold, drying winds is required. Frost pockets must be avoided.
Sun: Full sun is definitely preferred. Slight shade may be accepted, but growth may be leggy and flowering reduced.
Soil: Soil can be of any texture but must be well-drained. Highly fertile, heavy, wet and poorly drained soils must be avoided. Poor, stony, well-drained soils are favoured.

TABLE 4.23 SPECIES AND CULTIVARS OF *CHOISYA*

Name	Height	Spread	Leaf shape	Leaf colour	Flower timing and colour	Remarks
C. ternata (Mexican orange blossom)	1.5–2.5m (5–8ft)	1.5–2.5m (5–8ft}	Trifoliate	Dark-green	Spring, early summer and frequently in autumn. White	The hardiest and most tolerant of this selection
C. ternata SUNDANCE ('Lich') AGM	1.5–2.5m (5–8ft)	1.5–2.5m (5–8ft)	Trifoliate	Bright-yellow	Spring and early summer. White	Foliage greens in shade
C. × *dewitteana* APPLE BLOSSOM ('Pmoore09'PBR)	1.2m (4ft)	1.2m (4ft)	3–5, slender leaflets	Green	Spring. Pink in bud, opening to white with pink on reverse	
C. × *dewitteana* 'Aztec Gold'PBR	1–1.5m (3–5ft)	1–1.5m (3–5ft)	3–5 slender leaflets	Bright-yellow	Spring and autumn. White	Best leaf colour in full sun
C. × *dewitteana* 'Aztec Pearl' AGM	1.5–2.5m (3–8ft)	1.5–2.5m (3–8ft)	3–5 slender leaflets	Green	Spring and autumn. White	
C. × *dewitteana* GOLDFINGERS ('Limo'PBR)	1.5–2.5m (3–5ft)	1.5–2.5m (3–5ft)	3–5 slender leaflets	Bright-yellow	Spring and autumn. White	Best leaf colour in full sun
C. × *dewitteana* SNOW FLURRIES ('Lisflurry')PBR	1.5–2.5m (3–5ft)	1.5–2.5m (3–5ft)	5 narrow finger-like leaflets	Green	Spring and autumn. White	
C. × *dewitteana* WHITE DAZZLER ('Londaz'PBR) AGM	0.5–1m (20–40in)	0.5–1m (20–40in)	5–7 narrow leaflets	Green	Spring and autumn. White	A free-flowering compact cultivar
C. GREENFINGERS ('Lissfing')	1.25–1.5m (4–5ft)	1–1.25m 4–5ft)	5–7 narrow leaflets	Green	Spring and autumn. White	Dense and compact growth
C. ROYAL LACE ('Pmoore06'PBR)	0.75m (30in)	0.75m (30in)	5–7 narrow leaflets	Lime-green to yellow	Spring and autumn. White	Growth denser and more compact than *C.* 'Aztec Gold'PBR

pH: A wide range of soil pH from acid to alkaline is accepted but established plants may eventually become chlorotic on very limey soils.

Garden Design Considerations

Cistuses are well placed in Mediterranean, Mediterranean-style, gravel and coastal gardens, where they experience full sun, good drainage, warm conditions and an absence of cold, drying winds. Here they provide effective ground-cover and bright summer flower colour. The release of essential oils from the leaves and scent from the flowers of rock roses in summer is strongly evocative of Mediterranean and seaside gardens.

Inland they can still provide effective evergreen ground-cover and summer flowers, but the site must be open and sunny, well drained and sheltered. If cold temperatures should occur during winter, then the casualty rate can be high. They will slowly recover from winter cold damage if this is not too severe but react badly to hard pruning. They are low maintenance and suitable for ground cover on dry, sunny banks and slopes, where they will tolerate well dry soil conditions.

Smaller species and cultivars such as *Cistus* × *laxus* 'Snow White' and *C.* × *dansereaui* 'Decumbens' AGM are suitable for inclusion in larger, sunny, well-drained rock gardens, raised beds and smaller gardens generally.

Cistus × *argenteus* 'Silver Pink' with *Rosa* × *odorata* 'Mutabilis'. A good combination, both in flowering time and flower colour.

Cistus × *argenteus* 'Silver Pink' is excellent for ground cover in sunny, well-drained positions, especially by the seaside.

Cistus × *purpureus*. A larger sun rose that requires a warm, sheltered spot, good drainage and shelter from cold winds.

Cistus × *purpureus*. The flowers may be tissue-like and individually only last a day, but boy do they deliver colour!

Whilst their summer flowers are bright and cheerful, their flowering season is not very long, and their appearance can be rather dull the rest of the year. They are perhaps better mixed with other sun-loving but deciduous shrubs to give a lighter more varied effect when not in flower.

TABLE 4.24 *CISTUS* SPECIES AND HYBRIDS

Name	Habit	Height	Spread	Foliage	Flowers	Remarks
C. × *argenteus* 'Peggy Sammons'	Bushy shrub	1m (3ft)	1m (3ft)	Grey-green, ovate and downy	Rose-pink with a yellow centre of stamens, 5cm (2in) across	After establishment may eventually become chlorotic on chalky soils
C. × *argenteus* 'Silver Pink'	Mounded, bushy shrub	0.9m (2¾ft)	0.9m (2¾ft)	Grey-green	Silvery pink with a golden centre of stamens	
C. × *cyprius* AGM (common gum cistus)	Bushy shrub	1m (3ft)	1m (3ft)	Dark-green, shiny lanceolate leaves. Grey on underside	White, with dark-red blotch at base of each petal	
C. × *cyprius* var. *ellipticus* 'Elma' AGM	Bushy shrub	1–1.5m (3–5ft)	1–1.5m (3–5ft)	Dark-green, shiny lanceolate leaves	White, unblotched, 10cm (4in) across	After establishment may eventually become chlorotic on chalky soils
C. × *dansereaui* 'Decumbens' AGM	Low-growing and spreading shrub	0.5–1m (1¾–3ft)	1–1.5m (3–5ft)	Narrow leaves which are sticky to the touch	White with a crimson blotch at base of petals, 6cm (2¼in) across	
C. × *dansereaui* 'Jenkyn Place' AGM	Bushy shrub	1–1.5m (3–5ft)	1–1.5m (3–5ft)	Dark-green and narrow	White with a maroon blotch at base of petals, 7cm (2¾in) across	
C. × *hybridus* (syn. *corbariensis*)	Small, rounded shrub	0.9m (2¾ft)	0.9m (2¾ft)	Leaves with a wavy margin	White with a yellow centre of stamens	After establishment may eventually become chlorotic on chalky soils
C. ladanifer (common gum cistus)	Loose, open habit	1.5–2.5m (5–8ft)	1–1.5m (3–5ft)	Dark-green upper surface, underside grey. Leaf sticky to the touch and narrow	White with dark-red blotch at base of petals, 10cm (4in) across	After establishment may eventually become chlorotic on chalky soils
C. laurifolius (laurel-leaf cistus)	Open habit	1.5–2.5m (5–8ft)	1.5–2.5m (5–8ft)	Dark-green, sticky to the touch and ovate	White with a yellow centre of stamens, 8cm (3in) across	
C. × *laxus* 'Snow White' AGM	Bushy shrub	0.5–1m (1¾–3ft)	0.5–1m (1¾–3ft)	Green leaves sticky to the touch	White with a yellow centre of anthers	After establishment may eventually become chlorotic on chalky soils
C. × *lenis* 'Grayswood Pink' AGM	Spreading shrub	0.5–1m (1¾–3ft)	1–1.5m (3–5ft)	Grey-green, ovate	Light-pink with a yellow centre of anthers 2.5 cm across	After establishment may eventually become chlorotic on chalky soils
C. pulverulentus 'Sunset' AGM	Low-spreading shrub	0.1–0.5m (4–20in)	0.5–1m (20–39in)	Grey-green and wavy	Rose-pink with a yellow centre of stamens	After establishment may eventually become chlorotic on chalky soils
C. × *purpureus* AGM	Bushy shrub	0.5–1m (1¾–3ft)	0.5–1m (1¾–3ft)	Dull green, narrow and slightly wavy edged	Purple-pink with a deep red blotch at the base of each petal, 8cm (3in) across	
C. 'Snow Fire' AGM	Dense, rounded shrub	0.5–1m (1¾–3ft)	1–1.5m (3–5ft)	Lanceolate	White with a maroon blotch at the base of each petal, 6cm (2¼in) across	After establishment may eventually become chlorotic on chalky soils

CLERODENDRUM

A genus of mainly tropical trees, shrubs and climbers that is frequently used for ornamental purposes in tropical gardens and in heated glasshouses and conservatories in colder climbers.

Three species are hardy – one is recommended here.

Growing Requirements

Hardiness: USDA 7b/8a. RHS H5.
Aspect and shelter: Requires both a sheltered, preferably south-facing position.
Sun: Full sun or part shade.
Soil: Soil can be any texture, but well drained and preferably high organic content.
pH: Tolerant of a wide pH range.

Clerodendrum trichotomum var. *fargesii* AGM (Harlequin Glory Bower)
A vigorous, deciduous, large shrub from western China. When constrained to a single stem, it forms a very attractive, small tree.
Height: 5–6m (16–20ft).
Spread: 5–6m (16–20ft).
Leaves: Large, opposite, ovate leaves 10–22cm (4–9in) long and 5–12cm (2–5in) wide emerge in late spring.
Flowers: From pink buds, very fragrant flowers appear in late summer. The flower has a

Clerodendrum trichotomum var. *fargesii* can be trained to make a fine, small tree.

Clerodendrum trichotomum var. *fargesii*. The late summer, scented flowers are followed by bright-blue berries, each backed by a persistent dark-red calyx.

Clerodendrum trichotomum var. *fargesii* 'Carnival' AGM. A shrub to provide a summer garden with summer leaf colour and late-summer, scented flowers, followed by unusually coloured fruits.

Clerodendrum trichotomum var. *fargesii* 'Carnival' AGM. A seemingly subtropical mix of colour and scent.

white corolla, green sepals and a red calyx.
Fruit: This is followed by a vivid metallic-blue berry, forming a unique colourful combination with the persistent red, five-pointed calyx. Flowers and berries are freely produced. This display continues into the autumn. Unfortunately, the foliage has a most unpleasant smell, but only when touched or brushed against.

Clerodendrum trichotomum var. *fargesii* 'Carnival' AGM

A less vigorous, variegated cultivar, perhaps better treated as a large shrub, rather than a very small tree.
Height: 2.5–4m (7–13ft).
Width: 1.5–2.5m (5–8ft).
Leaves: Green shade by darker green and margined with yellow. Later the margins mature to a creamy-yellow.
Flowers and fruit: A little less free-flowering and fruiting, but still a colourful addition.

Garden Design Considerations

An unusual, distinctive, small garden tree providing late season scent, flower and eye-catching fruit colour. Good as an isolated specimen or in an informal cottage-style garden.

Clerodendrum trichotomum var. *fargesii* 'Carnival' provides a lot of summer and early autumn interest, adding a subtropical feel to a bed or border designed for a late season colourful display.

COPROSMA

A genus of about 100 species of evergreen shrubs or small trees found mainly in New Zealand, but also the East Indies, New Guinea and Chile.

These plants are dioecious and fruits are formed only if male and female plants are present together. It should be noted that most of the cultivars mentioned in Table 4.25 are male or female clones, and fruiting will only occur if the appropriate male or female plants are also present. The flowers are wind-pollinated and the fruit is a non-poisonous berry. The fruit is usually small in the case of the shrubby species and is usually bright orange but may be dark-red or blue.

Horticulturally, these evergreen shrubs are cultivated for their attractively variegated leaves, which may change in colour according to the season, cooler or cold conditions promoting the intensification of colour.

Potential cultivators must note that all the species and cultivars are only rated USDA 9b/10a. RHS H3. In the milder and warmer parts of the USA, these plants may be very attractive, permanent shrubs and important in garden design, providing structure as evergreens and contributing vivid, sometimes changeable, leaf colour.

Growing Requirements

Hardiness: USDA 9b/10a. RHS H3.
Aspect and shelter: In most of the UK they can be grown in a conservatory or greenhouse that is frost-free. The plants may be grown outside in the UK during the summer months. A sheltered, warm, sunny site is required. Leaf colour will intensify as cooler early autumn is experienced outside, but do not let the plants experience frost.
Sun: In the milder and warmer parts of the USA, they may be grown in full sun or part shade. In the UK, full sun is required in the winter and filtered sunlight in the summer. Good ventilation should always be maintained.
Soil and pH: The soil should be well drained and have a pH that is acid or neutral. The plants should be grown in acidic, loam-based potting compost with extra grit to ensure very good drainage. In drier areas, irrigation may be required in the summer, although they are drought-resistant. *Coprosma* is resistant to saltwater conditions and this may make it well suited to coastal plantings. A mulch of well-rotted leaf mould or composted pine needles, 5–7cm (2–2½in) deep every spring is beneficial.

TABLE 4.25 *COPROSMA* SPECIES AND CULTIVARS

Name	Habit	Height	Spread	Leaves	Summer colour	Autumn / winter colour
C. 'Beatson's Gold' (f/v)*	Compact, rounded shrub	1.5m (5ft)	1.5m (5ft)	Small, ovate leaves	Variegated with a gold centre and edged with lime and a darker green	
C. 'Evening Glow'PBR (f/v)	Small, compact shrub	1.5m (5ft)	1.5m (5ft)	Small, shiny leaves	Variegated with a green margin and central yellow	Becomes tinted with orange pink
C. 'Fire Burst'PBR (f/v)	Compact shrub	1.5m (5ft)	1.5m (5ft)	Glossy 2–3cm (1–1.5in) long	Variegated with cream	Becomes bronzy red in winter
C. 'Inferno'PBR (f/v)	Upright habit	1.5m (5ft)	1.5m (5ft)	Glossy	Variegated with green and lime-green. Edged orange-red	Variegated with green and lime-green. Edged bronzy-red and red in winter
C. 'Rainbow Surprise'PBR (v)	Bushy, compact shrub	1.5–2.5m (5–8ft)	1.5–2.5m (5–8ft)	Glossy	Variegated with green centre, edged with lime-green, cream, pink and red.	Colour intensifies
C. repens (mirror plant, looking-glass plant)	Upright or prostrate shrub			Deep-green, glossy foliage. Leaves 2–8cm (1–3in) long	Green	Green. Habit of growth depends on growing situation
C. repens 'Pacific Dawn'		1.2m (4ft)	1.5m (5ft)	Glossy	Variegated with green and red-copper	Variegated. Intensifies to deep maroon-purple
C. repens PACIFIC NIGHT ('Hutpac'PBR) (m)	Upright, small shrub	1.m (3ft)	1.5m (5ft)	Narrow leaves, glossy	Purple-red foliage	
C. repens PACIFIC SUNSET ('Jwncopps') (m/v)	Small shrub	1m (3ft)	1m (3ft)	Shiny leaves	Centre of leaf red with irregular brown margins	
C. repens 'Pina Colada'PBR (v)	Compact	1.5m (5ft)	1.5m (5ft)	Thick and glossy with a wavy edge	Variegated with gold and yellow	Red and orange
C. repens 'Tequila Sunrise'	Upright and spreading.	1m (3ft)	0.8m (2ft 8in)	Small and glossy.	Variegated with orange and green.	Variegated with burgundy and brown in winter.
C. 'Scarlet O'Hara'	A small shrub			Bright-green splashes on young growth	Red orange leaves with deep red brown in centre	Burgundy and brown in autumn

* v, variegated; f, female clone; m, male clone.

Adequate watering coupled with monthly liquid feeding should be carried out in summer, but plants kept drier and not fed during the winter.

Garden Design Considerations

Potential cultivators must note that all the species and cultivars are only rated USDA 9b/10a RHS H3.

In the milder and warmer parts of the USA, these plants are very attractive, permanent shrubs. Over much of this vast area, the plants will grow well in sun or dappled shade, but in hot parts of the country, where temperatures regularly go above 100°F (38°C), these shrubs will require some shade and irrigation. In addition to the normal roles of giving permanent structure and furnishing a garden in winter, their contribution of leaf colour is very important. They exhibit a range of leaf colours that changes with the seasons, generally becoming more intense in the cooler parts of the year. Their moderate height makes them well suited for use as low hedges either left unclipped and informal in shape or regularly trimmed in spring to give a more formal regular shape.

Coprosma 'Evening Glow'. The foliage is contrasted against a 'blue-leaved' conifer.

Coprosma 'Evening Glow'. The picture shows the complex leaf variegation.

As *Coprosma* consists of relatively small and compact evergreen shrubs with bright leaf colour, these plants are also very good for planting in tubs, planters and pots on patios and around outside sitting areas.

Whilst intolerant of cold, they are tough in other respects. They tolerate wind, being previously adapted to this in their native New Zealand, and make good subjects for coastal planting, where their tolerance to saline conditions is also useful.

Coprosma 'Fire Burst'PBR. A compact cultivar photographed in early summer. Like all *Coprosma*, the leaf colours change with the seasons and temperature.

Coprosma 'Pacific Dawn'. This forms an upright, spreading shrub.

Coprosma PACIFIC SUNSET ('Jwncopps') (m/v).

Coprosma 'Rainbow Surprise'PBR. This forms a compact, bushy shrub.

In the UK, the genus can only be considered as a hardy permanent shrub for outside garden planting in the mild West Country and coastal gardens, where frost and low temperatures are not regularly experienced. In the rest of the UK, it may survive with wall shelter, a good microclimate or if provided with some artificial protection, but it is better considered as a long-lasting conservatory or greenhouse plant. It may be grown outside during the summer months and makes an excellent small shrub in pots and planters, but must be brought into protected cultivation before the winter or the gardener should be prepared to lose it in a cold winter.

CORNUS

With the common names of dogwood and cornel, this is a genus with a very wide distribution that includes North America, Europe, Asia and China. Members include trees, shrubs and herbaceous plants. Nearly all are deciduous, and the leaves are opposite, with the exceptions of *Cornus alternifolia* and *Cornus controversa*.

In cultivation, all species can be grown as shrubs, but some will naturally, and others can be encouraged to, form small trees, some of which are compact and ideal for the smaller garden.

Hardiness: USDA 6a-1. RHS H7. This is a general hardiness rating for the genus. Note there are some differences with some species, and these are mentioned, where appropriate.

Cornus alba (Red-Barked Dogwood)

This species occurs in Siberia, down to North-Eastern China and into North Korea. A very hardy, deciduous shrub, producing a dense thicket of growth. An upright mass of suckering stems. Very vigorous and spreading.
Height: 2–3m (6.5–10ft).
Spread: 2m+ (6½ft+).
Leaves: Plain-green, 5–11cm (2–4.5in) long, turning dark-red in autumn before falling.
Stems: Upright, young stems produced in the last growing season are red, becoming duller with age.
Flowers: White and small in flat heads in early summer, not showy.
Fruits: White or pale-blue from autumn into early winter.

Growing Requirements

An exceptionally hardy and tough shrub.
Hardiness: USDA 6a-1. RHS H7.
Aspect: Any orientation except due north is acceptable.
Shelter and sun: Full sun to part shade is tolerated and no particular shelter is required.
Soil: The plant will grow well in almost any soil, but calcareous are not favoured because they tend to be too dry in summer. Well-drained soil is appreciated, as are soils that are moisture-retentive in summer. Short-term waterlogging in winter is tolerated and the plant relishes a damp site.
pH: Soil pH should be acid or neutral.
Pruning: When grown for coloured winter-stem effect, the following is best practice. Allow newly planted to grow freely and establish for one to three years. Then, in late winter, or as late as mid-spring, coppice (also known as stooling) the plants. This involves cutting all stems down to 5–7cm (2–3in) above the ground. Take care to make the pruning cuts horizontally just above an opposite pair of buds. This prevents excessive dieback from this severe pruning. Then apply a balanced fertiliser at 70g/m^2 (2oz/yd^2). Try to avoid the fertiliser falling on to the cut stem surfaces, as this can cause burning. Following this, mulch with 3–7cm (2–3in) of garden compost or horse manure. This will stimulate strong, rapid, new stem growth in the following season, which will have the brightest stem colour for the following winter. This feeding and mulching may have to be added to by irrigation during dry summer conditions to ensure maximum new stem growth.

TABLE 4.26 *CORNUS ALBA* AND CULTIVARS

Name	Habit	Height	Spread	Leaves	Stems	Remarks
C. alba	Produces a dense thicket of upright stems	3m (10ft)	3m (10ft)	Plain-green becoming dark-red in autumn	Young stems dark-red	Very vigorous, will grow and spread in both wet and drier soils
C. alba 'Aurea' AGM	Upright stems	1.5–2.5m (5–8ft)	1.5–2.5m (5–8ft)	Soft-yellow leaves	Reddish young stems	
C. alba BATON ROUGE ('Minbat'PBR)	Upright, suckering	1.5–2m (5–6½ft)	1.5–2m (5–6½ft)	Dark-green leaves. Dark-purple autumn leaf colour	Bright-scarlet stems	Less vigorous than 'Sibirica', but the brightest red
C. alba 'Elegantissima' AGM	Upright, suckering	3m (10ft)	3m (10ft)	Grey-green and marbled with white	Young stems a good red	A strong growing selection
C. alba IVORY HALO ('Regnzam')	Upright stems, suckering	3m (10ft)	3m (10ft)	White edge to leaf	Red young stems	A vigorous selection
C. alba 'Kesselringii'	Produces a mass of upright stems	1.5–2.5m (5–8ft)	1.5–2.5m (5–8ft)	Plain-green becoming dull-purple in autumn	Dark-red shoots become almost black in winter	Moderately vigorous
C. alba 'Sibirica' AGM (syn. 'Westonbirt')	Upright stems	1.5–2m (5–6½ft)	1.5–2.5m (5–8ft)	Plain-green, dark-red autumn leaf colour	Young stems very bright-red	Less vigorous than the straight species, but still quite vigorous enough
C. alba 'Sibirica Variegata' AGM	Upright stems	1.5–2m (5–6ft)	1.5–2.5m (5–6½ft)	Irregular margin of creamy-white	Red young stems	
Cornus alba 'Spaethii' AGM	Upright stems	1.2–1.8m (4–6ft)	1.2–1.8m (4–6ft)	Majority of leaf bright-yellow. Autumn colour variable	Young stems red, but less so than 'Sibirica'	Leaves do not burn in sun. Leaf colour does not fade later in season

Pruning in late winter before bud burst used to be recommended for fear of causing damage due to sap loss if carried out later. This involves some loss of stem colour at the end of winter and the beginning of spring. Further practical experience now informs the gardener that pruning can be delayed until mid-spring with no detrimental effect on the plants.

This hard pruning can be done annually but must be followed by feeding and mulching if the vigour of the plants is to be maintained. If growing conditions are not optimal, e.g. shaded, then this hard pruning should be done less frequently, giving the plant time to recover before hard pruning again.

Coppicing in this manner involves the loss of any function the plant may have as a barrier or hedge. This can be avoided by cutting only one-third of the stems hard back annually. This is essentially a compromise, since the barrier or hedge effect will be maintained and some new growth stimulated, but not as much as when complete coppicing is carried out.

A final option is to leave the plants unpruned. The plants will grow to the size indicated in Table 4.26. As such they will not be compact and not well suited to the smaller garden. They will, however, flower, be attractive to pollinating insects and produce white fruit that may also have a slight bluish colouring. This fruit also acts as a food source for birds in autumn and early winter.

Cornus alba produces white flowers in late spring followed in autumn by white berries, often with a touch of blue.

Cornus alba and *Euonymus fortunei* 'Silver Queen'. A pleasing autumnal colour combination. After leaf fall, the remaining red stems of the *Cornus* will show against the evergreen leaves of the *Euonymus*.

Cornus alba 'Aurea'. Bright-yellow foliage colour is maintained throughout the growing season. Winter-stem colour is red, but not as bright as other cultivars.

Cornus alba 'Aurea'. This bright-yellow leaf colour is excellent for planning leaf-colour combinations with other shrubs of different leaf colours.

Cornus alba BATON ROUGE ('Minbat'PBR). Coppiced to produce winter red-stem effect.

Cornus alba BATON ROUGE ('Minbat'PBR). The cultivar with the reddest stem colour.

Cornus alba 'Elegantissima'. A clone with moderate vigour and fairly muted winter-stem colour.

Cornus alba 'Elegantissima' showing the white and green variegation.

Cornus alba 'Sibirica' coppiced for winter-stem colour and forming a strong plant association with a base planting of *Hedera helix* 'Glacier'.

Cornus alba 'Sibirica' with very good, red winter-stem colour.

Cornus alba 'Sibirica' well paired with *Cornus sericea* 'Flaviramea' in a corner of a garden planned for winter colour and interest.

Cornus alternifolia (Pagoda Dogwood, Green Osier)

A deciduous tree or shrub from North-Eastern America, where it experiences periods of intense cold in winter.

Height and habit: This species can grow as a small tree with a height of up to 6.5m (20ft) or as a shrub with several erect stems. Branches are produced that are almost tiered horizontally, giving a semi-tabulate habit, giving rise to one of its common names, pagoda dogwood.

Leaves: The plant is deciduous, its leaves being alternately arranged, oval or ovate, tapered at both ends and 5–12.5cm (2–5in) long and 2.5–4cm (1–1½in) wide. Good autumn leaf colour is sometimes produced.

Flowers: Small, white flowers are produced in flattened cymes in early summer and followed by small, black fruit with a bluish bloom.

Growing Requirements

Hardiness: USDA 6a/7a. RHS H6.

Aspect and shelter: Any orientation is accepted providing that it is not north-facing and shelter is required from cold northerly winds.

Sun: Part shade is preferred, and a hot sunny position should be avoided.

Soil: A soil of any texture is acceptable, providing that it is well drained.

pH: A pH that is either acid or neutral is best, but slightly alkaline soil may be tolerated if it contains a lot of organic matter.

Pruning: Limited formative pruning may be required, as the plant's habit is a most important feature. If the plant is required to form a small tree, then it will be necessary to gradually remove the lower limbs to form a trunk. Any competing leader shoots should be removed. Pruning should be done between autumn and early spring and hard pruning must be avoided, as this will not be tolerated and may well kill the plant.

Cornus canadensis (syn. *Chamaepericlymenum canadense*) (Creeping Dogwood, Canadian Bunchberry)

This species comes from North America, Northern Asia and Greenland. Technically it is an herbaceous perennial, as its shoots are regrown from ground level annually; it is generally regarded and treated here as a shrub.

TABLE 4.27 *CORNUS ALTERNIFOLIA* AND CULTIVARS

Name	Habit	Height	Spread	Leaves	Flowers	Remarks
C. alternifolia	Tiered	6m (20ft)		Plain-green. Autumn colour reddish-purple	White, small	
C. alternifolia 'Argentea' AGM	Tiered	3m (10ft)	2.5m (8ft)	Green, margined with white. Autumn colour reddish-purple	White, small	Slow-growing
C. GOLDEN SHADOWS 'Wstackman'[PBR]	Tiered	1.8–2.4m (6–8ft)	1.5–2m (5–6½ft)	Broad cream margin around an irregular central green area	White, small	Slow-growing

TABLE 4.28 *CORNUS CANADENSIS*

Name	Habit	Height	Spread	Leaves	Flowers	Fruit
C. canadensis	Forms a low-growing mat over soil surface	10–15cm (4–6in)	0.5–1m (20–39in)	Obovate. In whorl-like arrangement at top of stem. Red autumn leaf colour	Head of tiny flowers surrounded by prominent white bracts	Red berry-like drupes

Cornus alternifolia 'Argentea'. The picture shows a good specimen with a well-developed semi-tabulate habit.

Cornus alternifolia 'Argentea'. Close-up of the rather distorted, variegated foliage.

Cornus canadensis. A very different plant to other cornus shrubs. In acidic ground, this plant makes delicate ground-cover, having white heads of 'flowers' in spring and red fruit in the autumn.

Height: A dwarf, rhizomatous, herbaceous perennial with new shoots produced from ground level annually. These vertical shoots are 10–15cm (4–6in) tall.
Spread: Spread by rhizomes is slow but, eventually, a low carpet of growth is produced.
Leaves: Closely spaced, alternate, obovate, green leaves appear like a whorl at the top of each stem.
Flowers: In early summer, each stem is topped by a cyme of tiny green flowers forming a dense head. The head of flowers is surrounded by four comparatively large, white bracts, each bract being 1.5 cm (0.5in) long and 1.1 cm (0.5in) wide.
Fruit: Later, the flowers are often followed by heads of red berry-like drupes, each fruit the size of a pea.

Growing Requirements

Hardiness: USDA RHS H7.
Aspect and shelter: A site of any aspect is acceptable and may be sheltered or exposed.
Sun: Part shade is best, since the plant is naturally a woodlander.
Soil: The soil can be of any texture but must be well drained and moisture-retentive.
pH: Soil pH should be acid or neutral, but not alkaline.

Cornus controversa (Wedding Cake Tree)

A deciduous tree that can grow 12–15m (40–50ft) tall. From China, Japan, Taiwan and the Himalayas, with a very distinctive habit.

Height and spread: This species grows as a tree 15m (50ft) high. It produces near-horizontal branches, a closely bunched group of new branches being produced towards the end of each growing season. This produces, over time, a very distinctive, tiered habit of branching.
Leaves: The leaves are alternate, green, oval or ovate and 7.6–15cm (3–6in) long by 5–7.5cm (2–3in) wide. In autumn, they often turn purple-red before falling.
Flowers and fruit: Small, white, star-shaped, scented flowers are produced in early summer. The flowers are gathered in 18cm (7in) across flattened cymes. The flowers are followed by blue-black fruits, each 0.6cm (¼in) in diameter.

TABLE 4.29 *CORNUS CONTROVERSA* AND CULTIVARS

Name	Habit	Height	Spread	Leaves	Flowers	Remarks
C. controversa	Tiered, tabulate	9–15m (30–50ft)		Oval or ovate, green. Red-purple autumn leaf colour	Small, white flowers in large, flat cymes	
C. controversa 'Variegata' AGM	Tiered, tabulate	4–8m (13–26ft)	4–8m (13–26ft)	Narrowly ovate, green, but broadly margined with cream	Small, white flowers in large, flat cymes	Slow-growing
C. controversa 'Pagoda'	Tiered, tabulate	15m (50ft)	15m (50ft)	Oval or ovate, green. Red-purple autumn leaf colour	Small, white flowers in large, flat cymes	Selected for a good regular tiered habit, but too large for a compact or small garden

Cornus controversa. The plain-green species grows into a medium-sized tree but does show the tabulate habit.

Cornus controversa. Attractive autumn leaf colour.

Cornus controversa 'Variegata'. An exceptionally well-managed, young plant. This shows the very distinct tabulate habit that can be achieved by careful selection, training and pruning of a young plant.

Cornus controversa 'Variegata'. A plant that has been allowed to develop a more relaxed semi-tabulate habit. The picture was taken early in the season whilst the foliage was still expanding.

Cornus controversa 'Variegata'. A picture taken in mid-summer when the foliage had fully developed.

Growing Requirements

Hardiness: USDA RHS H5.
Aspect and shelter: A site of any orientation is acceptable, although some shelter from cold winds should be provided.
Sun: Full sun or part shade is necessary.
Soil: The soil can be of any texture but must be well drained and moisture-retentive in summer.
pH: The soil should be of acid or neutral pH; avoid calcareous soils.
Pruning: Keep pruning to a minimum, but some formative pruning may be necessary during development to limit the growth of competing leaders, and to maintain the very desirable and distinctive tiered tabulate habit.

Cornus florida 'Sunset'. Photographed in late summer.

Cornus florida 'Sunset'. Photographed in autumn. Autumn leaf colour is now fully developed!

Cornus florida (Flowering Dogwood, American Box)

A deciduous, small tree or large shrub found from North-Eastern USA to Northern Mexico. In cultivation it grows 3–6m (10–20ft) high but can get considerably taller in the wild.
Height: 4–8m (13–26ft).
Spread: 4–8m (13–26ft).
Leaves: Green, opposite, oval or ovate, 8–16cm (3–6in) long, 4–8cm (1.5–3in) wide. Colouring to red and orange in autumn.
Flowers and fruit: The small flowers are produced in a crowded headed subtended by four white bracts. The bracts are formed in the autumn, protect the flower during the winter and expand early in late spring. Once expanded, each bract is 4–6cm (1.5–2in) long and reflexed back from the true flowers, forming with the three other bracts a corolla-like structure often referred to as the flower. In late summer, each flowerhead develops into a rounded, orange-red fruit.

Cornus florida 'Sunset'. Showing the variegated foliage and the beginnings of autumn colour.

Cornus florida 'Sunset'. Detail of the autumn foliage colour. Note how the colour matches the pattern of the original variegation.

Growing Requirements

Hardiness: USDA RHS H5.
Aspect and shelter: A site of any aspect, except north-facing, is suitable. In exposed positions some shelter is necessary. In the UK, the plant has proved susceptible to late spring frosts and the growth produced in the growing season does not ripen well due to the lack of warm, humid, autumnal weather, as experienced in its natural habitat. Protection from other overhead trees may help avoid damage from late spring frosts followed by rapid thawing by the sun. This is another feature of the UK climate not common in its native environment.
Sun: Full sun or part shade is suitable, ideally with protection from early morning sun.
Soil: The soil can be of any texture but must be well drained and moisture-retentive in summer. The plant is not tolerant of drought in dry summer weather.
pH: The soil pH should be acidic or neutral, i.e. up to pH 7, but not above, with plenty of organic material, but avoid alkaline, thin, chalky soils. Organic mulch applied in spring or early summer will help by keeping the roots cool and limiting soil moisture loss.

TABLE 4.30 *CORNUS FLORIDA* AND CULTIVARS

Name	Habit	Height	Spread	Leaves	Bracts
C. florida	Small tree or spreading shrub	4–8m (13–26ft)	4–8m (13–26ft)	Green, oval or ovate. Red and orange autumn colour	White
C. florida f. *rubra*	Small tree or spreading shrub	4–8m (13–26ft)	4–8m (13–26ft)	Green, oval or ovate. Red and orange autumn colour	Rosyred, but can be variable
C. florida 'Cherokee Chief' AGM	Spreading shrub	3m (10ft)	4m (12ft)	Green, oval or ovate. Red and orange autumn colour	Deep rosyred
C. florida 'Cherokee Princess'	Spreading shrub	3m (10ft)	4m (12ft)	Green, oval or ovate. Red and orange autumn colour	Large, white bracts 6cm (2in) long
C. florida 'Holman's Gold'	Spreading shrub	3m (10ft)	4m (12ft)	Green leaves margined with yellow. Turning red-purple in autumn	White bracts
C. florida 'Rainbow' AGM	Compact, upright growth	2.5–4m (8–13ft)	1.5–2.5m (5–8ft)	Green leaves with yellow margins. Leaves purple-red with red margins in autumn	White bracts
C. florida 'Sunset'	Compact, upright growth	3m (10ft)	4m (12ft)	Young leaves pink, later green with broad, irregular, yellow margin. Autumn leaf colour pink, red and purple	Pinkish white bracts

Cornus kousa (Szechuan Strawberry)

A deciduous, small tree or shrub growing up to 5m (20ft) high in the wild. Native to Japan, Korea and Central China. In China, the varietas *Cornus kousa* var. *chinensis* is found. This is stronger growing, forms a larger plant and can have larger bracts than the species.

Height: Growing up to 5m (20ft) in the wild.

Leaves: Ovate, with a slender point, 3.8–7.6cm (1½–3in) long and 1.9–7.6cm (¾–1¾in) wide with an undulating margin.

Flowers and fruit: Flowering occurs in early summer with very small flowers produced in a compact, rounded inflorescence in the centre of much more conspicuous bracts. These four white bracts immediately subtend the flowers and are ovate, pointed at their distal end, 2.5–5cm (1–2in) long and 1–1.9cm ½–¾in) wide. In late summer, the inflorescence develops into an attractive deep-pink, strawberry-like fruit cluster, which is edible raw or cooked.

Cornus kousa. This species usually grows into a large shrub or small tree.

Cornus kousa var. *chinensis*. The picture shows the white bracts surrounding the central mass of very small flowers.

Cornus kousa var. chinensis. This varietas is usually more vigorous and larger growing than the ordinary species and will grow into a medium-sized tree.

Cornus kousa var. *chinensis*. Most forms and cultivars of this species regularly develop rich autumn leaf colour.

Cornus kousa 'Madame Butterfly' is a distinct cultivar in which the tips of the bracts are turned upwards, giving the impression of butterflies.

Cornus kousa 'Miss Satomi'. This cultivar develops a broad, elegant, structured shape.

Cornus kousa 'Miss Satomi' has broad, deep-pink bracts.

Cornus kousa var. *chinensis* 'Wisley Cream' bearing the red, spherical fruits often developed by *Cornus kousa* and its cultivars. They are attractive and edible. Unfortunately, the outer rind, although edible, is tough and bitter-tasting, therefore, it is usually removed, and the inner flesh is full of inconvenient seeds.

TABLE 4.31 *CORNUS KOUSA* VARIETAS AND CULTIVARS

Name	Habit	Height	Spread	Leaves	Bracts	Remarks
C. kousa	Bushy	4–8m (13–26ft)	4–8m (13–26ft)	Ovate with a slender tip at the distal end, 3.8–7.6cm (1½–3in) long, 1.9–7.6cm (¾–1¾in) wide. Autumn colour reddish-purple	White, 2.5–5cm (1–2in) long, 1–1.9cm (½–¾in) wide	Can form a small tree or large shrub
C. kousa var. *chinensis* AGM	Upright	6–10m (20–33ft)	4–8m (13–26ft)	Autumn colour reds and oranges	White 4–6cm (1½–2in) long	Free flowering
C. kousa 'China Girl'	Bushy	5m (16ft)	7m (23ft)	Leaves become crimson-purple in autumn	White developing to pink	A vigorous selection
C. kousa 'Madame Butterfly'	Bushy	4–8m (13–26ft)	4–8m (13–26ft)	Leaves purple-red in autumn	White bracts with pink tips. Bracts turn upwards looking like butterflies	
C. kousa 'Miss Satomi' AGM	Broadly bushy	5m (16ft)	5m (16ft)	Leaves become deep red in autumn	Broad, deep-pink bracts	
C. kousa 'Schmetterling'	A compact cultivar	2.5–4m (8–13ft)	2.5–4m (8–13ft)		Broad, white bracts	
C. kousa 'Summer Fun' AGM	A compact cultivar	2.5–4m (8–13ft)	2.5–4m (8–13ft)	Grey-green leaves with irregular, broad, cream margin. Leaves become flushed with pink in autumn	White bracts	
C. kousa var. chinensis 'Wisley Cream' AGM	Bushy, but slightly more erect	4–8m 13–26ft)	2.5–4m (8–13ft)	Leaves become purple-red in autumn	White bracts	Free-flowering
C. kousa 'Wolf Eyes' AGM	Bushy	4–8m (13–26ft)	4–8m (13–26ft)	Leaves with cream margins that become pink in autumn	White bracts	

Growing Requirements

Hardiness: USDA 6b/7a. RHS H6.
Aspect and shelter: A site with any orientation is acceptable. *Cornus kousa* is better adapted to the UK climate than *Cornus florida*. It requires less sun to ripen its wood and flower well, it flowers slightly later and avoids most late spring frost damage and is less soil pH-sensitive.
Sun: Full sun or part shade is desired and some shelter from strong winds should be provided.
Soil: The soil can be of any texture but must be well drained, high in organic matter and moisture-retentive in summer. Avoid thin, dry, calcareous soils. Poor drainage or summer drought will result in poor growth or death of the plant.
pH: A wide pH range is tolerated, but acid or neutral is best.
Pruning: If it is desired to form a small tree, rather than a shrub, formative pruning by removing unwanted lower branches can be done whilst the plant is dormant in winter or early spring.

Cornus kousa 'Wulf Eyes'. This grows into a distinctive, large shrub.

Cornus kousa 'Wulf Eyes' showing the broad, white, marginal variegation on each leaf and the white bracts this unusual cultivar produces.

Cornus mas (Cornelian Cherry, European Cornel, Edible Dogwood, Sorbet, Cornel)

This species is native to Central and Southern Europe; it is not native to, but has been cultivated in, Britain for centuries. The wood is very hard and dense, sinking in water. Historically, it was used for weapons and still finds a use in tool handles and parts for machines. In the Caucasus, Central Asia and Germany, it has been, and is still, grown as a fruit crop used in numerous types of preserve and drinks.
Height: Growing up to 8m (25ft) tall in the wild.

Leaves: Opposite, dark-green, ovate 4–10cm (1½–4in) long and 2–4cm (1–1½in) wide, becoming reddish-purple in autumn.
Flowers: Produced on bare branches in late winter and early spring before the leaves emerge. Each flower is small 0.5cm (¼in) in diameter, yellow, born in dense, short-stemmed umbels. They are prominent because of the lack of any other flowers at this time and appear as a haze of yellow stars over an otherwise bare shrub.
Fruit: The fruit is a red and shiny, oblong drupe 2cm (1in) long and 1.5cm (¾in) in diameter, juicy with a single seed. When ripe in late summer or early autumn it is edible with a sharp taste. The fruit is not

TABLE 4.32 *CORNUS MAS* AND CULTIVARS

Name	Habit	Height	Spread	Flowers	Leaves	Fruit	Remarks
C. mas AGM	Can be a tree or shrub. Densely branched	2.5–6m (8–21ft)	2.5–4m (8–13ft)	Very early. Compact umbels of small, yellow, star-like flowers	Dark-green, produced after flowering. Reddish-purple autumn colour	Oblong, at first green, then red, finally black	One of the first shrubs to flower in late winter
C. mas 'Aurea' AGM	Bushy	2.5m (8ft)	2.5m (8ft)	As above	Bright-yellow in spring, later yellow-green	Few fruits produced	Grow in semi-shade or leaves may burn. Red and pink tints in autumn
C. mas 'Golden Glory' AGM	Bushy	2.5–4m (8–13ft)	2.5–4m (8–13ft)	Profuse flowering			
C. mas 'Jolico' AGM	Strong growing	5m (17ft)	5m (17ft)	As above		Heavy crops claimed	A German cultivar claimed to produce masses of red fruit that are three times normal size*
C. mas 'Variegata' AGM	Relatively small and compact	1.5–2.5m (5–8ft)	1.5–2.5m (5–8ft)	As above	Bold irregular margin of white	Few fruits produced	

* Several Eastern European cultivars, including 'Pancharevski', 'Kazanlushki' and 'Shumenski' are now becoming available. They have been selected to produce heavy crops of fruit, 50kg (110lb) per established plant being claimed. Each cultivar requires cross-pollination with another different cultivar to crop well. Two Greek cultivars for crop production have also been patented *Cornus mas* 'Ntoulia 1'PBR and *Cornus mas* 'Ntoulia 2'PBR. As far as the author is aware, none of these cultivars have yet been trialled under UK or other countries' growing conditions.

Cornus mas. Here, seen flowering very early in spring.

Cornus mas. The tiny, star-like flowers being worked by a bee.

Cornus mas 'Aurea'. This cultivar is slightly less vigorous than the plain species but does produce very bright-yellow foliage, which is very colourful in late spring and summer.

Cornus mas 'Aurea'. The bright-yellow foliage may burn a little in strong sunlight.

Cornus mas 'Aurea' in fruit. In Eastern Europe this is a valuable crop.

Cornus mas 'Variegata'. A good, clear variegation. In autumn, the white areas of the leaf turn pink before falling.

freely produced in the UK, but much heavier crops are produced where the climate is more continental, with a clear transition from winter cold to warmer spring and summer conditions.

Growing Requirements

Hardiness: USDA 6b/7a. RHS H6.
Aspect: A site of any orientation is acceptable. To obtain good crops of fruit it is recommended to grow two different cultivars together to ensure cross-pollination.
Shelter and sun: Sun or part shade is required, but shelter is not specifically necessary.
Soil: A soil of any texture is suitable, providing that it is well drained.
pH: The plant is notably tolerant of a very wide pH range, growing equally well on acidic and alkaline soils. It will thrive on thin, chalky, alkaline soils with a high pH.
Pruning: The plant may be grown as a shrub or small tree. To encourage a young plant to develop into a tree, a single stem should be selected and other basal branches removed. As this single stem develops into the trunk of the tree, it may be necessary to remove lower competing side-shoots.

Cornus sanguinea (Common Dogwood)

This species is native to Europe, including the UK. In the UK, it is frequently found as a component of hedgerows where its partly red-coloured, last season's growth can sometimes be clearly distinguished during winter.
Height: A deciduous shrub with an upright habit of growth achieving a height of 1.8–3.5m (6–12ft) in the wild.
Habit: Bushy, young growth green, but reddened on exposed side, bark later becoming greyish.
Leaves: Green ovate 10cm (4in) long and 2–3cm (¾–1¾in) wide, turning red in the autumn, giving rise to the species Latin specific name.
Flowers: White, small, smelling strongly, in cymes in early summer.
Fruit: Spherical about 0.6cm (¼in) in diameter and black, tasting bitter. Wood very hard.

Growing Requirements

Hardiness: USDA 6b/7a. RHS H6.
Aspect: A site of any aspect is acceptable.
Shelter and sun: Full sun or part shade is required, but special provision of shelter is not necessary.
Soil: A soil of almost any texture is accepted, good drainage is appreciated, but this shrub will also prosper where drainage is poor.
pH: The plant is not specific in its soil pH requirement, tolerating a very wide range from acidic to alkaline soils.
Pruning: For best winter-stem colour, this shrub should be coppiced to encourage maximum vigorous, annual, vegetative growth, which has the most vivid stem colour. For details of coppicing please *see* Growing Requirements for *Cornus alba*.

TABLE 4.33 *CORNUS SANGUINEA* AND CULTIVARS

Name	Habit	Height	Spread	Stems	Leaves	Remarks
C. sanguinea	Suckering, upright	2.5–3m (8–10ft)	1.5–2.5m (5–8ft)	Young stems partly red	Green, becoming red in autumn	Beneficial for pollinating insects when left to grow naturally
C. sanguinea 'Anny' (syn. Winter Flame)	Suckering, upright	2m (6ft)	2m (6ft)	Young stems orange, red and yellow	Green, becoming orange-yellow	Grow in full sun for best winter-stem colour
C. sanguinea 'Anny's Winter Orange' AGM	Suckering, upright	1.5–2m (5–6ft)	1.5–2m (5–6ft)	Young stems orange, red and yellow	Green, becoming buttery yellow in autumn	Arguably considered the best cultivar for winter-stem colour
C. sanguinea 'Magic Flame' AGM	Suckering, upright	1.5–2.5m (5–8ft)	1.5–2.5m (5–8ft)	Young stems orange, red and yellow	Green, becoming yellow, orange, and red in autumn	
C. sanguinea 'Midwinter Fire'	Suckering, upright. More vigorous than C. 'Winter Beauty'	2m (6ft)	2m (6ft)	Young stems orange, red and yellow	Green, becoming orange and yellow in autumn	Grow in full sun for best winter-stem colour
C. sanguinea 'Winter Beauty'	Suckering, upright	1.5–2m (5–6ft)	1.5–2m (5–6ft)	Young stems orange and yellow and yellow-red in upper length of stem	Pale yellow-green, becoming orange and yellow in autumn	

Cornus sanguinea growing in a countryside hedge. A dull-red stem and dark-red autumnal foliage is noticeable. By selection, a range of much more colourful cultivars have been introduced.

Cornus sanguinea 'Anny's Winter Orange'. A very strongly coloured cultivar.

Cornus sanguinea 'Anny's Winter Orange.' A close up of the intensely coloured stems.

Cornus sanguinea 'Midwinter Fire'. A more vigorous cultivar. Here plants have been coppiced in the foreground and pollarded at the back to display even more stem colour.

Cornus sanguinea 'Midwinter Fire'. A brightly coloured mass of winter stems.

Cornus sanguinea 'Winter Beauty'. A selection slightly less vigorous than *C.* 'Midwinter Fire'. In common with other selections, this cultivar also produces good autumn leaf colour.

Cornus sanguinea. Close-up of autumn leaf colour. On close inspection you can see that the coloured stems are waiting to be fully revealed after leaf fall.

Cornus sanguinea 'Winter Beauty' showing yellow stem colour at the base of the stems, changing to orange-red at the top.

Cornus sericea (syn. *C. stolonifera*) (Red Osier Dogwood)

This species is native to North America growing right across the continent. A very strong-growing, deciduous, suckering shrub, spreading underground by stolons. It is close to the Asiatic *Cornus alba* but separated by its longer, pointed leaves and stoloniferous nature.
Height: Capable of obtaining 2.5m (8ft) in the wild.
Habit: Upright, spreading forming a dense thicket of stems.
Leaves: Opposite, ovate to lanceolate with long, tapered points, 12cm (5in) long 2.5–5cm (1–2½in) wide, upper surface green, lower glaucous.
Flowers: Small, white in cymes 5cm (2in) across produced in late spring and early summer.
Fruit: White spherical 0.6cm (½in) in diameter.

Growing Requirements

Hardiness: USDA 6a-1. RHS H7.
Aspect: A site of any orientation is acceptable.
Shelter and sun: Sun or part shade is required but no shelter is necessary.
Soil: The soil can be of any texture. Drainage should ideally be good but damp sites with poor drainage will be tolerated.
pH: A wide pH range from acid to alkaline is accepted. A very tough shrub that may be selected where almost any other will fail.
Pruning: For best winter-stem colour, stooling or coppicing should be done every, or every other, year. For details of recommended method please *see* under Growing Requirements for *Cornus alba*.

TABLE 4.34 *CORNUS SERICEA* AND CULTIVARS

Name	Habit	Height	Spread	Leaves	Stems	Remarks
C. sericea	Suckering, upright	2.5m (8ft)	Indeterminate	Green	Dark-reddish	Rampant spreader
C. 'Bud's Yellow'	Suckering, upright	1.5–2.5m (5–8ft)	1.5–2.5m (5–8ft)	Green	Yellow-green	
C. 'Cardinal'	Suckering, upright	1.5–2.5m (5–8ft)	2.5–4m (8–13ft)	Some autumn leaf colour	Red	
C. 'Flaviramea' AGM	Suckering, upright	1.5–2.5m (5–8ft)	2.5–4m (8–13ft)	Orange-yellow autumn colour	Greenish yellow	
C. 'Hedgerows Gold' AGM	Suckering upright	1.5–2.5m (5–8ft)	2.5–4m (8–13ft)	Leaves have a green centre with an irregular yellow margin that later becomes creamy-white. Red and yellow autumn leaf colour	Red	
C. KELSEY'S GOLD ('Rosco') AGM	Suckering, upright	0.1–0.5m (4–20in)	1–1.5m (3–5ft)	Leaves yellow-green in growing season and autumn	Yellow-green with red tips	A dwarf cultivar
C. 'White Gold'	Suckering, upright	1.5–2.5m (5–8ft)	1.5–2.5m (3–5ft)	Leaves have a green centre with an irregular creamy-white margin. Yellow autumn leaf colour	Yellow stems	

Cornus sericea 'Bud's Yellow' here well displayed in a winter garden.

A contrast in autumn leaf colours between the yellow of *Cornus sericea* 'Flaviramea' and those of *Cornus alba* BATON ROUGE ('Minbat'[PBR]). This contrast catches the eye and is repeated in the contrast between winter-stem colours.

Cornus sericea 'Bud's Yellow'. The young stems do have a bright-yellow colour. Ideal for contrasting against red-stemmed cornus.

Cornus sericea 'Flaviramea', expertly coppiced and displaying their olive-green stem colour.

Cornus sericea 'Cardinal', a softer red than those found in cultivars of *Cornus alba* and *Cornus sanguinea*.

Garden Design Considerations

From a horticulturalist's point of view, *Cornus* is a very mixed genus. Plants grouped together by virtue of a similar floral pattern have widely divergent uses in the garden and landscape. Care has to be taken; to borrow a phrase from Beth Chatto, one must 'select the right plant for the right place'.

Small Trees for the Modern and Compact Garden

Cornus alternifolia and *Cornus controversa* are plants that fall into this category. *Cornus alternifolia* is the smaller, reaching a maximum of 6m (20ft) in the plain-green species, but less in its variegated forms. It is often also seen as a shrub or multi-stemmed tree. When positioned where its outline and effect, from its relatively fine-textured, variegated foliage, can be clearly seen from the mid-distance, it has a unique, almost mist-like appearance; closer to, it contrasts against the foliage of almost any other plant.

Cornus controversa is larger in overall height, reaching up to (50ft) in the largest of the plain-green forms, but only 8m (26ft) in the variegated forms. Whilst *C. alternifolia* has a tendency towards tiered branching, this is more strongly developed in *C. controversa* and this, together with larger leaves, gives an even more distinct habit. In both species, once the tree-like shape is established, the head of the tree can be raised. Such a tree will occupy a smaller area of ground, a valuable feature in a compact garden, and allows space for underplanting where space is less limited. A well-grown tree, whilst taking time to develop, is a magnificent plant, always gaining a favourable comment from the discerning gardener or designer.

Both plants are difficult to propagate and slow-growing; this ensures that the nursery demands, and deserves, a high price but they are worth every penny!

Cornus kousa and its more vigorous form *Cornus kousa* var. *chinensis* can be formatively pruned into tree-like forms. They have a relatively upright, branching habit, tiered branching is less marked and their effect is more informal and less stylised. Coupled with this, they have the added attractions of showy summer bracts, curious fruits and excellent autumn leaf colour.

In contrast, *Cornus kousa* 'Miss Satomii', with its gracefully layered branches, can unfortunately hardly be considered compact, but when covered with soft-pink bracts in summer, it is a joy to behold!

Shrubs to Provide 'Flower' Colour, Summer and Autumn Foliage Colour

Here again *Cornus* provides an abundance, the bracts of *Cornus kousa* having already been mentioned.

If acid or neutral soil and some shelter can be offered, then *Cornus florida* and its cultivars can produce bract colour in white, pink or red. Their autumn leaf colour is legendary and cultivars such as *Cornus florida* 'Rainbow' live up to their name, having brightly variegated foliage in the growing season and vivid autumn leaf colour. *Cornus* 'Rainbow', a compact cultivar, is also very suitable for where space is limited.

Ground Cover

The trees and shrubs mentioned require some form of ground cover to clothe the soil beneath them. Here, *Cornus canadensis* excels. It grows best in part shade and does require acidic, damp soil. Providing these criteria can be met, this plant will spread slowly over the surface, creating a mat of dwarf shoots, each topped with a neat whorl of white bracts surrounding a central button of true flowers, which, in a favourable season, are followed by red fruits and finally autumn leaf colour.

Less demanding in its growing requirements, *Cornus sericea* KELSEY'S GOLD ('Rosco') AGM' is a dwarf, suckering shrub only growing to a maximum of 50cm (20in) tall and could be called upon to perform a similar role at the front and towards the base of larger shrubs.

Winter-stem colour and the Winter Garden

Recent years have seen many public and botanical gardens in the UK seek to capitalise on our now milder winters by attracting visitors not only in the traditional periods of spring, summer and autumn, but also winter.

In turn, private gardeners have also realised the greater opportunity for appreciating their gardens in winter. *Cornus alba* and *Cornus sericea* plus their cultivars have featured strongly in this planting. Both species, when coppiced, respond by producing vividly coloured young stems. *Cornus alba* 'Siberica' has long been famed for this attribute. The newer *Cornus alba* BATON ROUGE ('Minbat'[PBR]) has even brighter red colour, is slightly less vigorous and very well-tailored for the smaller garden. There are also cultivars that provide very good summer leaf colour. These are, however, quite large shrubs, even when coppiced. *Cornus alba* 'Spaethii' is arguably the best with clear, yellow foliage throughout the growing season, continuing into the autumn followed by red stem colour.

More recently, the coloured stems of *Cornus sanguinea* selections have 'hit' the gardening world, providing another range of stem colour. Of these, *Cornus sanguinea* 'Anny's Winter Orange' is the most suitable where space is limited, since stem colour is excellent and it is the least vigorous and most compact. Regular coppicing can be done, but there is also the opportunity to take a more relaxed approach, allowing the plants to develop taller branches, whilst removing a few every year to retain plant vigour and stem colour. This gives a taller, slightly less intense and more subtle effect.

Cornus sericea is one of the toughest garden shrubs, the plain species being a rapid spreader and only suited to the largest estates. Careful selection has led to cultivars with moderated vigour and, whilst admittedly limited, a useful range of stem colours. The cultivar *Cornus sericea* 'Hedgerows Gold' AGM is worthy of special note not only having good red winter-stems, but bold yellow leaf variegation.

These winter-stem specialists can be grouped together to form a planting or garden for dazzling winter colour. Incorporating pollarded forms of *Salix* (willow) and controllable forms of coloured stem *Rubus* (brambles) will increase the colour range and contrasts obtainable. The site for such a planting must be where direct winter sun will play upon them to highlight these colours. Mixing the various groupings so that colour contrasts can be planned and to obtain the exciting 'zebra striping effect' when the wind blows is essential.

Hedging, Space Dividers and Barriers

Many species of *Cornus* lend themselves to these functions, *C. alba*, *C. sericea* and *C. sanguinea* being particularly suitable.

If a site provides the opportunity, planting such a winter border near open water, where the colours and patterns can be reflected, is a chance not to be missed! This range of *Cornus* and *Salix* luxuriate in damp conditions and grow to perfection. Planted in a line, or less formally across or around an area, they will, if left to their own devices, rapidly make an informal hedge. This provides a visual block, a serviceable barrier plus attractive foliage and flowers in the summer. The plants respond well to clipping and a more formal shape can easily be achieved.

Where soil and planting conditions are challenging, *Cornus* plants may also be very helpful. In contrast to most shrubs, they tolerate poor drainage and will succeed where most other shrubs will fail. If used with thoughtfulness and given adequate maintenance, they will create a pleasant vista where the alternative may have been much less desirable.

Japanese Style and Minimalist Gardens

Several *Cornus* species are native to Japan and grow well there. The need for striking formalised shapes and branching is well met by the tiered habit of both *C. alternifolia* and *C. controversa*. The show of white bracts and vivid autumn leaf colour is much loved by the Japanese and associates very well with autumn shades of the famed Japanese maples.

Landscape and stroll gardens are two styles of Japanese gardens, but Zen and minimalist are others. Here the adage 'less is more' is never truer. Each plant has to bring a clear statement to the design. The distinctive habits of *C. alternifolia*, *C. controversa* and *C. kousa* are again appropriate. The same strictures apply to the minimalist garden, where each plant must gain its entry by virtue of its distinctiveness.

Wildlife Gardens

With the onset of climate change and diversity loss, the need to plant for the benefit of wildlife has never been more urgent. Here again *Cornus* can help. Left to grow freely, *Cornus* flowers profusely, even if not in a visually significant manner, and provides a rich nectar and pollen source for pollinators and fruit for birds and other herbivores in autumn. Species such as *Cornus alba* and *Cornus sericea* and their less flamboyant cultivars, if left to grow naturally in a wildlife garden, would give a backing for flowering, smaller, herbaceous and annual plants typical of the wildlife garden and give some necessary structure and winter shelter.

With all these potential applications for the gardener and designer what a truly versatile genus *Cornus* is!

COROKIA

A small genus native to New Zealand. It consists of evergreen shrubs with alternate entire leaves and small, yellow, star-like flowers consisting of five petals and five stamens.

Growing Requirements

Hardiness: USDA 8b/9a. RHS H4.
Aspect: A south- or west-facing site is required. This genus is well adapted to New Zealand's windy climate. It grows well in coastal situations where the winds may be strong, but not cold or drying.
Shelter: A warm, sheltered position away from cold, drying winds is necessary.
Sun: Full sun is best, but part shade is tolerated.
Soil: Provided the soil is well drained, it can be of any texture.
pH: A very wide range of soil pH is accepted.

Garden Design Considerations

Corokia is a genus with a range of unusual characteristics, which the garden designer and gardener would do well to employ. The shape and habit of *Corokia cotoneaster* with its mass of intermeshed branches is in strong contrast to any other plant. This will add a contrasting form to any border. This is added to by the sudden appearance of bright-yellow, star-like flowers over an already unusual shrub, adding another element of surprise to those unfamiliar to this genus.

The small size of the plants within the genus makes them all suitable for inclusion in small, compact gardens, including courtyards, patios and container-gardening. In a formal setting, they can be left to grow in their own free-ranging manner to contrast strongly with the other regularly shaped or clipped forms, or they can be clipped to conform. Indeed, they accept clipping so well that they form very good low hedges and are a good substitute for the disease-susceptible *Buxus sempervirens* and its smaller form *Buxus sempervirens* 'Suffruticosa'.

The strong leaf and fruit colours seen in the *Corokia* × *virgata* cultivars can be employed to give colour contrast and to enliven the autumn and winter garden. If this were not enough, the group is well adapted to grow in gardens with a Mediterranean climate and those with a mild but windy coastal environment. In a cold garden, the use of *Corokia* is restricted, but it can be happy and very attractive nestling in the warmth and shelter at the base of a south-facing wall.

Corokia cotoneaster. A shrub with a unique overall form and bound to illicit interest when seen for the first time.

Corokia cotoneaster. The small, star-like flowers are produced in early summer.

TABLE 4.35 *COROKIA COTONEASTER*, *COROKIA* × *VIRGATA* AND SELECTIONS OF *COROKIA* × *VIRGATA*

Name	Habit	Height	Spread	Foliage	Flowers	Fruit	Remarks
C. cotoneaster (wire-netting bush)	A mass of twiggy, intermeshed branches	1.5–2.5m (5–8ft)	1.5–2.5m (5–8ft)	Small, narrow, dark-green leaves	Small, yellow, star-like flowers in spring and early summer	Red, conspicuous	Has a unique, sparse, angular look
C. × *virgata* (hybrid wire-netting bush)	Upright, twisted, silvery stems	1.5–2.5m (5–8ft)	1.5–2.5m (5–8ft)	Small, leathery green, silvery underside	Small, yellow, star-like flowers in spring and early summer	Bright orange	Similar, but slightly less contorted appearance than *C. cotoneaster. Corokia* × *virgata* and its selections below all tolerate hard pruning and are very suitable for low hedging
C. × *virgata* 'Frosted Chocolate'	Upright, twisted, silvery stems	1.2m (4ft)	1.5–2.5m (5–8ft)	Bronze-green, becoming chocolate coloured in winter	Small, yellow, star-like flowers in spring and early summer	Red fruit	As for *C.* × *virgata*
C. × *virgata* 'Red Wonder'	Upright, twisted, silvery stems	1.5–2.5m (5–8ft)	1.5–2.5m (5–8ft)	Dark-green, silvery underside	Small, yellow, star-like flowers in spring and early summer	Red fruit	As for *C.* × *virgata*
C. × *virgata* 'Sunsplash'	Upright, twisted, silvery stems	1.2–1.5m (4–5ft)	1.2–1.5m (4–5ft)	Cream and green variegated top surface	Small, yellow, star-like flowers in spring and early summer	Orange fruits	As for *C.* × *virgata*
C. × *virgata* 'Yellow Wonder'	Upright, twisted, silvery stems	1.2–1.8m (4–6ft)	1.5m (5ft)	Bronze-green	Small, yellow, star-like flowers in spring and early summer	Yellow fruits	As for *C.* × *virgata*

CORONILLA

A genus of eight species of which only one is hardy. The genus has the following distinguishing features: pinnate leaves, umbels of pea-shaped flowers and long pods that are separated into one-seeded compartments.

Growing Requirements

Hardiness: USDA9b/10a. RHS H3.
Aspect and shelter: A sheltered south- or east-facing position is required. The plant is a little tender and in cold areas, protection from cold winds and heavy frosts may be necessary.
Sun: Full sun.

Coronilla valentina subsp. *glauca*. This plant is luxuriating and flowering profusely in a warm, sheltered position at the base of a south-facing wall.

Coronilla valentina subsp. *glauca*. The bright-yellow fragrant flowers are produced mainly in the spring, but also sporadically throughout the year.

TABLE 4.36 *CORONILLA VALENTINA*, A SUBSP. AND CULTIVAR

Name	Habit	Height	Spread	Foliage	Flowers	Remarks
C. valentina	Upright, bushy	1.3m (4ft)	1.3m (4ft)	Pinnate, green	Bright-yellow, sweetly scented	Very good conservatory plant
C. valentina subsp. *glauca*	Bushy	1m (3ft)	1m (3ft)	Pinnate, leaves, glaucous and with fewer leaflets than species	Bright-yellow. Mainly in spring, but often again in summer	H4
C. valentina subsp. *glauca* 'Citrina' AGM	Bushy	1m (3ft)	1m (3ft)	Pinnate, leaves, glaucous and with fewer leaflets than species	Pale lemon-yellow, scented	H4
C. valentina subsp. *glauca* 'Variegata'	Bushy	1m (3ft)	1m (3ft)	Pinnate, but attractively variegated with creamy-yellow	Bright-yellow	H3

Soil: Moderately fertile soil of any texture, but well drained is necessary.
pH: Soil of a very wide pH range is satisfactory.
Pruning: After a few years, the plants do tend to become rather leggy. This can be corrected by pruning in early spring. This will be followed by fresh, new, vigorous growth.

Garden Design Considerations

Except in very favoured sites, *Coronilla valentina* is best confined to a frost-free conservatory, where it is a delight, flowering freely in spring and filling the space with its fragrant scent.

Coronilla valentina subsp. *glauca* is more compact, slightly hardier and frequently found luxuriating at the base of a warm, sheltered, south-facing wall. Here it flowers profusely starting in late winter, through spring and often again in summer, with vivid-yellow flowers and the same very pleasant scent.

Coronilla valentina subsp. *glauca* 'Citrina' has paler lemon-yellow scented flowers and is preferred by many. *Coronilla valentina* subsp. *glauca* 'Variegata' is attractively variegated, slightly more upright in habit and produces the same bright-yellow flowers, but, unfortunately, is less hardy than the two other subsp. variants and is best kept in a conservatory or placed outside for the summer only.

CORYLOPSIS

A genus of trees and shrubs from the western Himalayas, China and Japan. All, except one species are too large for inclusion here.

Corylopsis pauciflora AGM

A small, densely branched, deciduous shrub. The plant is slow-growing, but eventually reaches 1–1.5m (4–5ft) tall and a spread of 1.5–2m (5–6½ft). In early spring, masses of primrose-yellow, cowslip-scented flowers on short, drooping racemes are produced.

Following flowering, the leaves appear. Whilst emerging, these have a pink coloration, but upon expansion, they turn green and become ovate, 4–6cm (1½–2½in) long. During the summer, the shrub takes a backstage position, becoming a neat mound of green leaves. In autumn, the leaves briefly turn yellow-brown before falling, leaving a finely branched outline.

Corylopsis pauciflora flowering in early spring.

Corylopsis pauciflora. A close up of the small, pale-yellow flowers.

Corylopsis pauciflora. A young plant producing new leaves after flowering. This plant is benefiting from growing in a site with dappled shade.

Corylopsis pauciflora. New leaves have a faint red coloration, which fades to leave a light-green leaf.

Growing Requirements

Hardiness: USDA RHS H5.
Aspect and shelter: Any except north-facing.
Sun: Essentially a woodland shrub requiring part shade.
Soil: The soil should be well drained, but moisture-retentive in summer.
pH: An acid pH.
An annual spring mulch of well-rotted pine needles or leaf mould will be beneficial.

Garden Design Considerations

Providing the correct growing conditions can be provided, this is an excellent shrub for the spring garden, offering both colour and scent. It is best planted by the edge of woodland or a path where the delicate flowers and scent can be fully appreciated.

COTINUS

A genus of six or seven large shrubs and small trees found distributed in Europe, Asia, USA and Mexico. It is typified by alternate simple leaves and large panicles of small flowers, which develop, when spent, a smoke-like appearance leading to their common name of smoke bushes. Formerly in the genus *Rhus*, but now separated on the basis of leaf shape; *Cotinus* has simple leaves, but those of *Rhus* are pinnate.

Growing Requirements

Hardiness: USDA 7b/8a. RHS H5.
Aspect: A site with any orientation is suitable.
Shelter: No particular provision for shelter is necessary.
Sun: Full sun or part shade. Full sun will produce the most vivid autumn leaf colour. Care needs to be taken with yellow-leaved cultivars whose foliage may burn if placed in full sun and then experience moisture shortage in summer.
Soil: A soil of any texture is satisfactory, but the drainage must be good.
pH: A wide soil pH range is accepted.

Note on Maintenance

All the plants in Table 4.37 respond to hard pruning in late winter. The previous season's growth should be cut back to just above its starting point. The remaining branches then form a framework upon which the new growth can spring. This will initially reduce the overall size of the shrub. The plant will then respond during the growing season by producing vigorous, new growth with larger leaves than normal. This can produce

Cotinus coggygria. Seen here as the large shrub it grows into if left to grow naturally. It is tolerant of a wide range of growing conditions, including those that are hot and dry.

Cotinus coggygria. Showing the strong autumn leaf colour that this species develops every autumn.

Cotinus coggygria GOLDEN SPIRIT ('Ancot'[PBR]). In contrast to *Cotinus coggygria*, this cultivar needs a soil that does not dry out in summer if the foliage is not to burn.

Cotinus coggygria 'Royal Purple'. Growing as large as the original species, this cultivar produces purple leaves during the growing season and then delivers good red/orange autumn leaf colour.

Cotinus 'Flame'. A large, vigorous plant, not suitable for a small or compact garden. The large, green leaves turn a brilliant red in autumn.

Cotinus 'Grace'. Another large, vigorous plant. Both *C.* 'Flame' and *C.* 'Grace' produce very vigorous annual extension growth if hard-pruned in the previous winter. Such growth in autumn looks spectacular, but there must be adequate space.

Cotinus 'Ruby Glow'. A shrub that is much smaller than the two previous cultivars. Purple leaves are produced through the growing season. The numerous inflorescences give a good 'smoke bush' effect. In autumn orange and red leaf colour is produced.

Cotinus 'Ruby Glow'. Note the inflorescences are the same colour as the leaves.

a spectacular effect on a well-established plant. This hard pruning can be repeated annually, providing an application of fertiliser follows the pruning.

In the cases of *Cotinus* 'Grace' and *Cotinus* 'Flame', this new growth can be well over 1m (3ft) long. On a recently planted shrub this can look rather odd, but on an established larger plant the effect is spectacular.

Garden Design Considerations

Select your smoke bush with care! As can be seen, there has been much breeding and selection done within this genus. The dwarf and compact cultivars listed in Table 4.37 are excellent for inclusion in a patio planting scheme or in containers. The larger and more vigorous cultivars are definitely not suitable for the small or compact garden.

Correctly chosen, they will add excellent foliage colour to any planting. A range of colours is available during the growing season and most finish with a very bright display of orange and red autumn leaf colour. The purple and yellow-leaved cultivars can be planted adjacently to give strong colour contrasts during the growing season or the yellow-leaved forms could be paired against the smaller, red-leaved cultivars of *Physocarpus*, such as *Physocarpus opulifolius* LITTLE ANGEL ('Hoogi16'[PBR]).

TABLE 4.37 *COTINUS COGGYGRIA* AND CULTIVARS

Name	Habit	Height	Spread	Foliage	Flowers	Remarks
C. coggygria	Large vigorous shrub	4–8m (13–26ft)	4–8m (13–26ft)	Rounded green, gives good autumn leaf colour	Pale-pink inflorescence 15–20cm (5¾–7¾in) long in summer being grey-coloured in autumn	Venetian sumach
C. coggygria GOLDEN LADY ('Mincojaug'[PBR])	Medium-sized, bushy	1.5–2.5m (5–8ft)	2.5–4m (8–15ft)	Rounded, yellow flushed orange in spring, later yellow, by mid-summer lime-green, orange in autumn	Pink flowers becoming silvery grey in fruit	
C. coggygria GOLDEN SPIRIT ('Ancot'[PBR]) AGM	Large, bushy shrub	4–8m (13–26ft)	4–8m (13–26ft)	Rounded, at first coral-red, then yellow, bronze-yellow in autumn	Green flowers becoming grey in fruit	Needs a moisture-retentive soil if foliage is not to burn in summer
C. coggygria 'Lilia'[PBR]	Dwarf, bushy	1.2m (4ft)	1.2m (4ft)	Deep-purple, becoming orange and red in autumn	Pink flowers, grey in fruit	
C. coggygria 'Royal Purple' AGM	Large, vigorous shrub	4–8m (13–26ft)	4–8m (13–26ft)	Initially deep-purple. Becoming redder and translucent in autumn	Pink flowers becoming grey in fruit	
C. coggygria SMOKEY JOE ('Lisjo'[PBR])	Compact. Bushy	1.5–2.5m (5–8ft)	1.5–2.5m (5–8ft)	Green leaves	Purplish pink flowers becoming grey in fruit	
C. coggygria 'Young Lady'[PBR] AGM	Dwarf	1–1.5m (3–5ft)	1–1.5m (3–5ft)	Green leaves	Many pink flowers becoming grey in fruit	
Cotinus DUSKY MAIDEN ('Londus'[PBR])	Small, bushy	1.5–2.5m (5–8ft)	1–1.5m (3–5ft)	Undulating leaves of dark bluish green and purple. Become orange and red in autumn	Yellow flowers on purple stems becoming grey in fruit	
Cotinus 'Flame' AGM	Vigorous, large shrub	4–8m (13–26ft)	4–8m 13–26ft)	Large, green leaves, each 12cm (5in) long. Become orange-red in autumn	Large inflorescence of pink	
Cotinus 'Grace'	Vigorous, large shrub	4–8m (13–26ft)	4–8m (13–26ft)	Large, purple leaves, each 10cm (4in) long	Large inflorescence of deep-pink	
Cotinus 'Ruby Glow'	Small, bushy	1.8m (6ft)	1.8m (6ft)	Purple leaves become vibrant orange-red in autumn	Many inflorescences of pink flowers becoming grey in fruit	

COTONEASTER

A large and horticulturally important genus containing several hundred species. It comprises trees and shrubs, which can be evergreen, semi-evergreen or deciduous. Small, white or rose-tinted, cup-shaped, generally inconspicuous flowers appear in early summer. They are very attractive to bees and are good food plants, the many species and cultivars providing a succession of forage for bumblebees and honeybees. The resultant pollination frequently results in profuse red, orange, yellow or black fruits.

The main aesthetic qualities that cotoneasters bring to the garden are vivid autumn leaf and berry displays.

As an aid to selection, the shrub-like species are separated into low-growing forms suited to ground cover and taller growing shrubs. A few of the larger species, and those specially propagated, are listed under Trees.

Growing Requirements

As a genus the plants are non-demanding in their requirements, but a little attention to detail for some will make for better results.

Hardiness: All are USDA 6b/7a, RHS H6, except *Cotoneaster horizontalis*, which is hardier at USDA 6a-1, RHS H7.

Aspect: Most will accept a site of any orientation.

Shelter: Shelter is not required by deciduous plants, but evergreens require shelter from cold, drying winds in winter.

Sun: Full sun or part shade. Deciduous species will fruit more profusely in full sun, whereas evergreens prefer partial shade.

Soil: A reasonably fertile soil but it must be well drained. Many are resistant to summer drought.

pH: Any pH is fine.

Cotoneaster atropurpureus 'Variegatus' flowing over the sloping ground gracefully.

TABLE 4.38 LOW-GROWING COTONEASTERS FOR GROUND COVER

Name	Habit	Height	Spread	Leaves	D/E*	Flowers/fruit	Remarks
C. adpressus AGM (creeping cotoneaster)	Prostrate	0.1–0.5m (8–18in)	1.5–2.5m (5–8ft)	1cm (½in) long, rounded leaves with wavy edges. Turn red in autumn	D	White flowers followed by red berries	Growth follows the contours of the ground
C. atropurpureus 'Variegatus' AGM	Low and spreading	0.1–0.5m (8–18in)	0.5–1m 18in–3ft)	1cm (½in) long, pointed leaves with white margins that become tinted pink in autumn	D	Pink flowers followed by red berries	Growth form similar to *C. horizontalis* (in the table for taller growing species), but less vigorous
C. dammeri AGM (syn. *C. humifusus, C. humifusa*)	Prostrate and spreading	0.1–0.5m (8–18in)	1.5–2.5m (5–8ft)	Rounded leaves.1.5–3cm (½–¾in)	E	White followed by red berries	
C. procumbens 'Queen of Carpets' AGM	Prostrate	0.1–0.5m (8–18in)	1.5–2.5m (5–8ft)	1cm (½in) long and glossy	E	White followed by red berries	Excellent ground cover
C. salicifolius 'Gnom' AGM	Low and arching	0.1–0.5m (8–18in)	1.5–2.5m (5–8ft)	Glossy narrow leaves 10cm (4in) long	E	White followed by red berries	Specific name indicates willow-like leaves
C. × *suecicus* 'Coral Beauty' AGM	Arching and very dense	1m (3ft)	2m (6ft)	Ovate-elliptic, glossy 2cm (¾in) long	E	White followed by orange-red berries	Many fruits produced
C. × *suecicus* Juliette' AGM	Mat-forming	0.1–0.5m (8–18in)	1.5–2.5m (5–8ft)	Green, grey-green and creamy-white variegated	E	White followed by red berries	Numerous fruits
C. × *suecicus* 'Skogholm' AGM	Arching	0.6m (2ft)	3m (10ft)	Dark-green 2cm (¾in) long	E	White followed by small red berries	Few fruits produced

* D = deciduous; E = evergreen.

Cotoneaster atropurpureus 'Variegatus'. Note the pattern of the variegation and pointed distal end to the leaf.

Cotoneaster dammeri. Young plants in pots showing their spreading habit.

Cotoneaster dammeri. After initially spreading, the growth then thickens to cover the ground.

Cotoneaster × *suecicus* 'Juliette' growth 'flowing' over a low wall. This cultivar also makes an attractive, small standard when grafted on to an upright stem of another cotoneaster.

Cotoneaster × *suecicus* 'Juliette'. Note the pattern of the variegation and rounded end to the leaf.

TABLE 4.39 TALLER GROWING COTONEASTERS

Name	Habit	Height	Spread	Leaves	D/E*	Flowers/fruits	Remarks
C. conspicuus 'Decorus' AGM	Bushy and mound-forming	1–1.5m (3–5ft)	1.5–2.5m (5–8ft)	0.5–2cm (¼–¾in) long, elliptic, dark-green and glossy	E	White, followed by profuse red berries.	Shelter from cold, drying winds.
*C. franchetii***	Bushy	2.5–4m (8–13ft)	2.5–4m 8–13ft)	3.5cm (1½in) long. Glossy grey on upper surface, white underneath	E or semi-E	White followed by red berries	
*C. horizontalis*** (herring bone cotoneaster, wall cotoneaster)	Spreading, flat fans of growth	0.5–1.5m (18in–5ft)	1–1.5m (3–5ft)	1cm (½in) long. Glossy, dark-green. Bright-red in autumn	D	White followed by red berries that persist through winter	H7
C. lacteus AGM	Large, bushy shrub	2.5–4m (8–13ft)	4–2.5m (8–13ft)	Oval, 6cm (2½in) long. Dark-green, but nearly white on lower surface due to tiny hairs	E	Large clusters of white flowers followed by many small red berries	Too large to be considered a compact shrub but included here for use only as a hedging plant
*C. microphyllus*** (small-leaved cotoneaster)	Stiff, upright shoots	1m (3ft)	1.5m (5ft)	1cm (½in) long, narrow, dark-green and glossy	E	White followed by a few red berries	
C. nanshan	Arching and spreading	1m (3ft)	2m (6ft)	Small, rounded and glossy. Red and orange in autumn.	D	Rose-pink followed by orange-red berries 1–1.3cm (½in) across	Good autumn colour from fruit and leaves
*C. simonsii***	Upright and bushy	1.5–2.5m (5–8ft)	1.5–2.5m (5–8ft)	1.5–2.5cm (¾–1in) long	Semi-E, glossy, small leaves	White followed by red berries	

* D = deciduous; E = evergreen.

** *C. franchetii, C. horizontalis, C. microphyllus* and *C. simonsii* are all listed on Schedule 9 of the UK Wildlife and Countryside Act 1981 as invasive non-native species, under which they are plants that may not be planted or otherwise caused to grow in the wild. If not controlled, these species will self-seed and spread in the wild environment to the detriment of the native flora. In several areas they have become a problem and need to be eradicated.

Garden Design Considerations

All the cotoneasters in Table 4.38 have the potential to make very effective, low-maintenance ground-cover providing all weeds, particularly those that are perennial, are eradicated before planting. There are differences in habit and growth rate, but all will rapidly establish and make a dense covering.

Deciduous species will produce more fruit in sunny positions and have the advantage of providing autumn leaf colour. Evergreen species generally do better in semi-shade and give continuous leaf cover but may require some shelter against cold, drying, winter winds. All have some tolerance to dry conditions, and this can be employed to advantage when covering steep banks and slopes that tend to dry out in summer.

Those that have a completely prostrate habit, such as *Cotoneaster adpressus* and *Cotoneaster dammeri*, are useful in softening the edges of exposed brickwork, their growth flowing over dwarf walls, such as those of raised beds.

Cotoneaster atropurpureus 'Variegatus' tends to produce growth in a single, flat plane and can be used as a very effective wall shrub on shaded walls. A system of horizontal wires should be used to train the plant against a wall or fence.

A few, such as *Cotoneaster × suecicus* 'Juliette', can be trained to form very decorative small trees and are mentioned under Trees.

Taller Growing Cotoneasters

Here, a considerable number of species and cultivars could be included, but only those that are relatively small or compact are listed in Table 4.39.

In run-down urban areas, where any plant life may be sparse, the presence of these species may, in fact, be of benefit to local ecosystems, since when flowering they are very attractive, providing a good food source for bees, while birds eat the berries in winter.

Garden Design Considerations

All the plants in Table 4.39 are widely planted. They are particularly useful in covering banks and slopes that may be dry in summer and otherwise difficult to establish cover on. In a shrubbery or mixed border, *C. conspicuus* 'Decorus' and *C. horizontalis* are particularly worthy of inclusion by virtue of their distinctive habits of growth and displays of autumn fruit and, in the case of *C. horizontalis*, fruit and autumn foliage colour.

Cotoneaster conspicuus 'Decorus', *C. horizontalis* and *C. microphyllus* are good as wall shrubs, tolerating well the dry conditions often found at the base of a wall or fence and growing against either of any orientation, although fruiting will be reduced in shade and *C. conspicuus* 'Decorus' must be sheltered from cold, drying winds.

The following can be used for hedging and maintained at the range of height indicated:

Cotoneaster conspicuus 'Decorus'. Several plants clipped to form a series of rounded shapes.

Cotoneaster conspicuus 'Decorus'. A closer picture showing the overall textural effect achieved.

Cotoneaster franchetii. The natural irregular arching habit.

Cotoneaster franchetii. Foliage and fruit.

Cotoneaster horizontalis. Here, forming a hedge backed by a wrought iron fence.

Cotoneaster horizontalis. Herring-bone pattern of the lateral branches and fruit.

Cotoneaster lacteus. A large plant with an irregular, semi-arching habit.

Cotoneaster lacteus. Showing the abundant fruit produced and leaves, which are milky white beneath, hence the specific name.

- *C. simonsii* hedge 1–1.5m (3–5ft)
- *C. franchetii* hedge 1–2.5m (3–8ft)
- *C. lacteus* hedge 1.5–2.2m (5–7¼ft)

In addition to those grown as compact shrubs, some can be propagated and trained, or top-worked on to pre-grown vertical stems, to produce small or very small trees that are ideal for inclusion where space is limited.

TREE COTONEASTERS

Cotoneaster 'Cornubia' AGM (syn. *C.* × *watereri* 'Cornubia', syn. *C. frigidus* 'Cornubia')

Left to grow naturally, this makes a large shrub with a height of 4–8m (13–26ft) and a spread of 4–8m (13–26ft). When grown as a small tree, just one stem of a young plant is first tied up vertically to a cane and other basal side-shoots removed.

This single stem is then taken up to 4–6m (13–19ft) and, once the desired height is reached, this main shoot is released and allowed to grow naturally. Whilst the main stem is growing up vertically, laterals are allowed to develop. Both the main shoot and the laterals develop a graceful, arching habit. Once fully developed as for sale, the plant will have a central stem extending up to 6m (19ft) with lateral branches appearing up this stem.

At planting, the young tree should be staked, with the stake extending up to about one-third the height of the tree. This will prevent the tree from blowing over but allow the stem to flex and develop the normal tapered shape of the trunk. After one or two growing seasons, the tree should be capable of supporting itself and the stake should be released and removed at the start of the subsequent growing season.

This cotoneaster is semi-evergreen; during the winter some leaves may develop a bronze coloration and if the weather is very cold, some or all the leaves may drop off. In spring, new leaves will be produced.

The leaves are dark-green, narrowly elliptical and 12cm (5in) long. In early summer, many small, white flowers are produced, followed in the autumn by a profusion of large, 7mm (¼in) diameter red fruits. These are well displayed by the dark foliage and last into the winter.

As a tree, the overall dimensions will be height 4–6m (13–19ft) and spread 3–4m (10–13ft).

Cotoneaster 'Exburiensis' (syn. *C. salicifolius* 'Exburiensis')

Left to grow naturally, this vigorous plant makes a large shrub with a height of 5m (16ft) and spread of 5m (16ft).

To produce a single-stemmed tree, the same production process, planting and staking as explained for *C.* 'Cornubia' is followed.

This plant is evergreen. The leaves are mid-green, narrowly lance-shaped and 12cm (4¾in) long. In early summer, many small, white flowers are produced, followed in autumn by masses of pale-yellow berries, each 5mm (¼in) across. Once exposed to cold, these fruits become tinged with pink.

As a tree, the overall dimension will be height 4–5m (13–16ft) and spread 2–3m (6½–10ft).

Cotoneaster 'Hybridus Pendulus' (syn. *C. salicifolius* 'Pendulus', syn. *C.* × *watereri* 'Pendulus')

When left to grow naturally, this evergreen cultivar makes a vigorous, spreading, prostrate shrub with a height of 0.1–0.5m (¼–1½ft) and a spread of 1.5–2.5m (5–8ft). Along with several other cotoneaster species and cultivars described under Shrubs it is often used for very effective ground cover.

To produce a single-stemmed tree, the same process as described under *C.* 'Cornubia' may be followed or the tree may be produced by top-working. In this case, the height of the tree is determined by the height of the stem used for the graft. The main stem and laterals, when left to grow naturally, develop a pendulous, graceful habit.

The leaves are dark-green and elliptical. Small, white flowers are produced in early summer and these are followed by profuse red berries.

If taken up to or top-worked at 2m (6½ft), then overall eventual dimensions will be 2–3m (6½–10ft) tall with a width of 1.5–2m (5–8ft).

Cotoneaster × *suecicus* 'Coral Beauty' and *Cotoneaster* × *suecicus* 'Juliette' AGM

Both these cultivars can be top-worked to form attractive, very small, weeping trees. The vertical stem on to which they are grafted is usually about 85cm (33in) tall. The tree produced will have an eventual height of 1.25m (4ft) and a spread of 1.25m (4ft).

If allowed to grow naturally, they are low-growing shrubs and fully described under Cotoneasters for ground cover.

Garden Design Considerations

All the cultivars shown above have been selected to form striking individual specimens when produced as described. In the garden they all lend themselves for this purpose and make a striking focal point when in fruit.

The three slightly larger trees are intended for small gardens, but where overall space is not too limited. The two *C.* × *suecicus* cultivars are excellent for the same purposes, but where space is very limited, e.g. tiny patio gardens, a small 'secret' garden within a larger garden or in raised beds.

Cotoneaster 'Hybridus Pendulus'. A plant trained to form a small tree.

Cotoneaster 'Hybridus Pendulus'. Detail of leaves and fruit.

DANAE

A monotypic genus containing only *Danae racemosa* (Alexandrian laurel). This species is native to South-West Asia and Northern Iran. It is found growing as an understorey shrub beneath deciduous trees.

Growing Requirements

Hardiness: USDA7b/8a. RHS H5.
Aspect: The planting site can have any orientation.
Shelter: Shelter is required from cold, drying winds.
Sun: Full sun to part shade.
Soil: The soil can be of any texture providing that it is well drained.
pH: A wide pH range is tolerated.

Danae racemosa. A small, evergreen, fruiting shrub. Useful for dry, shaded areas.

Danae racemosa. What appear to be leaves are, in fact, modified stems called phyllodes. The fruit are more abundantly produced after a warm summer.

Danae racemosa AGM

Height: 1–1.2m (3–4ft).
Spread: 1–1.2m (3–4ft).
Habit: A small, evergreen, elegantly arching, suckering, slow-growing shrub.
Leaves: Botanically interesting, since what appears as its leaves are, in fact, flattened stems known as phyllodes. These are dark-green and shiny.
Flowers: Tiny, insignificant, greenish-white flowers are produced in spring.
Fruit: The flowers are followed by bright-red fruits in autumn after a warm summer. This does seem to be becoming more of a regular occurrence. The berries persist into early winter.

Garden Design Considerations

A useful, unusual, under-utilized, small, evergreen shrub for deciduous part-shade, even dry shade once established. In a small garden it can provide small-scale, ever-green, attractive ground-cover in difficult to plant, dry, shaded areas. The plant has a low-maintenance requirement, just needing the pruning out of old, discoloured shoots at their base in spring, just as new replacement shoots are appearing.

DAPHNE

A genus of small, beautiful shrubs, which can be deciduous or evergreen, and most have fragrant flowers. The leaves are always simple and entire.

In this genus, the perianth consists of only one floral whorl. The flowers are tubular at their base, opening at the mouth to form four lobes.

All parts of these plants are highly toxic, and the sap may irritate the skin.

Their range of application in the garden is very varied and the selection here has been restricted to those in general cultivation and of a compact nature. Only a few of the very compact and dwarf cultivars suitable for alpine troughs, crevice or rock gardens have been included, just to whet the appetite!

Growing Requirements

Hardiness: As a genus it varies between USDA8b/9a or 7b/8a and RHS H4 or H5. It cannot be considered reliably hardy and cold, wind-swept situations should be avoided.
Aspect and shelter: It doesn't like a northerly orientation and requires a sheltered site.
Soil: It requires good well-drained soil, but this must also be moisture-retentive in summer. Dryness at the roots in summer will result in poor growth and can cause plant death. On sandy soils, organic matter should be incorporated before planting and followed by annual mulching. Both *Daphne laureola* and *Daphne mezereum* are tolerant of heavier soils.
Pruning: Any form of pruning often results in dieback and should be avoided unless absolutely necessary.

This is a group about which it is difficult to generalize. Several have the reputation of being 'difficult' and it is suspected that, despite being shrubs, some may be naturally short-lived. Some species are known to be susceptible to virus infection, especially *Daphne mezereum*, and a complex of several viruses may accumulate and ultimately kill the plant.

They hate root disturbance and planting should be done with care and only in the spring under good growing conditions. Transplanting of established plants is better not undertaken.

Soil pH and light-level requirements vary from plant to plant, and information is given for each plant listed.

Having said all this, most gardeners consider it well worthwhile to pander to each plant's set of requirements, but they are not plants for the general landscape.

Daphne bholua

A species from the Himalayas. It can be deciduous or evergreen. Hardiness and leaf retention depend upon the height at which the original plant was collected.
Height: 1.5–2.5m (5–8ft).
Spread: 1–1.5m (3–5ft).
Habit: An upright, suckering shrub.
Flowering: Occurs in late winter and early spring. Deep-pink in bud, opening light-pink to white, variable from plant to plant. Flowers sweetly scented.

Growing Requirements

Hardiness: USDA 8b/9a RHS H4.
Aspect: South- or south-west-facing.
Shelter: Sheltered from cold winds.
Sun: Sun or part shade.
Soil: Soil good, well drained, moisture-retentive. If sandy, add organic matter.
pH: Moderately alkaline to mildly acidic.

Daphne cneorum (Garland Flower or Rose Daphne)

This small, prostrate, evergreen shrub is found in the mountains of central and southern Europe. Growing only to a height of 20cm (8in), it produces dense clusters of fragrant, pink flowers in spring and early summer. The cultivar *Daphne cneorum* 'Exima' AGM, said to have slightly larger flowers and leaves than the species, is the plant commonly found in cultivation and is the plant described here.

Growing Requirements

Hardiness: USDA 7b/8a. RHS H5.
Aspect: A site of any aspect is acceptable except north-facing.
Shelter: The planting position should be sheltered.
Sun: Full sun or only part shade is needed.
Soil: The soil can be any texture and must be well drained, but moisture-retentive in the summer.
pH: Neutral or alkaline.

This plant has a reputation for being 'difficult'. It hates being transplanted. Ideally, the young plants should be grown in loam-based compost, planted with care not to damage or disturb the roots into the soil described above. The plant should be kept just moist during establishment. Plants may be naturally short-lived, but top-dressing with an organic and gritty mix to eventually layer the prostrate stems may help to prolong life.

Daphne bholua, a white-flowered form. This species, when collected from the wild, has been found to be very variable in hardiness and flower colour.

Daphne bholua 'Jacqueline Postil'. An upright-growing, hardy, evergreen cultivar flowering in late winter and early spring.

Daphne bholua 'Jacqueline Postil'. The soft-pink, heavily scented flowers are produced in profusion.

TABLE 4.40 *DAPHNE BHOLUA* VAR. *GLACIALIS* 'GURKHA' AGM

Habit	Height	Spread	D/E*	Leaves	Flowers	Remarks
Upright, suckering shrub	2m+ (6ft+)	1–1.5m (3–5ft)	D	Oblanceolate	Occurs in late winter and early spring, after leaves have fallen	H4. The hardiest cultivar. Buds pink, opening white with a little pink

* D = deciduous; E = evergreen.

TABLE 4.41 *DAPHNE BHOLUA* 'JACQUELINE POSTILL' AGM

Habit	Height	Spread	D/E*	Leaves	Flowers	Remarks
Upright, suckering shrub	3m+ (10ft+)	1.5–2m (3–5ft)	Semi-E. Leaves drop after flowering, providing a backing for the flowers	Oblanceolate	Occurs in late winter and early spring, before leaves fall	H4. Buds dark-pink, opening pink

* D = deciduous; E = evergreen.

Daphne gemmata 'Royal Crown'

Daphne gemmata is small, compact shrub found in south-western China. *Daphne gemmata* has unscented flowers, but those of *Daphne gemmata* 'Royal Crown' are scented.

Only marginally hardy in the UK. May require winter protection, including frost protection, and may be better grown in a cool greenhouse.

Growing Requirements

Hardiness: USDA8b/9a. RHS H4.
Aspect: For outside cultivation, any orientation except north-facing is tolerated.
Shelter and sun: The site must be sheltered and in dappled or part shade.
Soil: The soil can be of any texture but must be well drained and should have high organic matter content.
pH: A wide soil pH range is accepted.
In the colder parts of the UK, cultivation in a pot is preferable, the plant being placed outside in a shaded position for the summer months but brought into a cool greenhouse before cold or frosty conditions develop.

Daphne laureola (Spurge Laurel)

Found in south and west Europe, including the British Isles. Usually grows in shade on calcareous soils.

Growing Requirements

Hardiness. USDA 7b/8a. RHS H5.
Aspect: Any orientation but avoid a hot, south-facing situation.
Shelter: Only slight shelter is necessary.
Sun: Will grow well in sun, half shade or full shade, but very useful for heavy shade.
Soil: Any that is well drained.
pH: Tolerant of a wide range, including high pH.

Daphne laureola subsp. *philippi*. A useful, small, evergreen shrub that is content to grow in dry shade if given care at, and after, planting.

Daphne laureola subsp. *philippi*. This is a compact form of the species. The flowers cannot be considered exciting, but this small, evergreen shrub is appropriate on the rock garden where it helps to furnish this feature in the winter and can act as a backdrop for small spring bulbs, such as *Galanthus*.

TABLE 4.42 *DAPHNE CNEORUM* 'EXIMA'

Name	Habit	Height	Spread	E/D*	Leaves	Flowers	Remarks
D. cneorum 'Exima' AGM	Small, prostrate, forming a mat of growth	0.1–0.5m (6–15in)	0.5–1m (1½–3ft)	E	Narrow, oblanceolate, 15mm (¾in) long, 5mm (¼in) wide	Flower buds red, opening to give bright-pink, fragrant flowers	Flowers late spring into early summer. Flowering very profuse

* D = deciduous; E = evergreen.

TABLE 4.43 *DAPHNE GEMMATA* 'ROYAL CROWN'

Name	Habit	Height	Spread	E/D*	Leaves	Flowers	Remarks
D. gemmata 'Royal Crown'	Compact and bushy	75cm (30in)	75cm (30in)	D	Dark-green	Late spring and early summer. Heads of scented yellow flowers. Red berries in autumn	Where hardy, good for the woodland or shaded garden. In a pot, good on a patio or near an entrance to scent the area when in flower

* D = deciduous; E = evergreen.

TABLE 4.44 *DAPHNE LAUREOLA* AND SUBSP.

Name	Habit	Height	Spread	E/D*	Leaves	Flowers	Remarks
D. laureola	Small, bushy shrub	0.5–1m (1½–3ft)	1–1.5m (3–5ft)	E	Thick, fleshy polished, obovate 8cm (3in) long	Occurs in late winter and early spring	Flowers lightly fragrant, yellowy-green in clusters beneath leaves. Followed by ovoid, black fruits
D. laureola subsp. *philippi* (dwarf spurge laurel)	Compact, small, semi-prostrate shrub	0.1–0.5m (4in–1½ft)	0.1–1.5m (4in–5ft)	E	Thick, fleshy, polished, obovate 6cm (2½n) long	Occurs in late winter and early spring	Flowers lightly fragrant, yellowy-green in clusters, slightly smaller than those of *D. laureola*. Followed by black, ovoid fruits

* D = deciduous; E = evergreen.

Daphne mezereum (Mezereon)

A small, deciduous shrub found wild in western Asia, right across Russia and Europe. A rare British native that usually grows on calcareous soils.

Growing Requirements

Hardiness: USDA6b/7a. RHS H6.
Aspect: Any orientation is acceptable, except cold north-facing.
Shelter: Some shelter is required.
Sun: Full sun or just part shade is needed.
Soil: The soil can be of any texture but must be well drained. Organic matter should be incorporated before planting. A cool root-run that does not dry out in summer is needed and mulching in spring is recommended.
pH: A wide pH range is accepted including chalky soils.

Daphne mezereum f. alba. A choice, small, deciduous, late winter-flowering shrub, producing white, scented blooms.

TABLE 4.45 *DAPHNE MEZEREUM*, PLUS A FORMA AND A VARIETAS

Name	Habit	Height	Spread	D/E*	Leaves	Flowers	Remarks
D. mezereum	Rounded bush	1–1.5m (3–5ft)	0.5–1m (20–40in	D	Oblanceolate, dull grey-green, short-stalked. Often poor due to virus infection	Late winter, early spring, before leaf emergence	Small, purple-red, highly scented. Clustered on one-year old growth. Red fruit
D. mezereum f. *alba*	More upright bush	1–1.5m (3–5ft)	0.5–1m (20–40in)	D	Oblanceolate, dull grey-green, short-stalked. Often poor due to virus infection	Late winter, early spring, before leaf emergence	White, highly scented. Yellow translucent fruit
D. mezereum var. *rubra*	Rounded bush	1–1.5m (3–5ft)	0.5–1m (20–40in)	D	Oblanceolate, dull grey-green, short-stalked. Often poor due to virus infection	Late winter, early spring, before leaf emergence	Deep red-purple, highly scented. Red fruit

* E = evergreen; D = deciduous.

TABLE 4.46 *DAPHNE ODORA* AND CULTIVARS

Name	Habit	Height	Spread	D/E*	Leaves	Flowers	Remarks
D. odora	Bushy	1–1.5m (3–5ft)	1–1.5m (3–5ft)	E	Dark-green, simple	Late winter, early spring. Buds pink, opening light-pink. Scented	Avoid a hot south-facing site
D. odora 'Aureomarginata'	Bushy	1–1.5m (3–5ft)	1–1.5m (3–5ft)	E	Dark-green, edged with cream. 7cm (2½in) long, 2cm (¾in) wide	Late winter, early spring. Buds pink, opening light-pink. Scented	
D. odora 'Mae-jima'	Bushy	1–1.5m (3–5ft)	1–1.5m (3–5ft)	E	Dark-green with broad yellow margin	Late winter, early spring. Buds purplish pink, opening light-pink. Scented	
D. odora Marianni 'Rogbret'	Bushy	1–1.5m (3–5ft)	1–1.5m (3–5ft)	E	Dark-green with broad yellow margin	Late winter, early spring. Buds purple, opening lavender pink. Scented	

* E = evergreen; D = deciduous.

Daphne odora

A small, evergreen shrub coming from China and long cultivated in Japan. The species is not common in cultivation, the variegated forms described here being much more widely planted. The variegated forms are credited as being slightly hardier, but their popularity compared to the species is probably due to a brighter overall appearance in the garden.

Growing Requirements

Hardiness: USDA RHS H4.
Aspect: A site of any orientation, except facing south, is suitable for *D. odora*, which does not enjoy excessive sun. The variegated cultivars need more sun and a site of any orientation is acceptable to them.
Shelter and sun: Full sun or part shade and a sheltered position are good for the growth of all.
Soil: Soil of any texture is accepted providing that it is well drained but adding organic matter to the soil prior to planting is beneficial
pH: A wide range, including alkaline conditions, are suitable.

Daphne PERFUME PRINCESS ('Dapjur01')

A hybrid *Daphne* created by crossing *Daphne odora* with *Daphne bholua* by the plant breeder Mark Jury. He and his family are also well known for introducing many fine *Magnolia* hybrids, several of which are mentioned under *Magnolia*.

Growing Requirements

Hardiness: USDA 7b/8a. RHS H5.
The same as for *D. odora* and *D. bholua*.

Daphne odora 'Aureomarginata'. A slow-growing, evergreen shrub flowering more profusely in a sunny position.

Daphne odora 'Aureomarginata'. Flowering in late winter and spring.

Daphne 'Mai-jima'. A small, evergreen shrub with a very bright, marginal leaf variegation.

Daphne 'Mai-jima'. The clusters of scented, pale-pink flowers are well set above the leaves.

TABLE 4.47 *DAPHNE* PERFUME PRINCESS ('DAPJUR01')

Name	Habit	Height	Spread	D/E*	Leaves	Flowers
D. PERFUME PRINCESS ('Dapjur01')	Bushy	1.5m (5ft)	1m (3ft)	E	Dark-green, lustrous ovate 5cm (2in) long 2cm (¾in) wide	Flowering from late winter to early spring in clusters, pink in bud opening white, pink at base. Scented

* E = evergreen; D = deciduous.

Daphne Perfume Princess ('Dapjur01[PBR]'). An evergreen shrub, potentially slightly larger than *D. odora* producing large clusters of scented flowers in late winter and early spring.

Daphne pontica (Pontic Daphne, Twin-Flowered Daphne)

A small, rounded, evergreen shrub from Bulgaria, northern Turkey and the Caucus.

Growing Requirements

Hardiness: USDA 7a/8a. RHS H5.
Aspect: A site of any aspect is accepted except a hot south-facing site.
Shelter: A little shelter from the coldest winds is required.
Sun: Any light level is acceptable from full sun, through various degrees of shade to heavy shade.
Soil: The soil can be of any texture, but the drainage must be good and a soil with a high organic content is enjoyed. Mulching in spring to retain soil moisture and keep a cool root run is recommended.
pH: Soil pH can be from acid to alkaline with calcareous soils being very acceptable providing organic matter is added at planting and maintained by annual mulching.

TABLE 4.48 *DAPHNE PONTICA*

Name	Habit	Height	Spread	D/E*	Leaves	Flowers	Remarks
D. pontica	Small, rounded, bushy, evergreen shrub	0.5–1m (1½–3ft)	1–1.5m (3–5ft)	E	Leathery, mid-green leaves. Sessile, obovate, 4.5–10cm (1¾–4in) long, 2–3.5cm (¾–1½in) wide	In clusters of usually twinned, yellow-green flowers, each 2cm (¾in) across	Flowers from spring to early summer. Elusive scent, strongest in mild weather. Fruit ovoid, black

* D = deciduous; E = evergreen.

Daphne pontica. A tough, evergreen shrub flowering from spring into summer, having a quiet charm of its own. Seedlings of this species may be used as rootstock for other daphnes.

Daphne pontica. The small, green flowers have an elusive scent.

Daphne × *rollsdorfii*

A hybrid *Daphne* made by crossing *D. petraea* with *D. sericea* Collina Group. They are a group of dwarf, evergreen shrubs of bushy and erect growth with clusters of fragrant flowers. The following two cultivars are selections from within this group.

Growing Requirements

Hardiness: USDA 7b/8a. RHS H5.
Aspect: A site facing north, south-east or west will suit this group of hybrids.
Shelter and sun: They require full sun or just part shade, plus shelter.

Soil: The soil can be of chalk, loam or sandy texture, but must be well drained and humus-rich. The roots of these plants need to be kept cool in the summer and an organic mulch is recommended.

TABLE 4.49 *DAPHNE* × *ROLLSDORFII* CULTIVARS

Name	Habit	Height	Spread	D/E*	Leaves	Flowers	Remarks
Daphne × *rollsdorfii* 'Arnold Cihlaz'	Small, mounded shrub, with erect stems	10–50cm (4–19½in	10–50cm (4–19½in	E	Oblong, dark glossy green, 25mm (1in) long 8mm (¼in) wide	Terminal clusters of pink flowers	In early summer and autumn
Daphne × *rollsdorfii* 'Wilhelm Schacht' AGM	Small, dense, rounded shrub	10–50cm (4–19½in	10–50cm (4–19½in	E	Oblong, dark glossy green, 25mm (1in) long, 8mm (¼in) wide	Purple-pink flowers in terminal clusters	In early summer and autumn

* D = deciduous; E = evergreen.

Daphne × *rollsdorfii* 'Arnold Cihlaz'. One of several cultivars selected from this cross. A gem for the rock garden.

Daphne × *rollsdorfii* 'Arnold Cihlaz'. Producing pink flowers in spring.

Daphne × *rollsdorfii* 'Wilhelm Schacht'. Another cultivar resulting from the same cross. Again, suitable for the rock garden.

Daphne × *rollsdorfii* 'Wilhelm Schacht'. Producing masses of deep-ink flowers in spring and often again later in the season.

Daphne tangutica AGM

Slow-growing, small, evergreen shrub from western China and the Himalayas.

Growing Requirements

Hardiness: USDA 7b/8a. RHS H5.
Aspect: Any orientation, except north-facing.
Shelter and sun: Full sun or part shade, plus some shelter.
Soil: Soil can be chalk, clay, loam or sand, but it must be well drained and preferably humus-rich.
pH: Neutral or nearly so.

Daphne × *transatlantica*

A hybrid *Daphne* created by crossing *Daphne caucasica* with *Daphne sericea* Collina Group. From this cross, several plants have been selected, named and propagated.

Growing Requirements

Hardiness: USDA 7b/8a. RHS H5.
Aspect: A site of any orientation, except facing north is acceptable.
Shelter: Some shelter should be provided.
Sun: Full sun to part shade, but not heavy shade is required.
Soil: The soil can be of any texture, must be well drained, but should not dry out in summer. The addition of organic matter is often beneficial.
pH: A wide range of soil pH is tolerated from slightly alkaline to slightly acid.

Garden Design Considerations

Here we are considering several medium-sized to very small shrubs with a wide range of potential applications in the garden.

The Winter Garden or Winter Border

The cultivars of *Daphne bholua* and *Daphne odora* should feature strongly. *Daphne bholua* 'Jaqueline Postil' could act as flowering backdrop, this cultivar being the best suited to the role because of its longer leaf retention. As feature plants within a border or larger planting, both this cultivar and *Daphne bholua* var. *glacialis* 'Gurkha' would be outstanding, providing a wonderful late-winter display of heavily scented flowers.

The smaller *Daphne odora* cultivars and *Daphne* PERFUME PRINCESS ('Dapjur01') would be better placed towards the front of such a planting because of their smaller stature. Their evergreen leaves help to provide a contrasting backdrop to items such as coloured stems contributed by other plants in the design.

Such winter-flowering, highly scented plants are excellent selections for deliberate placement near house doors leading outside and alongside garden paths, where their scent and colour can be readily appreciated.

The Garden In Spring and Early Summer

At this time of year, the garden should be full of colour and scent, much of which can be provided by judicious selection within the daphnes. Cultivars of *Daphne odora* and *Daphne* PERFUME PRINCESS ('Dapjur01') will continue to flower into early spring.

Daphne mezereum starts to flower in late winter and continues into early spring. It is the typical cottage garden daphne, having been lifted from the wild and placed in many such gardens in the previous two centuries. Such practice should be frowned upon and, indeed, is illegal today, but plants from cultivation can fulfil the same role. In a mixed border, a large rock garden or conifer and heather planting, it is a great stand-out plant, combining vivid flower colour and strong appealing scent.

A little later in mid-spring, the floral display is continued by *Daphne* × *transatlantica* ETERNAL FRAGRANCE ('Blafra'PBR) and *Daphne* × *transatlantica* PINK FRAGRANCE ('Blapink'PBR). These are extremely useful, long-flowering, dwarf shrubs that can be readily incorporated into a small border or planted densely to form a low evergreen hedge.

The Rock Garden

Of course, spring is the time of year associated with colour and interest in the rock garden and here again *Daphne* can make a wonderful contribution. *Daphne tangutica* and *Daphne tangutica* Retusa Group form small, dense bushes and are well placed in the larger rock garden and mixed border with sun or part shade to provide flower colour and scent.

A range of choice dwarf, some very slow-growing daphnes are available for the plantsman. Two *Daphne* × *rollsdorfii* cultivars have been detailed above to demonstrate something of what is available. These dwarf shrubs are only suitable for growing with other tiny alpine or rock garden plants in scree beds, specialist alpine beds, crevice beds, troughs and raised beds, being in danger of being overgrown and overwhelmed if mixed with other vigorous plants. *Daphne cneorum* 'Exima' with its prostrate stems forming a low mat of growth is another very beautiful dwarf shrub well suited to the range of alpine situations.

Cultivars of *Daphne odora* and many other small daphnes can be included in the rock garden, raised bed or deep container. Containers of adequate depth, minimum 30cm (1ft), are required, as it should be remembered these plants have an aversion to the roots becoming dry or too warm during the summer.

The Shaded or Woodland Garden

Here daphnes can be important components. Several favour part shade and organic rich soil, as mentioned in their entries above.

Daphne laureola and *Daphne pontica* are shade specialists, being perfectly content to grow even in dry shade, providing extra organic matter is incorporated prior to planting, and irrigation given to aid establishment. In this situation they supply much needed evergreen furnishing of the planting, and interesting if not vivid flowers in spring. Their foliage also helps as a contrast to highlight the many spring flowering bulbs that do well in deciduous shade such as *Galanthus* (snowdrops) and *Hyacinthus* (bluebells), *Erythronium* and *Trillium*.

Daphne laureola and the shorter, more compact *Daphne laureola* subsp. *philippi* also look well in the rock garden, contributing much needed evergreen leaves in the winter and a distinct habit of growth.

Late Summer Into Autumn

At this time, daphnes make a more muted contribution, but on the rock garden, *Daphne* × *rollsdorfii* cultivars and *Daphne tangutica* will often flower again.

Given a kind spring, *Daphne mezereum* may have red berries and *Daphne mezereum* f. *alba* presents a display of very attractive, yellow, translucent berries that can glow in the shafts of autumn sunlight.

Then, after a pause, the winter-time display starts again!

TABLE 4.50 *DAPHNE TANGUTICA* AND *DAPHNE TANGUTICA RETUSA* GROUP

Name	Habit	Height	Spread	D/E*	Leaves	Flowers	Remarks
D. tangutica AGM	Small bush	0.5–1m (1½–3ft)	0.5–1m (1½–3ft)	E	Leaves oblong, 8cm (3in) long	Flowers white on upper surface and rose-purple on underside of perianth. Scented	Flowers in spring and again in late summer. Red fruit
D. tangutica Retusa Group AGM	Slow-growing and dwarf. Dense, rounded shape	0.5–1m (1½–3ft)	0.5–1m (1½–3ft)	E	Shorter dark-green elliptic	Terminal clusters, pink in bud, opening white. Scented	Flowers in early summer and often again in autumn

* D = deciduous; E = evergreen.

TABLE 4.51 *DAPHNE* × *TRANSATLANTICA* CULTIVARS

Name	Habit	Height	Spread	D/E*	Leaves	Flowers	Remarks
D. × *transatlantica* ETERNAL FRAGRANCE ('Blafra'PBR)	Small, compact	0.5–1m (1½–3ft)	0.5–1m (1½–3ft)	Semi-evergreen	Small, dark-green, long and thin with round tip	White, pink flushed, fragrant	Very long flowering period from late spring to autumn
D. × *transatlantica* PINK FRAGRANCE ('Blapink'PBR)	Small, compact	0.9m (3ft)	0.9m (3ft)	Semi-evergreen	Small, dark-green, long and thin with round tip	Pink	Very long flowering period from late spring to autumn

* D = deciduous; E = evergreen.

× *transatlantica* ETERNAL FRAGRANCE ('Blafra'PBR). An evergreen shrub with a very long flowering period from late spring into early autumn.

× *transatlantica* ETERNAL FRAGRANCE ('Blafra'PBR). Producing small, scented, pale-pink flowers.

Daphne tangutica Pink Fragrance ('Blapink'PBR) essentially the same as ETERNAL FRAGRANCE but producing darker, pink, scented flowers.

ELAEAGNUS

Elaeagnus can be evergreen or deciduous trees or shrubs. Young growth is covered with silver or brownish scales, which gives a characteristic whitish or silvered appearance. The leaves are alternate. The flowers are small and silvery, each flower consisting outwardly of a cylinder, which expands into four lobes at the mouth, and are usually fragrant.

Growing Requirements

Hardiness: USDA 7b/8a. RHS H5.
Aspect: The planting site can face any direction except north.
Shelter: No special provision of shelter is required.
Sun: Full sun or part shade.
Soil: The soil can be of any texture, but it must be well drained.
pH: The genus has a wide tolerance to acid, neutral or alkaline soils, but plants may develop chlorosis in very alkaline conditions.

Garden Design Considerations

It could be argued that *Elaeagnus* is too large for small, compact gardens. To do so would be to lose some of the most brightly variegated shrubs available to gardeners.

Elaeagnus pungens 'Frederici' is slow-growing and makes the smallest shrub of those considered, but its variegation is erratic and not particularly bright. *Elaeagnus pungens* 'Maculata' is slightly faster growing and eventually does make a larger shrub, but its variegation is very bright. In winter, if set in a sunny position, it makes a valuable contribution and is well worthy of inclusion in any winter border or plant association.

Elaeagnus 'Quicksilver' is very different, being larger and deciduous. Its narrow foliage is very similar in appearance to *Pyrus salicifolia* 'Pendula' AGM (the weeping pear) and in the garden it can be employed in a similar way. Left to grow naturally, it makes a large, suckering shrub and can be used to provide a light backdrop to a large border, perhaps with other coloured leaf shrubs. Trained to form a small tree, it will provide height, foliage colour and light shade. As such it is suitable for a small garden or a garden room of a larger garden.

Elaeagnus × *submacrophylla* is a large, evergreen, spreading shrub. It is used extensively as low-maintenance ground-cover in landscaping schemes but is too large for small gardens, except possibly for dense, evergreen hedging, if space permits.

Elaeagnus × *macrophylla* 'Gilt Edge', which is potentially large, can, like all other *Elaeagnus*, be kept to a compact smaller size by regular clipping. The gardener is rewarded with a brightly variegated shrub of similar value to *E. pungens* 'Maculata'. *Elaeagnus* × *submacrophylla* 'Limelight', together with 'Vivleg'PBR, are faster growing than 'Gilt Edge', which is a feature not often wanted where space is limited.

Elaeagnus × *macrophylla* 'Gilt Edge' and *E. pungens* 'Maculata' can be employed to fulfil a wide range of garden functions, including formal and informal hedging, screening and ground cover over difficult areas, such as banks and slopes.

The thick, scaley leaves of *Elaeagnus* make it resistant to wind and very suitable for the provision of shelter in windswept gardens and inclusion in all coastal gardens.

TABLE 4.52 *ELAEAGNUS* CULTIVARS

Name	Habit	Height	Spread	Foliage	Flowers	Remarks
*E. pungens** 'Frederici'	Dense, bushy shrub	1.5m (5ft)	1.5–2.5m (5–8ft)	Evergreen. Narrow leaves irregularly marked with pale creamy-yellow	Small, silvery, fragrant. Autumn	Slow-growing
*E. pungens** 'Maculata'	Dense, bushy, with a few spines	1.5–1.8m (5–6ft)	2.5–4m (5–13ft)	Evergreen. Broad leaves, green with a large central splash of yellow	Small, silvery, fragrant. Autumn	Very bright variegation
E. 'Quicksilver'	Large shrub or small tree	2.5–4m (8–13ft)	2.5–4m (8–13ft)	Deciduous. Narrow, silvery leaves	Small, fragrant, yellow. Summer	Does sucker, but can be trained to be a single- or few-stemmed, small tree
E. × *submacrophylla* (syn. × *ebbingei*)	Large, spreading, dense-growing	2.5–4m (8–13ft)	2.5–4m (8–13ft)	Evergreen. Top surface dark silvery-green. Bottom surface silvery	Small, silvery, fragrant. Autumn. Followed by small, orange fruits	Flowers not produced until shrub is mature
E. × *submacrophylla* 'Gilt Edge' AGM		1.5–2m (5–6½ft)	2–3m (6½–10ft)	Evergreen. Large with top surface green with a broad yellow margin	Small, silvery, fragrant. Autumn	
E. × *submacrophylla* 'Limelight'		2–2.5m (6½–8ft)	2.5–4m (8–13ft)	Evergreen. Leaves green when young. Later develop a central area of yellow and pale-green	Small, silvery, fragrant. Autumn	Faster growing than 'Gilt Edge' and prone to reversion
E. × *submacrophylla* 'Viveleg'PBR	Fast-growing, rounded shrub	2–2.5m (6½–8ft)	2.5–4m (8–13ft)	Evergreen. Leaves large with a golden-yellow edge	Small, silvery, fragrant. Autumn	Similar to 'Gilt Edge', but more vigorous and faster growing

* *Elaeagnus pungens* has become an invasive species in south-eastern USA and should no longer be planted.

Elaeagnus pungens 'Frederici'. Foliage showing the lime and lime-yellow variegation.

Elaeagnus pungens 'Maculata'. Foliage showing the brighter lime and lime-yellow chorda × macrantha 'The Bride'. A close up of the pure white flowers. variegation. This cultivar is very effective in brightening any winter border.

Elaeagnus 'Quicksilver'. Here, this cultivar has been formed into a very graceful, small tree.

Elaeagnus 'Quicksilver'. The silver-grey foliage.

Elaeagnus × submacrophylla. The rather dull-grey foliage, plus unusually fruit.

Elaeagnus × submacrophylla 'Guilt Edge'. Seen here enlivening a small, wall-side border.

Elaeagnus × submacrophylla 'Guilt Edge'. The irregular, central green region and broad, golden margin is a very bright form of variegation.

ESCALLONIA

This genus consists of around forty species coming almost entirely from South America. With a single exception they consist of evergreen shrubs and trees that flower in summer and early autumn. Collectively they have alternate leaves and terminal racemes of five-petalled flowers.

Growing Requirements

Hardiness: USDA 8b/9a. RHS 4, in the main. Those with a rating of only RHS H3 are specifically noted in the table.
Aspect: It is best to avoid a north-facing site.
Shelter: All species and cultivars require shelter from cold, drying winds. They will, however, grow well where they experience mild but moist winds.
Sun: Full sun or part shade suits most, but yellow-variegated plants perform best in full sun.
Soil: Soil of any texture is accepted, but the soil must be well drained and retain adequate moisture in the summer.
pH: They will grow well in acidic, neutral and alkaline soils.

Several species and cultivars are too large for compact or small gardens. Only those that are suitable have been included in Table 4.53. All are evergreen but produce a new flush of leaves in spring.

Garden Design Considerations

The taller growing cultivars listed in Table 4.53 are excellent to form evergreen summer-flowering hedges or screens. These may be informal where the plants are left to grow naturally, or they may be clipped to give a more formal effect. Care should be taken not to site either feature where the plants may be subject to cold, drying winds. Escallonias are very useful plants for coastal gardens where they will grow well and resist the less-cold, less-drying, seaside winds.

All the cultivars listed are suitable for cottage-style and other informal plantings where their bushy, evergreen habit and long flowering season fits in well.

The showy compact cultivars *E.* PINK ELLE ('Lades')PBR, *E.* GOLDEN CARPET ('Alcaura')PBR and *E.* GLOWING EMBERS ('Lowat21')PBR are ideally suited to today's small or compact gardens, bringing a long season of flower and foliage colour, plus being of a size suitable for patio gardens and containers. These compact plants also find a place in the mixed borders of larger gardens, where their long season of colour is, of course, very welcome.

TABLE 4.53 *ESCALLONIA* CULTIVARS

Name	Habit	Height	Spread	Leaves	Flowers	Remarks
E. 'Apple Blossom' AGM	Bushy	1.5–2.5m (5–8ft)	1.5–2.5m (5–8ft)	Small dark-green	Pink and white. Summer early autumn	
E. 'Edinensis'	Bushy	1.5–2.5m (5–8ft)	1.5–2.5m (5–8ft)	Glossy leathery	Tubular pink. Summer early autumn	
E. GOLDEN CARPET ('Alcaura')PBR	Dwarf, compact	60cm (23½in)	0.5–1m (1½–3ft)	Yellow in spring, lime-green in summer	Tubular red, summer	
E. GLOWING EMBERS ('Lowat21')PBR	Bushy, compact	1.2–1m	1–1.5m (3–5ft)	Orange, red flush, then yellow in spring, followed by mid-green in summer	Dark-pink, summer	USDA RHS H3. Needs full sun
E. 'Iveyi' AGM	Bushy	1.5–2.5m (5–8ft)	1.5–2.5m (5–8ft)	Glossy dark-green	White fragrant. Summer early autumn	
E. PINK ELLE ('Lades')PBR	Bushy, compact	1–1.5m (3–5ft)	1–1.5m (3–5ft)	Dark-green	Large heads of pink flowers summer and early autumn	USDA RHS H3

Escallonia GLOWING EMBERS ('Lowat21'PBR). A cultivar that is compact and bushy in growth.

Escallonia GLOWING EMBERS ('Lowat21'PBR). New foliage in spring is reddish at the tips, developing into bright-yellow. By summer this has changed to a shiny green.

Escallonia PINK ELLE ('Lades'PBR). This cultivar has compact, bushy growth. The large heads of flowers are produced in summer and early autumn.

EXOCHORDA

A small genus of deciduous shrubs originating from China and Central China. In late spring into early summer, they produce racemes of spherical buds that open into white flowers, hence their common name of pearl bush.

Growing Requirements

Hardiness: USDA 6b/7a. RHS H6.
Aspect: A site facing any direction will be accepted.
Shelter: No special provision for shelter is required.
Sun: Full sun or part shade.
Soil: Providing that it is well drained a soil of any texture is satisfactory.
pH: A wide pH range is acceptable, but avoid shallow, alkaline soils, as chlorosis will develop.

Garden Design Considerations

These are an easily grown group of plants, providing very alkaline conditions are avoided. *Exochorda* × *macrantha* 'The Bride' provides a rich, if not long, display of white flowers. It is excellent for lifting the colours in any border but contributes little after flowering. It is very appropriate for a bridal garden. The cultivar *E.* × *macrantha* 'Niagara' is more compact and, therefore, better suited to where space is limited.

The more recently introduced *E.* MAGICAL SPRINGTIME ('Kolmasprit') is more compact than both of the *E.* × *macrantha* cultivars and has a longer period of flowering. In the modern compact garden, this would be the cultivar to select.

Exochorda serratifolia 'Snow White' as a large shrub could only be accommodated in at least a medium-sized garden. It can, however, be grown on a single stem to form a small tree. This will make a brief, but spectacular show in spring. It will also provide the valuable commodity of dappled shade in summer. As a tree, it will occupy little ground space and could easily be accommodated at the back of a border. A little shelter from the strongest winds would be a wise precaution to protect what may become rather a large head on a somewhat slender stem.

Exochorda × *macrantha* 'The Bride' in full flower in spring. The display is stunning, but rather short-lived.

Exochorda × *macrantha* 'The Bride'. Nominally a plant of the same cultivar supplied by the same nursery at the same time. There is obviously a wide variation in plants supplied as this cultivar, but they are all very attractive plants, even if of very varying habits.

MAGICAL SPRINGTIME ('Kolmasprit'). A plant that is more compact and upright-growing than *E.* × *macrantha* 'The Bride'.

Exochorda × *macrantha* 'The Bride'. A close up of the pure-white flowers.

Exochorda serratifolia 'Snow White'. Both the flowers and the entire plant are larger than the two other plants also mentioned. It can make a very attractive small tree.

Exochorda serratifolia 'Snow White' . The flowers are also attractive to butterflies.

TABLE 4.54 *EXOCHORDA* CULTIVARS

Name	Habit	Height	Spread	Foliage	Flowers	Remarks
E. macrantha 'The Bride' AGM	Medium-sized with an arching habit	1–5–2.5m (5–8ft)	1.5–2.5m (5–8ft)	Leaves simple, spoon-shaped, pale-green	White flowers, each 3cm (1in) across in short racemes	Flowers once in late spring and early summer
E. × *macrantha* 'Niagara'PBR	More compact than *E.* 'The Bride'	1.5m (5ft)	1.5m (5ft)	Oblong pale-green	White flowers, each 3cm (1in) across in short racemes	Flowers once in late spring and early summer
E. MAGICAL SPRINGTIME ('Kolmasprit')	Upright, branching	1–1.5m (3–5ft)	1–1.5m (3–5ft)	Simple-shaped, light-green	Profuse white flowers	Flowers on old and new wood. Flowers once in late spring and early summer and again in summer
E. serratifolia 'Snow White'	Tall shrub or small tree	4m (13ft)	3m (10ft)	Oblong, pale-green	Large, white flowers	Flowers once in late spring and early summer

FUCHSIA

A genus of about 100 species coming mainly from Central and South America, plus New Zealand. The flowers are produced on stalks from the leaf axils and are normally pendulous. The calyx is tubular at its base but separates into four segments. The corolla is formed from four petals that emerge from the base of the calyx tube. The flowers are usually showy and produced from summer into autumn.

The hardiness of plants within the genus is variable. Some may be evergreen, semi-evergreen or deciduous, depending on the climate where they are grown. Here species and cultivars have been selected that are compact and rated at least USDA 9b/10a RHS H3 or USDA 8b/9a RHS H4. In milder climates these plants may prove fully hardy, but in colder climates these plants may be killed to the ground by winter, and will normally sprout from ground level or above in the following spring. In these colder climates, planting the plant with the top of the root system 5cm (2in) below the soil level, and providing winter frost-protection in the form of a mulch around the base of the plant, may well prove beneficial.

Growing Requirements

Hardiness: USDA 9b/10a or 8b/9a. RHS H3 or H4.
Aspect: Avoid a north-facing aspect.
Shelter: The site should be sheltered, especially from cold winter winds.
Sun: Plant in full sun or part shade.
Soil: The soil can be of any texture but must be well drained and retain moisture in the summer.
pH: A wide pH range is acceptable.

Garden Design Considerations

Fuchsias are excellent for contributing colour to the summer and autumn garden.

The largest species and cultivars make very attractive hedges where the winters are mild enough. Medium-sized plants are very well suited to flowerbeds and borders and their informal habit allows them to fit well into cottage-garden plantings. Smaller cultivars are great for patio gardens, raised beds and containers. Here their long flowering season is a very valuable asset.

Fuchsias grow particularly luxuriantly in sheltered seaside gardens. Here, if the winters are mild, growth and flowering may be almost continuous, but cold winds must be avoided.

Fuchsia 'David'. A good, reliable, hardy cultivar.

Fuchsia 'David' providing summer and autumn flower colour.

Fuchsia 'Delta's Sarah'. An attractive flower but not a particularly hardy cultivar.

Fuchsia 'Display'. A free-flowering cultivar.

Fuchsia 'Lady Thumb'. A dwarf, hardy cultivar.

Fuchsia magellanica var. *gracilis* 'Versicolor'. A graceful plant that is surprisingly hardy.

Fuchsia magellanica var. *gracilis* 'Versicolor'. A cultivar that is garden-worthy because of its habit, variegated foliage and elegant, pendant flowers.

Fuchsia 'Mrs Popple'. An old, hardy cultivar.

Fuchsia 'Riccartonii'. A tall-growing, hardy cultivar.

Fuchsia 'Riccartonii'. In the mild south-western parts of the UK, widely used as very attractive hedging.

TABLE 4.55 SPECIES AND CULTIVARS OF *FUCHSIA*

Name	Habit	Height	Spread	Flower form	Petals	Sepals	Hardiness	Remarks
'Alice Hoffman' AGM	Upright	40–60cm (16–24in)	40–60cm (16–24in)	Semi-double	White	Rose-pink	USDA 8b/9a. RHSH4	
'David' AGM	Erect and bushy	90cm (36in)	100cm (40in)	Single	Purple	Cerise	USDA 8b/9a. RHSH4	
'Delta's Sarah' AGM		2m 6ft)		Single	Purple-blue	White	USDA 9b/10a RHSH3	
'Display' AGM	Small shrub	60–75cm (24–30in)	45–60cm (18–24in)	Single	Rose-pink	Carmine	USDA 8/u/9a RHSH4	
'Dollar Prinzessin' (syn. 'Dollar Princess')	Upright shrub	30–45cm (12–18in)	45–60cm (18–24in)	Double	Purple	Cerise	USDA 8/u/9a RHSH4	
'Genii' AGM	Dwarf	75–90cm (28–36in)	75–90cm (28–36in)	Single	Violet	Cerise	USDA 8/u/9a RHSH4	Lime-yellow leaves
'Hawkshead' AGM	Vigorous and upright	0.5–1m (20–40in)	0.1–0.5m (4–20in)	Single	White	White	USDA 8/u/9a RHSH4	
'Lady Boothby' AGM	Very vigorous	5m (17ft)	1m (3ft)	Single	Deep-purple	Crimson	USDA 8b/9a. RHS H4	A cultivar with very long internodes. Makes an unusual wall shrub
'Lady Thumb' AGM	Dwarf	15–30cm (6–12in)	15–30cm (6–12in)	Semi-double	Carmine-red	Rose-scarlet	USDA 8b/9a. RHS H4	
F. magellanica AGM	Erect	3m (10ft)	2–3m 6–10ft)	Single	Purple	Scarlet	USDA 8b/9a. RHS H4	
F. magellanica var. *gracilis* 'Versicolor' AGM	Slender and spreading	1–1.5m (40–60in)	1–1.5m (40–60in)	Single	Violet	Scarlet	USDA 8b/9a. RHS H4	Leaves rose tinted when young later grey-green
'Mrs. 'Popple' AGM	Upright, shrubby	1–1.1m (39–43in)	1–1.1m (39–43in)	Single	Violet	Scarlet	USDA 8b/9a. RHS H4	
'Riccartonii' AGM	Large upright shrub	2–3m (6–9ft)	1–2m (3–7ft)	Single	Dark-purple	Deep-scarlet	USDA 6b/7a RHS H6	Often used as a hedge in milder areas
'Tom Thumb' AGM	Dwarf	15–30cm (6–12in)	15–30cm (6–12in)	Single	Carmine- red	Rose-scarlet	USDA 8b/9a RHS H4	

× *HALIMIOCISTUS*

A bi-generic hybrid genus resulting from a cross between *Cistus* and *Halimium*. It consists of dwarf, evergreen, summer-flowering shrubs.

Growing Requirements

Hardiness: USDA8b/9a. RHSH4.
Aspect: A south- or west-facing site is required.
Shelter: Shelter from cold, drying winds is required.
Sun: Full sun is required.
Soil: This must be very well drained. Light sandy or chalky soil, not over-rich in nutrients, is preferred; avoid heavy clay.
pH: A pH range from acid through neutral to alkaline is accepted.

Garden Design Considerations

× *Halimiocistus* are great little shrubs for adding a splash of light flower colour in summer, providing that their demands for sun, shelter and excellent drainage can be met. Their small stature and desire for sun makes it eminently suitable for patios, containers, rock gardens and gravel gardens. South-facing banks, slopes, raised beds and the tops of walls, especially dry-stone walls, tend to be hot, dry and very well drained, potentially difficult places to plant, but ideal for this group.

TABLE 4.56 × *HALIMIOCISTUS*

Name	Habit	Height	Spread	Foliage	Flowers	Remarks
× *H. sahucii* AGM	Dense, low-growing, evergreen shrub	40cm (16in)	90cm (35in)	Dark-green, narrow leaves	White with a yellow centre of stamens, 2.5cm (1in) across	Flower from beginning to end of summer
× *H. sahucii* ICE DANCER ('Ebhals'[PBR])	Compact, low-growing, evergreen shrub	30cm (12in)	90cm (35in)	Leaves cream and green in summer, all-green in winter	White with a yellow centre of stamens	Flower from beginning to end of summer
× *H. wintonensis* AGM	Dwarf, evergreen shrub	15–37cm (6–15in)	100cm (40in)	Narrow, grey-green leaves	White with a basal zone of crimson 5cm (2in) across	Flower from beginning to end of summer
× *H. wintonensis* 'Merrist Wood Cream' AGM	Dwarf, evergreen shrub	30–60cm (12–24in)	60–90cm (24–36in)	Grey-green, elliptic leaves	Pale-yellow with a basal zone of maroon and yellow centre of stamens 5cm (2in) across	Flower from beginning to end of summer

HEPTACODIUM MICONIOIDES (SEVEN SON FLOWER OF ZHEJIANG)

This plant was first introduced from China in 1980. Often considered a large shrub, it can be grown as a small tree and, as such, is ideal for the smaller garden.

It gained its Chinese name from the apparent nature of the inflorescence. Each inflorescence consists of three branches of flower clusters appearing at two nodes, these being apparently topped by a single cluster. The terminal bud is not, in fact, a flower cluster, but a continuation of the inflorescence. This of no account to the Chinese, since seven is a lucky number!

A deciduous, very hardy plant that is easily restricted to a single stem, thus forming into a small, single-stemmed tree. Due to the plant's comparatively recent introduction, it is difficult to predict an ultimate size, but after ten years a tree 6m tall by 4m wide (20 by 13ft) is to be expected.

The plant has several attractive features to commend it. During the growing season it bears large, three-veined leaves that are retained into early winter. Small, white, scented flowers are produced in late summer and early autumn. Following flowering, the calyx of the flower enlarges and turns pinkish red, giving a second display. After leaf fall, the distinctive, pale, peeling bark is clearly visible, continuing the interest provided by this small tree into winter.

Heptacodium miconioides growing as a small tree.

Heptacodium miconioides producing white, scented flowers in late summer.

Heptacodium miconioides in autumn showing the persistent red calyx.

Heptacodium miconioides. The light-coloured, peeling bark is of interest, particularly in bright winter sunlight.

Growing Requirements

Hardiness: USDA 6a-1 RHS H7.
Aspect and shelter: A little shelter should be provided and a north-facing site avoided.
Sun: The tree is best in full sun or partial shade.
Soil and pH: A wide pH range is tolerated but should ideally be moisture-retentive and well drained.

Garden Design Considerations

A small tree that lends itself well to being an isolated specimen plant, having the attributes of being unusual, providing scent and flowers late in the growing season and continuing winter interest.

The tree could be included in an otherwise almost exclusively herbaceous border, converting this to a mixed border with a much-extended season of interest. The plant would also mix in well in an informal garden, where plant diversity is an asset.

HIBISCUS

A genus of about 670 species of herbs, shrubs and trees, found mainly in tropical and subtropical regions. The flower consists of five free petals that form a trumpet shape. Only *Hibiscus moscheutos* and *Hibiscus syriacus* are hardy.

Hibiscus moscheutos (rose of Sharon; USDA 8b/9a RHS H4), behaves as an herbaceous perennial and is best cut down to almost ground level in late autumn. It requires a warm, sunny position and can achieve a height of 1.5m (5ft) and spread of 1m (3ft) in one growing season. It produces large trumpet-shaped flowers, up to 20cm (8in) across in late summer in white or different shades of pink.

Hibiscus syriacus

This is a hardy, deciduous shrub introduced to Europe from Syria, but native to southern China. The species and its cultivars tend to have a multi-stemmed, upright habit.

Growing Requirements

Hardiness: USDA 7b/8a. RHS H5.
Aspect: The planting site should face south or west. Avoid a north- or east-facing site.
Shelter: A sheltered, warm site is highly beneficial.
Sun: Full sun is necessary to obtain maximum flowering.
Soil: The soil can be of any texture but must be well drained.
pH: An alkaline or neutral pH is preferred and avoid a strongly acidic soil.

Cultivars of *Hibiscus syriacus*

Leaf emergence: This is always late, usually not until late spring.
Flowering: From late summer into autumn.

The cultivars in Table 4.57 are available as normal shrubs or top-worked on to an upright stem at about 1.5m (5ft) to produce a standard plant that is, in effect, a very small tree.

Cultivars of *Hibiscus syriacus:* the Chiffon Series

These are distinguished by their flower form, which is no longer in the form of a trumpet. The stamens have become modified to become petaloid, forming a central mass of what appears as smaller petals. Overall, the flower has a ruffled or chiffon-like effect and can be described as semi- or fully double.

TABLE 4.57 CULTIVARS OF *HIBISCUS SYRIACUS*

Name	Habit	Height	Spread	Flower
'Bredon Springs'	Upright, multi-stemmed	1.5–2.5m (5–8ft)	1–1.5m (3–5ft)	Very large, pale purple
'Diana' AGM	Upright, multi-stemmed	1.5–2.5m (5–8ft)	1–1.5m (3–5ft)	White, 12cm (5in) across
'Dorothy Crane'	Upright, multi-stemmed	2.5–4m (8–13ft)	1.5–2.5m (5–8ft)	White with a red centre
'Hamabo' AGM	Upright, multi-stemmed	1–1.15m (3–5ft)	1–1.5m (3–5ft)	Pink with a large red blotch on each petal, 12cm (5in) across
'Helene'	Upright, multi-stemmed	1.5–2.5m (5–8ft)	1–1.5m (3–5ft)	Single, white with dark-purple centre, 12cm (5in) across
'Oiseau Blue' AGM (syn. 'Blue Bird')	Upright, multi-stemmed	3m (12ft)	2m (6½ft)	Blue with a red/purple centre
'Red Heart' AGM	Upright, multi-stemmed	2.5m (8ft)	2.5m (8ft)	White with a red centre
'Woodbridge' AGM	Upright, multi-stemmed	1.5–2.5m (5–8ft)	1.5–2.5m (5–8ft)	Deer pink with a dark-red centre, 8cm (3in) across

TABLE 4.58 CULTIVARS OF *HIBISCUS SYRIACUS*: THE CHIFFON SERIES

Name	Habit	Height	Spread	Flower
LAVENDER CHIFFON ('Notwoodone'PBR)	Upright, multi-stemmed	1.5–2.5m (5–8ft)	1–1.5m (3–5ft)	Double, light-pink, 10cm (4in) across
MAGENTA CHIFFON ('Rwood5'PBR)	Upright, multi-stemmed	3m (10ft)	2m (6½ft)	Large, reddish-pink, ruffled effect
PINK CHIFFON ('Jwnwood4'PBR)	Upright, multi-stemmed	1.5–2.5m (5–8ft)	1–1.5m (3–5ft)	Large, pink with darker centre, ruffled effect
Starburst Chiffon ('Rwood6')	Multi-stemmed	1.5m (5ft)	1.75m (6ft)	Semi-double pink with deep-pink veining and white centre, ruffled effect

Garden Design Considerations

Hibiscus' main strength are their contribution of strong flower colour in late summer and early autumn, when this is often lacking from many other shrubs.

They vary in overall size, but several are suitable for the smaller garden, their upright habit meaning that they are economical in the use of space.

Those available as standards can function as small trees in compact gardens, making a very attractive feature in late summer. The size of the head of these standards can be kept small by hard pruning back to a basic framework in winter.

The fact that they are late into leaf in spring may be considered only a minor disadvantage.

Hibiscus syriacus 'Bredon Springs'. This cultivar produces numerous, very large flowers.

Hibiscus syriacus 'Dorothy Crane'.

Hibiscus syriacus 'Hamabo'. Here, grown as a small standard tree.

Hibiscus syriacus 'Hamabo'.

Hibiscus syriacus 'Oiseau Blue'. Here, grown as a small standard tree.

Hibiscus syriacus 'Oiseau Blue'.

Hibiscus syriacus 'Woodbridge'. Here, grown as a small standard tree.

Hibiscus syriacus 'Woodbridge'.

HYDRANGEA

A genus of about seventy species from Eastern Asia, South and North America. The genus consists of trees, shrubs and climbers, the majority of which are deciduous, but a few are evergreen.

The flowers are usually produced in the form of a flattened corymb, consisting of two types of flowers. In the centre of the inflorescence there are small, fertile flowers, which, whilst coloured, are usually visually insignificant. Around the central fertile flowers is a ring, or rings, of sterile flowers with coloured, conspicuous sepals.

Hydrangea macrophylla

This is a large group that may be of hybrid origin. They have large leaves and flower from mid to late summer. From a horticultural viewpoint they are split into two groups.

Growing Conditions

Hardiness: USDA7b/8a. RHS H5.
Aspect: The planting site can face in any direction.
Shelter: A sheltered site is required. In cold gardens, plants can be susceptible to late spring frost damage. In these gardens it is wise to plant in a semi-shaded situation where some protection will be afforded by the plants overhead. Also, try to avoid where the plants will be exposed to early morning sun. Another strategy is to delay pruning until late spring so that the remaining old flower heads offer some protection to the flower buds below them.
Sun: Full sun or part shade. Where the soil retains adequate moisture in summer, full sun will result in profuse flowering. If the soil retains inadequate moisture, then full sun will result in wilting and a poor display during flowering.
Soil: A soil of any texture is acceptable. The soil should contain plenty of organic matter to retain moisture during the summer and in winter the drainage must be good.
pH: A soil pH from acid through to alkaline conditions is acceptable, but shallow, alkaline soils should be avoided.
Pruning: This is best done in spring. Young plants need very little pruning, only removal of old flower heads, if present. For established plants, remove just a few shoots down to ground level to encourage new replacement growth. Remove old flower heads down to the largest buds, usually those immediately below. If the top buds have been frost damaged, prune to undamaged buds just below.

Flower Colour

In *Hydrangea macrophylla* the colour of the flowers can be white, blue, pink, red or purple. White flowers remain white, but the flower colour of other cultivars varies due to the presence of aluminum ions in the soil solution. In acid conditions, more aluminum ions are available in the soil solution and the flowers will be blue. In alkaline conditions, fewer aluminum ions are available and the flowers will be pink or red. The shade of colour a cultivar develops depends, to an extent, on the cultivar involved; for instance, pink cultivars tend to become light blue under acid conditions, but red cultivars tend to become purple under acid conditions.

'Bluing compound', containing aluminum sulphate, can be applied to the soil surface around the plants and this will affect the aluminum ion concentration and tend to blue the flowers.

The *Hortensia* Group (Mop-Heads)

The inflorescence consists of only sterile florets that form large, rounded heads.

TABLE 4.59 *HORTENSIA* GROUP CULTIVARS

Name	Habit	Height	Spread	Flower	Remarks
'All Summer Beauty'	Compact	1–1.5m (3–5ft)	1–1.5m (3–5ft)	Pale-blue on acid soil. Pink on alkaline soil	Flowers on old and new wood. This should give a longer season of flowering
'Ami Pasquier'	Compact	1–1.5m (3–5ft)	1–1.5m (3–5ft)	Pale-pink on acid ground. Deep crimson or purplish red on alkaline soil	
'Bela'PBR	Compact	1m (3ft)	1m (3ft)	Deep-blue on acid soil. Pink on alkaline soil	
'Cardinal Red'	Rounded and bushy	1.5m (5ft)	1.5m (5ft)	Red on alkaline soil	
'Deutschland'	Upright and spreading	1.2m (4ft)	1.5m (5ft)	Blue on acid soil. Pink on alkaline soil	
'Frau Nobuko'PBR (Lady Series)	Compact	1m (3ft)	1m (3ft)	Bicoloured. White with a pink stripe	
'French Bolero Blue'	Compact	1m (3ft)	1m (3ft)	Blue	Flowers on lateral and terminal buds
'French Bolero Pink'	Compact	1m (3ft)	1m (3ft)	Light-pink	Flowers on lateral and terminal buds
'Madame Emile Mouillère'		1.5–2.5m (5–8ft)	1–1.5m (3–5ft)	White, may turn pink in sun	
MAGICAL NOBELESSE ('Hokomano'PBR)	Compact	1m (3ft)	1m (3ft)	White with a small pink eye and green tips. As the flowers age they become green all over	

Hydrangea macrophylla 'Ami Pasquier'.

Hydrangea macrophylla 'Bela'PBR.

Hydrangea macrophylla 'Deutschland'.

Hydrangea macrophylla 'Frau Nobuko'PBR (Lady Series).

Hydrangea macrophylla 'French Bolero Pink'.

Hydrangea macrophylla MAGICAL NOBELESSE ('Hokomano'PBR).

The Lacecap Group

These produce an inflorescence consisting of a flattened corymb in the centre of which there are fertile flowers. Round the non-showy fertile flowers there is a ring of showy sterile flowers.

Garden Design Considerations

Hydrangea macrophylla cultivars provide excellent late summer and autumn flower colour and some also provide autumn leaf colour. New cultivars that produce their flowers on old and new wood are claimed to have an even longer season of flowering. Also relatively new are those that flower on terminal and lateral buds, which produce more but smaller inflorescences. This is preferred by some who consider that more, smaller heads of flower to be more attractive than fewer but larger heads.

They do have a weakness with regard to late spring frosts, which can damage the flower buds and are better avoided in gardens where this is a frequent problem or at least position the plant where this risk is minimized, e.g. in the lee of a south- or west-facing wall or fence. In contrast, they grow and flower exceptionally well in seaside gardens, where they do not generally experience this problem.

Both mophead and lacecap cultivars fit in well with informal and cottage gardens, where they provide the long season of flower colour required.

Where flower colour combinations are important in plant associations and border design, care needs to be taken regarding the influence of soil pH. Many gardeners have been disappointed when expecting blue flowers only to be presented by pink!

Where the soil does tend to dry out in late summer, they are better planted in part or dappled shade to prevent wilting and a poor display. Many cultivars perform best in partly shaded conditions and white or light-coloured cultivars are particularly attractive in these situations, their flowers reflecting the light.

The modern, compact cultivars now readily available are very good for container and large pot planting. Many will give a long season of flower colour, followed by autumn leaf colour and even their spent flower heads are considered by many to be attractive through the winter.

TABLE 4.60 LACECAP GROUP CULTIVARS

Name	Height	Spread	Colour	Remarks
'Doctor Jean Varnier'	1.2–1.5m (4–5ft)	1.2–1.5m (3–5ft)	Pale-lilac on acid soil. Pale-pink on alkaline soil	
'French Cancan' (Rendezvous Series)	1–1.5m (3–5ft)	1.2–1.5m (3–5ft)	Pale-pink fertile flowers. Pink and fringed sterile flowers	
'Klaveren'	0.9–1.2m (3–4ft)	1m (3ft)	Mauve-purple on acid soil. Pink on alkaline soil	Fringed sterile flowers. Good autumn leaf colour
'Lanarth White' AGM	1.2m (4ft)	1.2m (4ft)	White	
RUNAWAY BRIDE SNOW WHITE ('Ushyd0405'PBR)	1–1.5m (3–5ft)	1–1.5m (3–5ft)	White sometimes flushed pink.	Flowers on lateral and terminal buds. RHS Chelsea Plant of the Year 2018
'Veitchii' AGM	1.5m (5ft)	1–1.5m (3–5ft)	Blue fertile flowers. White sterile flowers	
'Zorro'PBR AGM	1–1.5m (3–5ft)	1–1.5m (3–5ft)	Blue on acid soil. Pink on alkaline soil	Distinct dark stems

Hydrangea macrophylla 'French Cancan' (Rendezvous Series).

Hydrangea macrophylla 'Klaveren'.

Hydrangea macrophylla 'Lanarth White' AGM.

Hydrangea macrophylla RUNAWAY BRIDE SNOW WHITE ('Ushyd0405').

Hydrangea macrophylla RUNAWAY BRIDE SNOW White ('Ushyd0405').

Hydrangea macrophylla 'Veitchii'.

Hydrangea macrophylla 'Zorro'PBR.

Hydrangea macrophylla 'Veitchii' planted on the edge of woodland. Here, the light-coloured flowers show up clearly.

Hydrangea paniculata

This deciduous shrub from China, Japan and Taiwan was introduced in 1861. The leaves can be 7.6 cm (3in) to 15cm (6in) long and produced in pairs or threes. Flowering occurs on the current year's growth and the size of the inflorescence depends on the vigour of the shoots. Dense, pyramidal panicles consisting of both small and fertile, and large, showy and infertile flowers are produced in late summer and early autumn. The flower colour when immature is green, upon opening this changes to white, then often develops pink-purple coloration and, finally, turns brown once finished.

Propagation is done by softwood cuttings taken in early summer or hardwood cuttings taken in the dormant season.

Growing Requirements

Hardiness: USDA 7b/8a. RHS 5.
Shelter: Little or no shelter is required, north- or east-facing sites being tolerated.
Sun: Full sun or part shade is required. In full sun, ensure adequate moisture is available at all times.
Soil: The plant requires fertile, well-drained and moisture-retentive soil. Addition of organic matter to the soil, prior to planting, is recommended mainly to aid moisture-retention. Dryness at the roots, especially at flowering time, will not be tolerated.
pH: A wide pH range is tolerated but avoid very thin alkaline soils.
Pruning and feeding: To obtain the largest flower panicles, the correct pruning technique, plus fertiliser application in spring, will be required. Plants grown in pots will require liquid feeding and copious watering through the growing season.

Hydrangea paniculata with branches pruned back in late winter to leave a framework.

Pruning

When left unpruned the species and more vigorous cultivars will grow into large shrubs or small trees, the size of the plants, and to some extent the size of the panicles, depend on growing conditions. This variation was clearly seen in wild plants by the author whilst in Japan.

Pruning consists of shortening the previous year's growth, consisting of the finished flowerhead and one-year old stem, before growth recommences in spring. The ultimate size of the plant depends on the vigour of the cultivar and how hard it is pruned. Hard pruning, leaving only 5cm (2in) to 7cm (3in) of the previous year's growth will result in strong growth with comparatively large panicles of flowers. If several new shoots grow out from each cut branch, these may be reduced to two or three, which will then produce even larger panicles. Less severe pruning will result in less vigorous new growth and more, but smaller panicles of flowers. By pruning each winter, the overall size of the shrub is kept limited over several years. This pruning should be accompanied by fertiliser application in spring, followed by mulching to help retain the winter's rain in the soil. Hard pruning, reduction in shoot number and feeding may eventually weaken the plant.

Hydrangea paniculata 'Unique' pruned to form a small tree.

Pruning less severely each winter will allow the shrub to develop a larger permanent framework on which a display of numerous but smaller panicles will be produced.

By careful pruning each winter, two or three upward growing shoots may be allowed to develop. These eventually produce a multi-stemmed, small tree. This effect can be enhanced by removal of any lower branches, sometimes referred to as 'crown lifting', to expose this vertical stem architecture. The head of the tree should not be pruned too hard, since, if very large panicles are produced, they will be very susceptible to wind damage.

A final variation is sometimes offered by nurserymen where a strong, single stem is produced to about 1.2m (4ft) and then a head developed on top, to produce 'a bush on a leg', otherwise known as a pre-grown standard plant. Such a plant is liable to wind damage. The stem needs to be securely staked, with the stake extending to and secured into the head of the plant. Some pruning of the head should be done, but not very hard as the large panicles subsequently produced would in turn be very liable to wind damage.

Cultivars Available

As a result of modern breeding and selection, the number of cultivars available has increased very rapidly in recent years.

The main difference to note is in overall size. Older cultivars tend to be larger, but newer varieties are smaller and more compact, specifically with the modern, small garden in mind.

Most cultivars produce white flowers that, as they age, develop pink tones to a greater or lesser degree. Some cultivars have more predominantly green flowers that age to white and may eventually assume light-pink tones.

Garden Design Considerations

This species and its cultivars are popular because a range in size is available, adding to their versatility. They also provide a vivid, reliable display of late season colour that changes over time.

They have a wide range of applications in the garden, which include:

- Pot-grown specimen plants. These can be small and compact or larger, if appropriate.
- Ideally suited to a woodland garden. They revel in dappled shade and the light-coloured flowers can appear to glow in twilight.
- Their informal shape makes them well suited to a mixed border or a cottage-garden style.
- They could be employed as a spectacular informal flowering hedge; the width of the hedge being controlled by regular annual pruning.
- The panicles of flower can be used as cut flowers, the green colour of the immature flowers being very popular with florists and flower arrangers. Finally, the finished panicles can be harvested and used in dried-flower arrangements.

TABLE 4.61 *HYDRANGEA PANICULATA* CULTIVARS

Cultivar	Height	Spread	Flower colour
BOBO ('Ilvobo')PBR	0.9m (3ft)	1.2m (4ft)	White → pink
DART'S LITTLE DOT ('Darlido') PBR	1.2m (4ft)	1.0–1.5m (3–5ft)	White → pink
DIAMOND ROUGE ('Rendia')PBR	1.5m (5ft)	1.5m (5ft)	White → pink → raspberry red
'Floribunda'	2.5 – 4m (8–13ft)	1.5–2.5m (5–8ft)	White → pink
FRAISE MELBA ('Renba')PBR	1.2m (4ft)	1.2m (4ft)	White → pink
'Grandiflora'	3m (10ft)	2m (6½ft)	White → pink
'Limelight'	1.5–2.0m (5–6½ft)	1m (3ft)	Green → white → pink
'Jane' (syn. H. paniculata LITTLE LIME)	0.75m (2½ft)	0.75m (2½ft)	Lime-green → creamy-white → blush pink
'Magical Moonlight'	2.5m (8ft)	2.5m (8ft)	Green → white
PASTELGREEN ('Rencolor')PBR	1.2m (4ft)	1m (3ft)	Green → white → pink → a mix of these colours
'Phantom'	1.5m (5ft)	2.5m (8ft)	White → pink
PINKY WINKY ('Dvppinky')PBR	1.5m (5ft)	1.5m (5ft)	White → pink
'Silver Dollar'	1.25m (4ft)	1.65m (5½ft)	White → pink
SUNDAE FRAISE ('Rensun')PBR	1.5m (5ft)	2.5m (8ft)	Green → white
'Unique'	1–1.5m (3–5ft)	2.5m (8ft)	White → pink
VANILLE FRAISE ('Renhy')PBR	1.5m (5ft)	2.5m (8ft)	White → pink
'Wim's Red'PBR	1.5m (5ft)	2m (6½ft)	White → pink → red

Hydrangea paniculata BOBO ('Ilvobo'PBR).

Hydrangea paniculata DART'S LITTLE DOT ('Darlido'PBR). A very dwarf cultivar.

Hydrangea paniculata DART'S LITTLE DOT ('Darlido'PBR).

Hydrangea paniculata 'Grandiflora'.

Hydrangea paniculata 'Limelight'.

Hydrangea paniculata 'Limelight'.

Hydrangea paniculata 'Jane' (syn. *H. paniculata* LITTLE LIME).

Hydrangea paniculata 'Jane' (syn. *H. paniculata* LITTLE LIME).

Hydrangea paniculata PHANTOM Flowers early in the season.

Hydrangea paniculata PHANTOM Flowers later in the season.

Hydrangea paniculata PINKY WINKY ('Dvppinky'PBR).

Hydrangea paniculata PINKY WINKY ('Dvppinky'PBR).

Hydrangea VANILLE FRAISE ('Renhy'PBR). A plant showing flowers that are new coloured white and those that are older coloured pink.

Hydrangea quercifolia (Oak-Leaved Hydrangea)

This species is native to south-east USA. A deciduous shrub with large, deeply lobed leaves that develop very good autumn leaf colour. The flowers consist of conical heads of white flowers. These are produced in late summer and, as the flowers age, they gradually change to shades of pink.

Several cultivars have been produced, some of which are more vigorous and larger than the species, whilst other are smaller and more compact.

Growing Requirements

Hardiness: USDA7b/8a. RHS H5.
Aspect: A planting site facing any direction is satisfactory.
Shelter: There is no particular need to provide shelter.
Sun: Full sun or part shade.
Soil: A soil of most textures is suitable, but avoid shallow, chalky soils. It must retain adequate moisture in the summer.
pH: A soil with an acid or neutral pH is suitable but avoid a high pH.

Garden Design Considerations

Essentially the same as for *Hydrangea macrophylla* cultivars, but with the emphasis on autumn leaf colour rather than flowers.

An additional way the larger *Hydrangea quercifolia* cultivars can be used is for ground cover. The rather spreading nature of these plants, plus their large leaves, which effectively shade all the soil beneath them, makes them well suited for this purpose.

The smaller, compact cultivars, such as *Hydrangea quercifolia* 'Pee Wee' and *H. quercifolia* 'Ruby Slippers', are very useful where space is limited, as in the small and compact gardens we are considering, and for use in containers and large pots.

TABLE 4.62 *HYDRANGEA QUERCIFOLIA* AND CULTIVARS

Name	Habit	Height	Spread	Foliage	Flowers	Remarks
H. quercifolia	Spreading	1.15m (4ft)	1.15–2.5m (4–8ft)	Large strongly lobed leaves. Develops good red autumn leaf colour	Conical heads of white flowers	
H. 'Alice'	Spreading	1.5–2.5m (5–8ft)	1.15–2.5m (4–8ft)	Good autumn red leaf colour	Large heads of white flowers that gradually turn pink	More vigorous and larger than species. USDA8b/9a. RHS H4
H. 'Burgundy'	Spreading	2m (6½ft)	2m (6½ft)	Very good autumn red leaf colour	Large cones of white flowers that gradually turn pink	Produces exceptionally good dark-red autumn leaf colour. More vigorous and larger than the species
H. 'Little Honey'[PBR]	Compact	1–1.5 (3–5ft)	1–1.5m (3–5ft)	Golden-yellow in spring. Green once mature. Red in autumn	Creamy-white upright panicles	Slower growing than other cultivars
H. 'Pee Wee'	Compact and low-growing	1–1.2m (3–4ft)	1m (3ft)	Good autumn red leaf colour	Broadly conical panicles of white flowers that gradually turn pink	
H. 'Ruby Slippers'	Small and compact	1m (3ft)	1.5m (5ft)	Leaves mahogany red in autumn	Large dense upright panicles of white flowers become dark-pink	Autumn leaf colour particularly good
H. SNOW QUEEN ('Flemygea') AGM	Spreading	2m (6½ft)	2m (6½ft)	Leaves in autumn become bronzed-red	Panicles of white flowers become pink	
H. SNOWFLAKE ('Brido') AGM	Spreading	1.5–2.5m (5–8ft)	1.5–2.5m (5–8ft)	Good autumn leaf colour	Panicles of white flowers. The sterile flowers produce extra white bracts giving a double-flowered appearance	

Hydrangea quercifolia complete plant.

Hydrangea quercifolia inflorescence.

Hydrangea quercifolia autumn leaf colour.

Hydrangea quercifolia SNOW QUEEN ('Flemygea') AGM.

Hydrangea quercifolia SNOW QUEEN ('Flemygea') AGM.

Hydrangea quercifolia SNOWFLAKE ('Brido') AGM.

Hydrangea quercifolia SNOWFLAKE ('Brido') AGM.

HYPERICUM (ST. JOHN'S WORT)

A genus containing trees, shrubs, herbaceous and annual plants that can be evergreen or deciduous. All have yellow flowers with prominent stamens that are often followed by seed capsules that may be coloured.

Here only the shrubs are considered. Collectively these all make good, yellow-flowering shrubs providing colour in summer and autumn.

Growing Requirements

They are all non-demanding shrubs of easy cultivation.

Hardiness: With a solitary exception all are USDA 7b/8a. RHS H5. *Hypericum* 'Rowallane' is USDA 8b/9a. RHS H4.

Aspect: The site for planting can be facing in any direction.

Shelter: Exposed or sheltered.

Sun: Sun or part shade.

Soil: Soil of any type will suffice, providing that it is well drained.

pH: A very wide range of soil pH is tolerated.

Garden Design Considerations

The main attributes of this genus to the gardener are that they continue flower colour from shrubs, into the summer; most other shrubs having finished. They have a low-maintenance requirement.

Hypericum calycinum has been used extensively for low-maintenance ground-cover, particularly in moderately shaded areas. The plant's popularity has decreased in recent years because it is too invasive for smaller gardens and because of the prevalence of *Hypericum* rust.

Hypericum × hidcoteense 'Hidcote' AGM is used extensively in the urban landscape and has become common, but it is still a very good choice for late summer and early autumn colour and well worthy of continued planting. *Hypericum kouytchense* is similar to *H. × hidcoteense* 'Hidcote' AGM but offers something a little different and less familiar, with its large flowers perched on the end of arching branches.

Where comparatively large areas, including sunny banks that tend to be dry during the summer, need to be planted with attractive but low-maintenance shrubs, both *Hypericum × hidcoteense* 'Hidcote' AGM and *Hypericum kouytchense* would be good choices since they are both tough and will flourish under these challenging conditions.

Hypericum 'Rowallane' AGM can perhaps claim to be the queen of this genus, bearing as it does large, golden-yellow cups of beautifully formed flowers on a shrub of distinctive habit. It is, however, the least hardy of those included and does need a well-sheltered, sunny position. If you have such a position, I recommend you go ahead. Plant and enjoy it.

Hypericum lancasteri and *Hypericum × dummeri* 'Peter Dummer' are both very well suited to today's smaller gardens offering a long season of flowering, low maintenance and not at all invasive. Small areas requiring ground cover would be well covered by either of these shrubs. These two small plants, along with the much smaller *Hypericum olympicum*, *Hypericum olympicum* 'Citrinum' and the very petite *Hypericum reptans*, are excellent for inclusion in a rock garden. They are of appropriate size and spread, and provide flower colour in summer and early autumn. Not a period when most rock garden and true alpine plants are in flower.

Finally, these smaller hypericums are also ideal for the currently very popular subjects of pot, raised bed and container-gardening. They provide a splash of summer colour, associate well with most summer bedding plants and small tender perennials and are low maintenance; who could ask for more?

Hypericum calycinum functioning as ground cover.

Hypericum calycinum.

Hypericum × *hidcoteense* 'Hidcote' AGM. A superb summer-flowering shrub.

Hypericum × *hidcoteense* 'Hidcote' AGM.

Hypericum × *inodorum* (Magical Series) MAGICAL BEAUTY ('Kolmbeau'PBR).

Hypericum × *inodorum* (Miracle Series) MIRACLE MARVEL ('Allmarvel'PBR).

Hypericum × *inodorum* (Magical Series) MAGICAL PUMPKIN ('Kolmapuki'PBR). Flowering and producing coloured fruit in summer. The fruit persists into the autumn.

Hypericum × *inodorum* (Magical Series) MAGICAL PUMPKIN ('Kolmapuki'PBR).

Hypericum kouytchense AGM. A medium-sized shrub with flowers produced at the end of arching branches.

Hypericum olympicum f. *uniflorum* 'Citrinum' AGM. Ideal for the rock garden or the top of a dry-stone wall.

Hypericum olympicum f. *uniflorum* 'Citrinum' AGM.

Hypericum 'Rowallane' AGM. This choice cultivar needs a warm, sheltered position.

Hypericum 'Rowallane' AGM.

TABLE 4.63 SPECIES AND CULTIVARS OF *HYPERICUM*

Name	Habit	Height	Spread	Flowers	Leaves	Remarks
H. calycinum (rose of Sharon, Aaron's beard)	Low-growing and spreading	0.6m (2ft)	Indefinite. Spreads by underground runners	Yellow, 10cm (4in) across. In summer and autumn	Evergreen ovate leaves in pairs.10cm (4in) long	Good ground cover, especially in shade, but can become invasive*
H. dummeri 'Peter Dummer' AGM	Bushy and compact, with reddish arching shoots	0.45m (1.6ft)	0.6m (2ft)	Yellow in summer, followed by red tinged fruits	Semi-evergreen. Bronzed when young, reddening again in autumn	
H. × *hidcoteense* 'Hidcote' (syn. *H.* 'Hidcote')AGM	Bushy	1–1.5m (3–5ft)	1.5m (5ft)	Yellow, 6cm (2½in) across from summer into early autumn	Semi-evergreen, leaves narrowly ovate	Very popular with commercial landscapers, but also an excellent garden plant
H. × *inodorum* 'Elstead'	Upright	1.5m (5ft)	1.5m (5ft)	Clusters of yellow flowers each 2.5cm (1in) across. Followed by bright-red fruits	Semi-evergreen ovate leaves	Often badly attacked by rust fungus*
H. × *inodorum* (Magical Series) MAGICAL UNIVERSE ('Kolmundi'PBR)	Upright	1m (3ft)	1m (3ft)	Yellow flowers summer followed by maroon fruits in late summer, early autumn	Semi-evergreen leaves 11cm (4¼in) long	Just one cultivar of the dwarf Magical Series. These are grown mainly for cut flower when in fruit. Rust-resistant
H. kouytchense AGM	Bushy, small shrub	1m (3ft)	1–1.5m (3–5ft)	6cm (2½in) across with long stamens, followed by reddish capsules	Semi-evergreen, ovate leaves	Produces arching stems with flowers held out horizontally
H. lancasteri	Small, spreading shrub.	0.5–1m (1½–3ft)	0.5–1m (1.6–3ft)	Golden-yellow 3–5.5 cm (1¼–2in) across. Summer to early autumn.	Paired leaves, each 6cm (2¼in) long	Good for small scale ground cover
H. × *moserianum* 'Tricolor'	Small, spreading shrub	0.1–0.5m (0.3–1½ft)	0.3–0.5m (1–1.6ft)	Yellow, small. Summer to early autumn	Semi-evergreen, leaves variegated with green, pink and cream	Needs a sheltered position
H. olympicum AGM	Very small shrub	0.25m (10in)	0.3m (1ft)	Star-shaped yellow 6cm (2½in) across	Deciduous. Small grey-green, pointed leaves 4cm (1.5in) long	
H. olympicum f. *uniflorum* 'Citrinum' AGM	Very small shrub	0.25m (10in)	0.3m (1ft)	Pale lemon-yellow flowers	Deciduous. Small grey-green, pointed leaves 4cm (1.5in) long	
H. reptans (creeping St. John's wort)	Prostrate, very small shrub	5cm (2in)	0.20m (8in)	Yellow 4.5cm (1.5in) across. Summer to early autumn	Deciduous, 0.6–2cm (¼–¾in.) long, crowding branches	
H. 'Rowallane' AGM	Upright, arching shrub	1.5–2.5m (5–8ft)	1m (3ft)	Golden-yellow 7cm (2½in) across. Summer	Semi-evergreen	USDA 8a/9b. RHS H4. Needs a sheltered position

* Indicates species or cultivars that may be badly affected by *Hypericum* rust (*Melampsora hypericorum*). Infection is seen as masses of bright orange spots on the leaves, which develop into leaf drop and dieback. It is more common in the south-west of the UK. Where this is a problem, pruning growth hard back in late winter may help to reduce the problem. New growth will be clear of the disease and it will take time for it to become apparent again.

ILEX

A large genus with over 500 species occurring in both hemispheres in temperate and tropical regions, which may be trees, shrubs or climbers.

The leaves are alternate. The flowers are small with four white or off-white petals. The genus is usually dioecious with male and female flowers being borne on separate plants. Only female plants bear fruit, which is technically a drupe but usually referred to as a berry.

Ilex aquifolium (Common Holly)

This is an evergreen tree or shrub native to Europe, including the UK. Left to grow naturally, it forms a dense, bushy, pyramidal tree. The leaves are glossy, dark-green, approximately 2.5–7.5cm (1–3in) long and 2–6cm (¾–2½ in) wide, but variable in size, outline and spines. Seedlings have very spiny leaves. As the plant ages the leaves produced gradually become less spiny. At the top of a holly tree, the leaves may be almost entirely spineless.

Where native, this species is excellent for nature conservation. Its flowers are a food source for insects and the fruit is an important food source for birds in winter. Where not native, *Ilex aquifolium* may become a noxious weed and the species should not be planted. This has occurred on the west coast of the USA, where it has invaded woodlands and shaded out native species.

Ilex aquifolium. Close-up of a male plant in flower. The anthers can be seen, but the ovary and stigma in each flower, which if fertilised would develop into a berry, are missing. This plant will never set any fruit but will serve as a good pollinator for female plants.

Growing Requirements

Hardiness: USDA 6b/7a. RHS H6.
Aspect: The planting site can face in any direction.
Shelter: No special provision needs to be made for shelter.
Sun: Full sun or part shade.
Soil: The soil can be of any texture but must be well drained.
pH: A soil pH ranging from alkaline to acidic will be accepted.

Plants should be sold either as root balled or container-grown, not bare root. Root-balled plants are best planted in early spring or autumn and kept well-watered whilst establishing.

Fruiting

Seed-raised plants of *Ilex aquifolium* will be either male or female. When planting several, as in a hedge, a mix of male and female plants is usually planted, and fruiting will be assured. If only planting one plant and fruiting is required, then the plant will need to be sexed before planting. This will mean waiting to see if fruits are produced.

Holly cultivars are vegetatively propagated clones and are all male or female. To ensure fruiting of a female plant, a male plant must be in the vicinity.

TABLE 4.64 *ILEX AQUIFOLIUM* AND CULTIVARS

Name	Habit	Height	Spread	Foliage	Male or female	Remarks
I. aquifolium AGM	Pyramidal, bushy, and dense-growing	12m+ (40ft+)	4–8m (13–26ft)	Dark-green, undulating, and spiny	Can be male or female	Fruit red
I. aquifolium 'Alaska'	Small, upright tree	8m (26ft)	3m (10ft)	Narrow, spiny or entire	Female	Red fruits. Slow-growing
I. aquifolium 'Argentea Marginata' AGM	Conical tree	12m (40ft)	4–8m (13–26ft)	Dark-green with a creamy margin	Female	Free-fruiting
I. aquifolium 'Argentea Marginata Pendula'	Small, weeping, dome-shaped tree	2–4m (6½–13ft)	2–4m (6½–13ft)	Dark-green with a creamy margin	Female	Red fruits. Train leader up to form a dome-shaped tree
I. aquifolium 'Bacciflava'	Small tree or shrub	3–4m (10–13ft)	3m (10ft)	Dark-green, not undulating, most entire	Female	Bright-yellow fruit
I. aquifolium 'J. C. van Tol' AGM	Upright tree	4–8m (13–26ft)	2.5–4m (8–13ft)	Dark-green almost spineless	Female, but will set fruit on its own	Large crops of red fruit
I. aquifolium 'Ferox'	Open-growing and bushy	4–8m (13–26ft)	2.5–4m (8–13ft)	Small, green leaves with many short spines on upper leaf surface	Male	Hedgehog holly. Slow-growing
I. aquifolium 'Ferox Argentea' AGM	Bushy shrub	4–8m (13–26ft)	2.5–4m (8–13ft)	Small, green leaves with a creamy margin and many short spines on upper surface	Male	Silver hedgehog holly. Slow-growing
I. aquifolium 'Pendula'	Small, weeping tree	4m (13ft)	2.5–4m (8–13ft)	Dark-green and spiny	Female	Slow-growing
I. aquifolium 'Silver Queen' AGM	A small, dense-growing tree	4–8m (13–26ft)	2.5–4m (8–13ft)	Dark-green, bordered white and marbled grey	Male	

Note: The naming of hollies can be confusing. For example, *Ilex aquifolium* 'Silver Queen' is a male clone and produces no fruit.

Ilex aquifolium 'Alaska' clipped into a cloud-pruned hedge. By cloud-pruning the hedge is developed into a more interesting feature.

Ilex aquifolium 'Argentea Marginata'. A good, bright variegation.

Ilex aquifolium 'Argentea Marginata Pendula'. Variegation and a pendular habit! Such a plant, when trained and pruned into a small, pendulous tree, makes a striking specimen.

Ilex aquifolium 'Bacciflava'. A yellow-berried form of the species.

Ilex aquifolium 'J. C. van Tol'. A female plant that is capable of self-pollination. Useful where there is no room to plant a female plus a male plant and there are no other male hollies in the vicinity to ensure good pollination.

Ilex aquifolium 'Pendula'. More quickly grown into a small tree than variegated cultivars, this too makes a good small-specimen tree.

Ilex aquifolium 'Pendula'. This is a female clone and can set a good crop of fruit.

TABLE 4.65 *ILEX × ALTACLERENSIS* AND CULTIVARS

Name	Habit	Height	Spread	Foliage	Male or female	Remarks
I. × altaclerensis	Can be between a medium-sized shrub to a tree	12m+ (40ft+)	4–8m (13–26ft)	Large, dark-green with few spines	Can be male or female	A variable hybrid of which there are many cultivars
I. × altaclerensis 'Belgica Aurea' AGM (syn. 'Silver Sentinel')	Erect-growing	8–12m (26–40ft)	2.5–4m (8–13ft)	Large, flat, 8–10cm (3–4in) long nearly spineless. Green with a white margin	Female	Red fruits
I. × altaclerensis 'Golden King' AGM	Bushy	6m (20ft)	5m (17ft)	Broad, almost spineless with an irregular white margin	Female	Fruits reddish-brown maturing to red
I. × altaclerensis 'Lawsoniana' AGM	Bushy	4–8m (13–26ft)	2.5–4m (8–13ft)	Large almost spineless. Green splashed with yellow and yellow-green	Female	Dull-red fruits
I. × altaclerensis 'Ripley Gold'	Bushy	4–8m (13–26ft)	2.5–4m (8–13ft)	Broad flat with a yellow margin	Female	Dull-red fruits. This cultivar is similar to *I. × altaclerensis* 'Lawsoniana'

Ilex × altaclerensis (Ilex aquifolium × Ilex perado) (Highclere Holly)
A large, vigorous, evergreen holly. Leaves usually larger than those of *Ilex aquifolium*, with normally only a few spines. Useful for tall hedging.
Hardiness: USDA 6b/7a. RHS H6.

Ilex × meserveae (Ilex aquifolium × Ilex rugosa) (Blue Holly)
Bred by in the USA by Mrs Kathleen Meserve to produce a group of hybrids that are hardier than *Ilex aquifolium* and *Ilex × altaclerensis*. These hybrids have characteristic blue-green foliage, and tend to be slower growing and more compact than *Ilex aquifolium*.
Hardiness: USDA6a-1. RHS H7.

Ilex crenata (Japanese Holly)

An evergreen, small-leaved holly from Japan. Fruits shiny and black. Growth variable.

Growing Requirements

Hardiness: USDA6b/7a. RHS H6.
Aspect: A planting site of any orientation is suitable.
Shelter: No special provision for plant shelter is necessary.
Sun: Full sun or part shade.
Soil: Providing the soil is well drained and holds adequate moisture in the summer a soil of any texture is acceptable.
pH: A very wide range of soil pH is accepted.

Ilex × altaclerensis 'Belgica Aurea'. This upright-growing cultivar makes a good, isolated specimen plant.

Ilex × altaclerensis 'Golden King'.

Ilex × meserveae Blue Angel ('Conang').

Ilex × meserveae Blue Angel ('Conang').

Ilex × meserveae 'Casanova'[PBR].

Ilex × meserveae 'Casanova'[PBR].

TABLE 4.66 *ILEX × MESERVEAE* AND CULTIVARS

Name	Habit	Height	Spread	Foliage	Male or female	Remarks
I. × meserveae	Bushy, medium-sized shrub			Blue-green, spiny	Can be male or female	Female plants bear bright-red berries
I. × meserveae BLUE ANGEL ('Conang')	Bushy	2.5–4m (8–13ft)	1.5–2.5m (5–8ft)	Blue-green, spiny. Each leaf 4cm (1½in) long	Female	Slow-growing. Red berries
I. × meserveae BLUE MAID ('Mesid')	Bushy	2.5–4m (8–13ft)	1.5–2.5m (5–8ft)	Blue-green, spiny	Female	Counterpart to Blue Prince. Many red fruits
I. × meserveae BLUE PRINCE ('Conablu') AGM	Bushy	2.5–4m (8–13ft)	2.5–4m (5–8ft)	Large, blue-green, spiny	Male	Purple-blue stems. Used to pollinate female cultivars
I. × meserveae BLUE PRINCESS ('Conapri') AGM	Bushy	2.5–4m (6–13ft)	1.5–2.5m (5–8ft)	Blue-green. Each leaf 6cm (2¼in) long	Female	Many bright-red fruits

TABLE 4.67 *ILEX CRENATA* AND CULTIVARS

Name	Habit	Height	Spread	Foliage	Male or female	Remarks
I. crenata	Shrub or small tree	4–8m (13–26ft)	1–1.5m (3–5ft)	Glossy, wavy, small 3cm (1in) long by 2cm (¾in)wide. Sometimes spiny	Can be male or female	Black berries on female plants
II. crenata 'Convexa' AGM	Bushy shrub	1.5–2.5m (5–8ft)	1–1.5m (3–5ft)	Small, dark, convex	Female	Black berries. Good for shaping into balls, pyramids, cloud pruned shrubs and hedging
I. crenata 'Fastigiata' AGM	Upright-growing. Forms a narrow column	2.5–4m (8–13ft)	1–1.5m (3–5ft)	Small, dark-green	Female	Slow-growing. Can be used to make a focal point
I. crenata 'Glory Gem'	Shrub or small tree	4–8m (13–26ft)	2.5–4m (8–13ft)	Glossy green	Male	Suitable for shaping into balls
I. crenata 'Green Hedger' AGM	Compact shrub	1–1.5m (3–5ft)	1–1.5m (3–5ft)	Small, dark-green, oval	Female	Black berries. Faster growing than *I. crenata* 'Convexa'; may be better for hedging

Ilex crenata trained and clipped to form a small tree.

Ilex crenata. New growth is lighter in colour.

Ilex crenata 'Shiro-furkurin' leaves grey-green with a creamy-white margin. This cultivar has a stable variegation and takes well to being clipped.

Ilex crenata as a Dwarf Hedge

Ilex crenata has been advocated for use as dwarf hedging and as an alternative to *Buxus sempervirens* and *Buxus sempervirens* 'Suffruticosa'. Judging from the plant specification above, it would appear that the species, or one of its cultivars, would be ideal; however, reports from gardeners and trials have been mixed. Problems have been experienced in establishing the plants with subsequent poor and slow growth. Clipping should be done in late spring just before growth starts, but the plants cannot be clipped so tightly as box and do not produce such a neatly defined, formal hedge.

In the USA, black root rot caused by the soil-borne fungus, *Thielaviopsis basicola*, has been very problematic.

It would appear that the search for a perfect replacement for box in hedging goes on.

Ilex verticillata (Winter Berry)

A dense-growing, multi-stemmed, deciduous shrub found in eastern North America.

Growing Requirements

Hardiness: USDA6b/7a. RHS H6.
Aspect: A site facing in any direction is suitable.
Shelter: Shelter does not specifically need to be provided.
Sun: Full sun or part shade is required.
Soil: Clay or loam soil is suitable but avoid calcareous soil. Drainage should be good, but poor drainage will be tolerated.
pH: This must be acid or neutral (pH3–6). Avoid a soil with a high pH.

To obtain berries, it is essential to have a male plant to act as a pollinator for the female plant. Different cultivars of winter berry flower over a period in spring. To obtain berries it is important to match early blooming male and female plants and likewise with late blooming cultivars.

TABLE 4.68 *ILEX VERTICILLATA* AND CULTIVARS

Name	Habit	Height	Spread	Male or female	Fruit	Remarks
I. verticillata	Multi-stemmed shrub	1.5–2.5m (5–8ft)	1.5–2.5m (5–8ft)	Male and female	Bright-red from autumn to spring	
I. verticillata male	Multi-stemmed shrub	1.5–2.5m (5–8ft)	1.5–2.5m (5–8ft)	Male		In the UK, plants that have been examined and are male are sold specifically to act as pollinators
I. verticillata 'Red Sprite'	Upright and rounded	1m (3ft)	1.2m (4ft)	Female	Bright-red	Early blooming
I. verticillata 'Winter Red'	Suckering	1.5–2.5m (5–8ft)	1.5–2.5m (5–8ft)	Female	Bright-red	Late-blooming. The most commonly available cultivar in the UK
I. verticillata 'Jim Dandy'	Multi-stemmed shrub	1–2m (3–6ft)	1.2–1.5m (4–5ft)	Male		Early blooming
I. verticillata 'Southern Gentleman'	Multi-stemmed shrub	2–2.5m (6–7ft)	2m (6ft)	Male		Late blooming

Garden Design Considerations

Of all the hollies, *Ilex aquifolium* (common holly) and its cultivars probably have the widest range of applications in garden and landscape design. Its dense, evergreen nature and positive reaction to pruning makes it eminently suitable for hedging. Such hedges provide shelter, privacy and a uniform backdrop. The species is resistant to wind damage and is useful in coastal gardens where tolerance of wind and provision of shelter for other plants are vitally important.

In winter, this evergreen species gives a balance to deciduous shrubs, plus structure and enclosure to the garden. Interplanted in a hedge with deciduous species that in juvenile, clipped form retain their foliage in winter, such as beech and hornbeam, it creates a very attractive pattern, perhaps analogous to the blue veining seen in Danish blue cheese. This is seen in the famous 'tapestry hedges' at Hidcote Manor Garden, an idea now seen replicated around the world.

Plants can be clipped into simple, pleasing shapes, such as columns, cubes and clouds. The 'four fat men', one vertical clipped feature in each corner of a knot garden, in Rosemary Very's garden in Gloucestershire are fondly remembered by many gardeners.

Grafting with holly has long been practised and top-worked standard hollies function very well as small trees in small, formal gardens and add height to plantings in larger gardens. Given time, pendulous cultivars can be trained to form spectacular, small, weeping, often umbrella-shaped trees.

Holly is traditionally linked with mistletoe at Christmas. The red berries add colour to the scene both outside and in the house; those with variegated foliage provide even more colour.

Holly (*Ilex aquifolium*) finds a welcome in the nature conservationist's garden. Where it is native, the flowers produce nectar for pollinating insects and the fruit valuable food for birds in winter. Several insects also use the foliage for food and shelter.

The cultivars of *Ilex* × *altaclerensis* carry out the same roles as common holly, but sometimes on a slightly larger scale, their greater vigour being employed to produce taller hedges. In today's smaller garden, they still have a very useful role competing with common holly cultivars for provision of the brightest variegated foliage.

The blue hollies have similar garden roles but are very appropriate in the smaller gardens of today. Here their slower growth and compactness makes them very useful. Where the climate is more challenging, their extra degree of hardiness broadens their range of application.

Ilex verticillata (winter berry) is more familiar to North American gardeners, where its vivid display of bright-red berries through autumn, winter and into early spring is much appreciated, provided an acid, humus rich perhaps boggy soil can be provided. Early and late blooming, and red and orange fruited cultivars are known. Providing that the fruiting and pollinating cultivars are correctly matched, they are very useful to add colour to the winter garden and landscape scene.

LAGERSTROEMIA

A genus of about fifty species. Originating from India, south-eastern Asia and northern Australia, it consists of trees and shrubs that may be evergreen or deciduous. The leaves are opposite and simple. Terminal panicles of flowers are produced ranging in colour from purple to white.

In the UK, the only *Lagerstroemia* species that is even marginally hardy is *Lagerstroemia indica*. This species requires a warm summer to initiate its flowers, but this is occurring increasingly frequently, and its cultivation is becoming more popular. Plant breeders have responded by producing and releasing several new cultivars and series of cultivars that are said to be hardier and very suitable for the UK climate.

Lagerstroemia indica (Common Crape Myrtle)

An upright-growing, deciduous tree or shrub. The leaves are simple, often bronzed or reddish when young but becoming dark-green. Autumn colour is often well developed, the leaves changing to shades of orange before falling. The flowers are produced in large panicles, the petals of which have a crepe paper-like appearance and texture. Flower colour varies from purple through shades of pink to white. In winter the exfoliating bark is revealed, which is an attractive feature of brown and contrasting grey patches.

Growing Requirements

Hardiness: USDA 9b/10a. RHS H3.
Aspect: The planting site should face south or west. Other orientations are very unlikely to be successful.
Shelter: A warm, sheltered site is required.
Sun: Full sun is required.
Soil: A soil of any texture is satisfactory, but good drainage is required.
pH: The plant is not pH-sensitive, and a wide pH range is satisfactory.

The base of a sheltered south-facing wall would be an ideal position. In autumn, a mulch of 5–7cm (2–2½in) of garden compost should be applied as protection against winter cold. If the plant is in a pot or container, then this should be moved to a sheltered spot and wrapping in horticultural fleece is recommended where low winter temperatures are possible. Plants in containers can also be moved into a temperate greenhouse for overwintering.

Modern cultivars appear to be developing into two groups: small and dwarf cultivars intended for container cultivation or planting out, and larger shrubs intended for open-ground planting. All have leaves that are reddish when young and developing, but with maturity become dark-green. Good autumn leaf colour, mainly in shades of orange is displayed before leaf fall.

Garden Design Considerations

With the coming of climate change, it appears there will be winners and losers in the world of plants. In the southern parts of the UK, it is becoming difficult to grow some Asiatic primulas that require cool summer conditions. Also, in southern UK, the climate is becoming more suited to *Lagerstroemia indica* and its cultivars. In warm, sheltered gardens growing *Lagerstroemia* is a viable proposition and a new and exciting prospect for gardeners. They represent a good option to provide vivid colour in summer and autumn, both as a carefully positioned shrub grown in the ground or as a container plant in all types of garden, including compact city, courtyard or patio gardens. Crape myrtles are also well suited to sheltered Mediterranean and coastal gardens, where low winter temperatures and cold winds are not encountered.

TABLE 4.69 *LAGERSTROEMIA INDICA* AND CULTIVARS

Name	Height	Spread	Flowers
L. indica AGM	4–8m (13–26ft)	4–8m (13–26ft)	Range from purple through shades of pink to white
L. indica 'Fuchsia d'Été'	4m (13ft)	2.5m (8ft)	Deep-pink
L. indica BERRY DAZZLE ('Gamad V1')	120cm (4ft)	70cm (2¼ft)	Purple/pink
L. indica MIMIE FUCHSIA ('Dablage01'PBR)	1m (3ft)	1.8m (6ft)	Pink
L. indica PETITE PINKIE ('Monkie')	1.5m (5ft)	1.5m (5ft)	Clear pink
L. indica RAPSODY IN PINK ('Whit V111')	4.5m (15ft)	3.5m (11½ft)	Light-pink
L. indica 'Rosea'	3–7m (10–23ft)	4.5m (15ft)	Deep-pink
L. indica VIOLET D'ÉTÉ ('Indyvio') (Indya chimes series)	2–4m (2½–13ft)	2.5m (8ft)	Violet-purple
L. indica 'Violet Fili'PBR (Fleming filigree series)	30–45cm (1–1½ft)	30–45cm (1–1½ft)	Violet
Members of the With Love Series			
L. indica WITH LOVE BABE ('Milaperl'PBR)	40cm (16in)	40cm (16in)	Pale-pink
L. indica WITH LOVE ETERNAL ('Milavio'PBR)	60cm (23in)	40cm (16in)	Light-pink on a compact panicle
L. indica WITH LOVE KISS ('Milarosso'PBR)	60cm (16in)	50cm (20in)	Light-pink. Large panicle

Lagerstroemia indica. Well positioned against a warm wall.

Lagerstroemia indica.

Lagerstroemia indica BERRY DAZZLE ('Gamad VI'). Showing autumn leaf colour.

Lagerstroemia indica VIOLET D'ÉTÉ ('Indyvio')

Lagerstroemia indica VIOLET D'ÉTÉ ('Indyvio')

Lagerstroemia indica RAPSODY IN PINK ('Whit V111').

Lagerstroemia indica (With Love Series) WITH LOVE KISS ('Milarosso'[PBR]).

Lagerstroemia indica (With Love Series) WITH LOVE KISS ('Milarosso'[PBR]) growing in a container and performing very well.

LEYCESTERIA

A small genus of six deciduous shrubs coming from the Himalayas and China.

Growing Requirements

Hardiness: USDA RHS H5 for the most part but note exceptions shown in Table 4.70.
Aspect: A site of any orientation is acceptable.
Shelter: A degree of shelter form severe cold will be appreciated.
Sun: Full sun or part shade is acceptable for all. Heavier shade will be tolerated by *Leycesteria formosa*.
Soil. Soil of any soil texture is tolerated, providing that it is well drained.
pH. Growth will be good over a wide soil pH range.

An Important Additional Feature of *Leycesteria formosa* and its Cultivars

The one-year-old stems of this plant are bright-green and stand out prominently in the winter garden. The plant needs hard annual pruning to remove old brown stems and stimulate new growth. Pruning is best done in the dormant period. This keeps the plant healthy, as older stems soon lose vigour and die and it has the benefit of enhancing the contribution made by this plant in the winter. *Leycesteria* 'Smoldering Embers' has young, reddish stems.

Garden Design Considerations

Leycesteria formosa (the pheasant berry) contributes flower and fruit colour to the garden in summer and early autumn. If hard pruned, the resultant new growth has vivid-green stem colour and this can be employed in the winter. It could be considered rather too large for what it contributes in all but larger gardens.

Leycesteria formosa GOLDEN LANTERNS ('Notbruce'[PBR]) provides strong yellow colour during the summer. It can be used alone as a focal point or contrasted with shrubs, such as the *Berberis thunbergii* cultivars with red foliage.

Leycesteria formosa LITTLE LANTERNS ('COD 01') is a new cultivar and will be very useful in small and compact gardens. It has the same attributes as *L.* Golden Lanterns but is considerably smaller. It is also of a size suitable for use in containers, pots, and patio gardens.

Leycesteria formosa GOLDEN LANTERNS ('Notbruce'[PBR]) in a large border designed to show off foliage colour, the *Leycesteria* making a bold central splash of colour.

Leycesteria formosa. Flowers and first fruit in summer.

TABLE 4.70 *LEYCESTERIA FORMOSA* CULTIVARS AND OTHER HYBRIDS

Name	Habit	Height	Spread	Foliage	Flowers	Remarks
L. formosa pheasant (berry, Himalayan honeysuckle)	Vigorous, upright	1.5–2.5m (5–8ft)	1.5–2.5m (5–8ft)	Mid-green, cordate with a drip tip	In pendulous racemes, 5–10cm (2–4in) long. Flowers white with prominent purple-red bracts	Flowers soon replaced by purple berries, the purple-red bracts remaining
L formosa 'Gold Leaf'	Vigorous, upright	1.8m (5–8ft)	1.5m (5–8ft)	Same shape and size as species. New growth tinged with crimson, later yellow-green	In pendulous racemes, 5–10cm (2–4in) long. Flowers white with prominent purple-red bracts	Flowers soon replaced by purple berries, the purple-red bracts remaining
Leycesteria formosa LITTLE LANTERNS ('CD CO1')	Compact, upright	1.25m (4ft)	1.25m (4ft)	Same shape and size as species. New growth lime-green with red shoots. In summer bright golden-yellow	In pendulous racemes, 5–10cm (2–4in) long. Flowers white with prominent purple-red bracts	Flowers soon replaced by purple berries, the purple-red bracts remaining
L. formosa GOLDEN LANTERNS ('Notbruce'PBR) AGM	Vigorous, upright	1.5–2.5m (5–8ft)	1.5–2.5m (5–8ft)	Same shape and size as species. New growth orange-red, then golden-yellow for rest of growing season	In pendulous racemes, 5–10cm (2–4in) long. Flowers white with prominent purple-red bracts	USDA 8b/9a. RHS H4. Flowers soon replaced by purple berries, the purple-red bracts remaining
L. formosa 'Purple Rain'		1.5–2.5m (5–8ft)	1.5–2.5m (5–8ft)	Same shape and size as species. Has a purple mid rib	Same as for *L. formosa*, but with darker purple more prominent bracts	
L. 'Smoldering Embers'		1.5–2.5m (5–8ft)	1.5–2.5m (5–8ft)	Ovate, 12 cm (4½in) long	In pendulous racemes, corolla pale-yellow with small reddish bracts	USDA 8b/9a. RHS H4

Leycesteria formosa LITTLE LANTERNS ('COD CO1'). New shoot tip.

Leycesteria formosa LITTLE LANTERNS ('COD CO1'). New tip and foliage below.

Leycesteria formosa GOLDEN LANTENS ('Notbruce'PBR). Plant in late summer. Foliage still yellow and flowers being produced.

Leycesteria formosa GOLDEN LANTERNS ('Notbruce'PBR) AGM.

LOROPETALUM

A small genus of four species of evergreen shrubs. It is separated from the closely related genus *Hamamelis* (witch hazels), partly on the basis of *Loropetalum* being evergreen and *Hamamelis* deciduous. They have four to six strap-shaped, wavy petals, each 1–2cm (¼–¾in) long.

In the UK, except in the mildest gardens, they may need winter protection during cold, wet weather in winter.

Growing Requirements

Hardiness: USDA 8a/9a. RHS H4.
Aspect: A south- or west-facing site should be selected.
Shelter: A sheltered site protected from cold winds and excessive winter wet should be provided.
Sun: Full sun or preferably part shade is required.
Soil: The soil can be of any texture, but plenty of organic matter should be incorporated prior to planting. It must be well drained but retain adequate moisture during the summer.
pH: Whilst not strictly calcifuge, a neutral to acid soil is preferred.

Garden Design Considerations

Oh, for a slightly warmer winter climate! This species and its cultivars are widely used in gardens and for landscaping in southern regions of China and Japan where the winters are slightly less severe than in most of the UK. Here it provides masses of spring flowers, often set against coloured foliage. In southern France, it also grows exceptionally well, as evidenced in the 'other garden' 'Serre de la Madone' of Major Lawrence Johnston the creator of Hidcote Manor Garden.

In slightly colder climates, it may be grown successfully outside in sheltered positions, away from cold winds and protected from excessive winter wet. Small or compact gardens in urban areas may well offer these conditions. Here, though, it may still need extra protection during the coldest winter spells.

They are well suited for use as container or patio plants, where they can receive individual winter protection or be moved into a cold greenhouse or conservatory during winter and then moved out to provide a good spring display and an attractive, often coloured, leaf shrub for the summer.

TABLE 4.71 CULTIVARS OF *LOROPETALUM CHINENSE*

Name	Habit	Height	Spread	Foliage	Flowers	Remarks
L. chinense CAROLINA MOONLIGHT ('Nci 002')	Bushy	1.5m (5ft)	1.5m (5ft)	Dark-green	Creamy-white. Late winter early spring	Strong contrast between foliage and flower colour
L. chinense var. *rubrum* 'Blush'	Bushy	2m (6½ft)	1.5m (5ft)	Bronze, red, may go green in summer	Pink	
L. chinense var. *rubrum* 'Burgundy'	Bushy	1.5m (5ft)	1.5–2.5m (5–8ft)	Rich red, may go olive-green in summer	Bright-pink	
L. chinense var. *rubrum* 'Chang Nian Hong'[PBR] (syn. 'Ever Red')	Bushy	1.5m (5ft)	1.5m (5ft)	Dark-reddish-purple	Red	
L. chinense var. *rubrum* 'Fire Dance'	Bushy	0.5–1m (20–40in)	0.5–1m (20–40in)	Dark-reddish-purple, may go green in summer	Bright-pink, slightly fragrant. Late winter early spring	
L. chinense var. *rubrum* 'Pearl'	Bushy	1.5m (5ft)	1.5m (5ft)	Dark-purple	Deep-pink	
L. chinense var. *rubrum* RUBY SNOW ('Iwai'[PBR])	Bushy	0.75m (2½ft)	1m (3¼ft)	Dark-purple	Creamy-white	Good contrast between leaves and flowers

Loropetalum chinense var. *rubrum* BLACK PEARL (Pearl'[PBR]). Plants for sale in late spring.

Loropetalum chinense var. *rubrum* BLACK PEARL (Pearl'[PBR]). Detail of foliage and flowers.

MAHONIA

A genus of about seventy species, nearly all of which are evergreen shrubs found in the Himalayas, Asia, North and Central America.

Plants in the genus *Mahonia* are separated from the closely related genus *Berberis* by their leaves. In *Mahonia*, the leaves are pinnate, whereas in *Berberis*, the leaves appear to be single and simple. The flowers of *Mahonia* are produced in terminal clusters of racemes. After flowering, blue/black berries are produced.

Growing Conditions

Hardiness: Most are USDA 7b/8a. RHS H5, but some are only USDA 8b/9a, RHS H4. Please see specific hardiness ratings in Tables 4.72 and 4.73.

Aspect: A site of any orientation is acceptable.

Shelter: Most do not specifically require a sheltered site, but those of a rating of USDA 8b/9b, RHS H4 do require shelter from cold and desiccating winds.

Sun: Here *Mahonia* does vary in its requirements and this is shown individually for each plant in Tables 4.72 and 4.73.

Soil: Soil of any texture is accepted, provided it is well drained. It is important that adequate moisture is retained in summer.

pH: A wide range of soil pH is tolerated, but with *M. nervosa*, a high pH is better avoided.

Garden Design Considerations

The smaller *Mahonia* species and cultivars have a wide range of possible applications. *Mahonia aquifolium* and its cultivars are excellent for low maintenance and attractive ground-cover. They contribute good flower colour in the spring and produce a dense, effective canopy of evergreen leaves that prevent weed intrusion. *Mahonia aquifolium* is sometimes regarded as a poor-quality shrub, but this is because it is often confined to the poorest and toughest garden situations, with no maintenance. Given a well-prepared site and a little annual maintenance, it is a fine shrub providing a good spring display and high-quality evergreen foliage. The toughness of this shrub should be well utilised, but not over-exploited.

Mahonia aquifolium 'Smaragd' has very dark, almost black foliage in winter and looks striking when planted against, or amongst, an evergreen, yellow-leaved plant, such as *Carex elata* 'Aurea' AGM.

Mahonia gracilipes may be planted when something unusual and interesting is required. The bicoloured nature of the flowers is unusual in mahonias, but the revelation of the reverse of the foliage, it is bright chalky-white, is always exciting.

The comparatively recent introduction of *Mahonia eurybractea* subsp. *gonpinensis* 'Soft Caress' caused something of a sensation as it is not prickly! For small-scale ground-cover and inclusion in large planters it is most desirable, producing small but attractive flowers and sought-after foliage. Mostly low-growing, it can tend to produce some taller shoots. If desired, these taller shoots can be pruned hard, down to ground level, if necessary, and such shoots will regrow producing several bushier shoots.

Mahonia nitida 'Cabaret' introduced new colour to *Mahonia*. The upright sprays of orange and yellow flowers really catch the eye. Its small size makes it eminently suitable for patio and pot culture.

Mahonia aquifolium. A very underrated spring-flowering, evergreen shrub.

Mahonia aquifolium 'Smaragd'. Here, the near-black foliage of the *Mahonia* contrasts well against the evergreen yellow foliage of the *Carex*.

Mahonia eurybractea subsp. *gonpinensis* 'Soft Caress'. A small, young, flowering plant.

Mahonia eurybractea subsp. *gonpinensis* 'Soft Caress'. Detail of the flowers and smooth foliage.

Mahonia eurybractea subsp. *gonpinensis* 'Soft Caress'. An older plant showing how its habit develops if left to its own devices.

Mahonia nervosa. A low-growing species good for ground cover in part shade.

TABLE 4.72 SMALLER *MAHONIA* SPECIES AND CULTIVARS

Name	Habit	Height	Spread	Foliage all leaves pinnate	Flowers	Flowering period and fruit	Sun	Remarks
M. aquifolium AGM	Low-spreading, suckering	0.5–1m (20–40in)	1–1.5m (40–60in)	Glossy, dark-green, prickly	Clusters of yellow flowers in terminal and lateral buds up stem	Spring. Blue/black. Will self-seed	Full sun to full shade	Oregon grape, mountain grape. USDA7b/8a. RHS H5
M. aquifolium 'Apollo' AGM	Low-spreading, suckering	0.5–1m (20–40in)	1.1.5m (40–60in)	Glossy, deep-green, can become purplish in cold weather	Abundant clusters of yellow flowers in terminal and lateral buds up stem	Spring. Blue/black	Full sun to full shade	USDA7b/8a. RHS H5
M. aquifolium 'Smaragd'	Compact, suckering	0.6m (2ft)	0.9m (3ft)	Very dark-green. Can become bronzed in winter	Clusters of yellow flowers in terminal and lateral buds up stem	Spring. Blue/black	Full sun to full shade	USDA7b/8a. RHS H5
M. eurybractea subsp. *gonpinensis* 'Soft Caress'	Upright	1–1.5m (40–60in)	1–1.5m (40–60in)	Dark-green. Narrow spine free leaflets	Upright, small yellow clusters. Honey scented	Late summer/autumn. Blue/black	Full sun to part shade	USDA 8b/9a. RHS H4. RHS Chelsea Flower Show Plant of the Year 2013
M. gracilipes	Small, suckering	1–1.5m (40–60in)	1.1.5m (40–60in)	5–9 well-separated, spiney leaflets. Green above, white beneath	Small sprays of flowers. Outer petals purple-red, inner petals creamy-white	Late summer/autumn. Fruit not freely produced	Part shade	USDA 7b/8a. RHS H5
M. nervosa	Compact, suckering	0.5–1m (20–40in)	0.5–1m (20–40cm)	9–10 leaflets	Dense clusters of flowers	Late spring/early summer Blue/black	Part to full shade	USDA 7b/8a. RHS H5. Requires shade and soil that does not dry out in summer
M. nitens 'Cabaret'[PBR] AGM	Upright	1.2m (4ft)	0.8m (32in)	Glossy with spines on leaflets	Upright clusters. Buds orange-red, flowers yellow. No scent	End of summer/autumn. Blue/black	Full sun to part shade	USDA 8b/9a. RHS H4. Needs shelter, particularly from cold, drying winds
M. × wagneri 'Pinnacle' AGM	Upright	1–1.5m (40–60cm)	1–1.5m (40–60cm)	Glossy. New growth can be copper tinted	Dense yellow clusters in both terminal and lateral buds.	Late winter/spring. Blue/black	Full sun to part shade	USDA 7b/8a. RHS H5

Garden Design Considerations

The larger *Mahonia* species and cultivars are often too large for small or compact gardens. They can, however, be kept shorter and bushier by pruning hard, about one-third of the tallest stems of established plants, every year just after flowering. Where space does permit, these larger plants are very effective in providing shelter and a background for smaller, perhaps more tender, plants. They also make very effective hedges and screens.

Mahonia × media 'Winter Sun' is much less spreading than other *M. × media* cultivars and has a very upright habit. This might be regarded as a rather leggy and gaunt shrub. In fact, this plant has strong architectural value, with a strong outline and, when viewed closely, displays a distinct corky bark. This is together with bold, lustrous foliage and a terrific display of flowers. As a single specimen, underplanted with smaller shrubs, this would make a notable feature in a small garden.

Mahonia oiwakensis subsp. *lomariifolia*. A large, rather tender shrub not suitable for small or compact gardens.

Mahonia oiwakensis subsp. *lomariifolia*.

Mahonia japonica. An excellent shrub for tall ground-cover.

TABLE 4.73 LARGER *MAHONIA* SPECIES AND CULTIVARS

Name	Habit	Height	Spread	Foliage	Flowers	Flowering period and fruit	Sun	Remarks
M. oiwakensis subsp. *lomariifolia* AGM	Large shrub. Very erect growing	2.5–4m (8–13ft)	1.5–2.5m (5–8ft)	Long leaves with many spiny leaflets	Racemes 25cm (10in) long. Bright-yellow, slightly scented flowers	Late autumn into winter. Blue/black	Full sun to part shade	USDA 8b/8a. RHS H4. Needs shelter especially from cold, drying winds
M. japonica	Medium-sized shrub	1–1.5m (3–5ft)	1.5–2.5m (5–8ft)	7–15 spiny leaflets	Fragrant, light-yellow flowers in long arching clusters	Late autumn to early spring	Full sun to part shade	USDA 7b/8a. RHS H5. *Mahonia japonica* × *M. oiwakensis* subsp. *lomariifolia* produced *M.* × *media* of which those shown below are selections
M. × *media* 'Buckland' AGM	Large erect growing	2.5–4m (8–13ft)	2.5–4m (5–8ft)	17–21 leaflets	Long spreading clusters	Late autumn/winter	Full sun to part shade	USDA 7b/8a. RHS H5
M. × *media* 'Charity'	Large, erect-growing	2.5–4m (8–13ft)	2.5–4m (8–13ft)	Up to 21 leaflets	Long spreading clusters	Late autumn/winter	Full sun to part shade	USDA 7b/8a. RHS H5
M. × *media* 'Lionel Fortescue'	Large, erect-growing	2.5–4m (8–13ft)	2.5–4m (8–13ft)	Each leaf 45 cm (17½in) long	Bright-yellow, upright clusters, up to 40 cm (16in) long	Late autumn/winter	Full sun to part shade	USDA 7b/8a. RHS H5
M. × *media* 'Winter Sun' AGM	Upright shrub	2.5–4m (8–13ft)	1.5–2.5m (5–8ft)	Long leaves	Bright-yellow, erect clusters	Late autumn/winter	Full sun to part shade	USDA 7b/8a. RHS H5. Much less spreading than other *M.* × *media* cultivars. Very profuse in flowering

Mahonia japonica. The sweetly scented flowers are produced in late winter.

Mahonia × *media* 'Charity'. A mature plant in full flower.

Mahonia × *media* 'Charity'. Arching racemes of flowers.

Mahonia × *media* 'Winter Sun'. The racemes of flowers are upright, as is the habit of the entire plant. Very good as an isolated specimen where space is limited.

MYRTUS

Formerly a genus of evergreen shrubs, the majority of which have been transferred to other genera, with only one species of horticultural importance, namely *Myrtus communis* (common myrtle) remaining.

Myrtus communis AGM (Common Myrtle)

Thought to be originally from western Asia, this bushy, evergreen shrub is commonly naturalised in southern Europe. Height is 1.5–2.5m (5–8ft) and spread 1.5–2.5m (5–8ft). The leaves are ovate, pointed, 2–5cm (1–2in) long and 1–1.5cm (½–¾in) wide, dark-green and glossy on top, paler beneath and fragrant when crushed. The flowers produced in late summer are white, bowl-shaped with prominent anthers and fragrant. These are followed by purple-black berries 1cm (½in) long.

In many cultures, it is associated with friendship and love and is considered an essential ingredient of wedding bouquets.

Growing Requirements

Hardiness: USDA 8b/9a. RHS H4.
Aspect: The aspect of the site should be south- or west-facing.
Shelter: To obtain flowering and fruiting, it is essential that a warm, sheltered site is provided. In cold areas, winter protection may be necessary.
Sun: Full sun.
Soil: The soil can be of any texture, but must be well drained, moisture-retentive in summer and fertile.
pH: Soil pH can be acid, neutral or alkaline.

Garden Design Considerations

These delightful small shrubs find a number of uses in the garden. The classic position is at the base of a warm, south- or west-facing wall where these shrubs will luxuriate and fill the air with late summer fragrance. They are also suitable for sunny flower borders, low hedging and screening in warm gardens, mild climatically favoured coastal gardens and sunny courtyards.

TABLE 4.74 *MYRTUS COMMUNIS*, SUBSP. AND CULTIVAR

Name	Habit	Height	Spread	Leaves	Flowers	Remarks
Myrtus communis AGM	Bushy	1.5–2.5m (5–8ft)	1.5–2.5m (5–8ft)	Dark-green, ovate	White, fragrant	
Myrtus communis subsp. *tarentina* AGM (Tarentum myrtle)	Compact and bushy	1.5m (5ft)	1.5m (5ft)	Narrowly oval, small, 1–2cm (½–¾in) long and 3mm (⅒in) wide	White with a pink tinge. White berries	A compact, free-flowering form
Myrtus communis 'Variegata'	Bushy	1.5–2.5m (5–8ft)	1.5–2.5m (5–8ft)	Leaves grey-green with a narrow creamy-white margin	White, fragrant. Purple-black berries	

Myrtus communis. The white, scented flowers and dark-green leaves.

Myrtus communis 'Variegata'. White, scented flowers and variegated leaves.

NANDINA

A monotypic genus containing only *Nandina domestica*. The plant comes from Central China and Japan where it has been cultivated in gardens for many centuries. In China, it was often planted either side of an entrance, its presence being said to ward off evil spirits. Today it is still planted in this position, but its presence is more a sign of welcome to visitors, sometimes known as the friendship plant.

Growing Requirements

Hardiness: USDA 7b/8a. RHS H5.
Aspect: A planting position of any aspect is acceptable, except north-facing.
Shelter: The planting site should be sheltered.
Sun: Full sun or just part shade is required.
Soil: The soil can be of any texture but must be well drained and retain some moisture in the summer.
pH: A pH range of acid through to alkaline is tolerated.

Nandina domestica (Sacred Bamboo, Heavenly Bamboo)

Not a bamboo, but an evergreen shrub producing upright, unbranched, bamboo-like stems. Height 1–1.5m (39–59in). Spread 1–1.5m (39–59in). Leaves large, tripinnate with lanceolate leaflets. Each leaf 50–100cm (20–39in) long and each leaflet lanceolate 4–11cm (2–4in) long and 1.5–3cm (1–1½in) wide. New leaves in spring purplish, changing to green as the season progresses. Old leaves may turn red before falling. Flowers white in panicles produced in summer, followed by red berries in the autumn. These fruits persist through the winter. Natural spread is by rhizomes and seed.
Toxicity: All parts are toxic to animals, especially birds, but it is not normally considered toxic to humans.
Invasiveness: *Nandina domestica* is considered invasive in North Carolina, Tennessee, Georgia and Florida. Purchase or continued cultivation of non-sterile varieties is discouraged in the south-east of the United States.

Nandina domestica. The species is potentially a larger plant than most of the cultivars.

Nandina domestica. Produces a crop of red fruit lasting throughout the winter.

Nandina domestica 'Fire Power'. This is a very compact plant.

Nandina domestica 'Fire Power'. Displays bright-red, slightly puckered leaves.

Nandina domestica 'Gulf Stream'. This cultivar has leaves of red and green during the winter.

Nandina domestica MAGICAL LEMON AND LIME 'Lemlin'PBR.

Nandina domestica 'Seika'PBR. The red foliage of this cultivar makes a strong contrast against the yellow foliage of *Pinus contorta* 'Chief Joseph'.

Nandina domestica 'Twilight'PBR. In winter, the foliage is green splashed with white. In spring, the new growth is pink.

Nandina domestica 'Seika'PBR. The finely cut, red foliage of this cultivar.

TABLE 4.75 *NANDINA DOMESTICA* AND CULTIVARS

Name	Habit	Height	Spread	New leaf colour spring	New leaf colour summer	Leaf colour autumn/ winter	Flowers	Fruits	Remarks
N. domestica	Upright	1–1.5m (3–5ft)	1–1.5m (3–5ft)	Purplish-green	Green	Green	White	Red	
N. BLUSH PINK 'Aka'[PBR]	Compact	0.9m (35in)	0.9m (35in)	Pink	Red	Deeper red eventually green		None	
N. BRIGHTLIGHT 'Shelten004'	Upright	1.5m (5ft)	50cm (1½ft)	Lime-yellow	Lime-yellow	Green	White	Few red fruits	Longer, more slender foliage
N. 'Fire Power'	Compact	0.6m (2ft)	0.45m (1½ft)	Red	Yellow-green	Red	White, few	Few red fruits	Leaves a little puckered
N. 'Gulf Stream'	Compact	1–1.5m (3–5ft)	0.5–1m (18in–3ft)	Bronze-red	Green with red and orange flushes	Shades of red and gold	White	Few red fruits	
N. Magical LEMON AND LIME 'Lemlin'[PBR]	Compact	0.6m (2ft)	0.6m (2ft)	Yellow-green	Lime-green	Green			
N. 'Richmond' AGM	Upright	2m (6ft)	1.5m (5ft)	Purplish	Green	Green	White	Many red fruits	
N. 'Seika'[PBR]	Compact	70cm (27in)	0.6m (2ft)	Fiery red	Green	Red	Few white	Red few	
N. 'Sunset'[PBR]		1–1.5m (3–5ft)	1–1.5m (3–5ft)	Coral orange	Green	Red			
N. 'Twilight'[PBR]		1–1.5m (3–5ft)	0.5–1m (1½–3ft)	Bright-pink	Green splashed with white	Green	White	Rarely any fruit	

Cultivars of *Nandina domestica*

Recent years have seen the introduction of numerous cultivars. They are deservedly popular, being compact, slow-spreading, evergreen shrubs providing vivid, changing leaf colour throughout the year.

The leaf colour of the coloured leaf cultivars starts with only the new foliage produced in the spring. This may change or fade as the season moves on to summer. In the autumn, the foliage colour may change again, and in winter may, but not always, become green.

Flowering and fruiting only occurs to much extent with the plain-green species and cultivars. In coloured leaf cultivars, flowering and fruiting is usually very little or entirely absent.

Nandina has very low-maintenance requirements, needing only the occasional removal of old stems and the collection of fallen old leaves.

Garden Design Considerations

Nandina domestica has a unique appearance; no other plant has the combination of tall cane-like stems and large, fine-textured, pinnate leaves. Its use in modern Chinese-style gardens is very appropriate, reflecting its use in the ancient scholar gardens of Suzhou and Wuxi in China. Today, Chinese gardeners and clients of garden designers may appreciate the plant's incorporation into contemporary gardens, remembering *Nandina*'s symbolic use as indicating friendship to visitors when planted either side of an entrance.

If the features of white flowers and red fruits in autumn and winter are intended to be part of a design, then *Nandina domestica* or *Nandina domestica* 'Richmond' AGM are the plants to select and to place in a sunny, warm sheltered position.

In the modern garden, *Nandina* is very appropriate, imparting a feeling of contemporary style. The cultivars bring leaf colours and fine leaf-texture to a design; their compact, very slow-spreading nature makes them ideal for the modern garden, where space is very often limited. The colour selected, which often changes with the seasons, is a matter for personal taste, but a range is available. As with the original species, the site should be warm, sunny and sheltered, whether in a small, mixed planting, raised bed, planter or large pot on a small patio.

OSMANTHUS

A small genus of about thirty species consisting of evergreen trees and shrubs, which are found in Asia, the islands of the Pacific and in North America. The distinguishing characteristics of the genus are, opposite leaves, flowers in axillary or terminal clusters, the flowers consisting of four petals, four sepals and two anthers. The flowers are usually fragrant.

Osmanthus fragrans is an extremely desirable species for the garden but is not hardy in north temperate climates. In China and Japan, it has been widely cultivated for centuries and there are many different cultivars. It grows to form a large, evergreen shrub with profuse, very fragrant flowers. The flowers are also used to make osmanthus tea.

Growing Requirements

Tough easily grown shrubs.

Hardiness: USDA 7b/8a. RHS H5 for species and hybrids listed in Tables 4.76–4.78

Aspect: A site facing in any direction is suitable.

Sun: Full sun or part shade is fully satisfactory.

Soil: Provided the soil is well drained it can be of any texture.

pH: A very wide pH range from alkaline to acidic is readily accepted.

Osmanthus × *burkwoodii*. Trained, pruned and clipped into a multi-stemmed, cloud-pruned shrub.

Osmanthus × *burkwoodii*. One of the clipped 'clouds' producing a few flowers.

Osmanthus × *burkwoodii*. When left to grow freely, each growth produces more of its white, sweetly scented flowers.

Osmanthus delavayi. A large shrub showing its naturally slightly pendant growth habit.

Osmanthus delavayi. A typical spray of its white, scented flowers.

Osmanthus heterophyllus 'Goshiki'. A slow-growing cultivar with compact growth.

Osmanthus heterophyllus 'Goshiki'. Showing a form of variegation.

Osmanthus heterophyllus 'Variegatus'. Showing the white, scented flowers and variegated foliage.

TABLE 4.76 *OSMANTHUS × BURKWOODII* AGM (*OSMANTHUS DECORUS × OSMANTHUS DELAVAYI*)

Name	Habit	Height	Spread	Foliage	Flowers	Remarks
O. × burkwoodii	Slow-growing forming a dense bush	2.5–4m (8–13ft)	2.5–4m (8–13ft)	Dark-green, ovate 2.5–3cm (1–1.5in) long	Small, white, highly scented flowers produced in late spring	More compact growing than *O. delavayi*

TABLE 4.77 *OSMANTHUS DELAVAYI* AGM

Name	Habit	Height	Spread	Foliage	Flowers	Remarks
O.delavayi	Slow-growing, but eventually makes a large shrub	2.4–4m (7½–13ft)	2.5–4m (8–13ft)	Small, glossy, serrated, dark-green leaves, 2.5cm (1in) long	Small, white, highly scented flowers produced in late spring, followed by blue/black fruits	More open arching habit than *O. × burkwoodii*
O. delavayi 'Pearly Gates'	A compact cultivar	1–1.5m (3–5ft)	0.5–1m (20–40in)	Oval, dark-green	Small, white, highly scented flowers in late spring	

TABLE 4.78 *OSMANTHUS HETEROPHYLLUS* (HOLLY OLIVE)

Name	Habit	Height	Spread	Foliage	Flowers	Remarks
O. heterophyllus	Dense-growing, holly-like shrub.	4–8m (13–26ft)	4–8m (13–26ft)	Leaves dark-green, variably shaped with spines, holly-like	White scented, produced in late summer. Followed by blue/black fruit	
O. heterophyllus 'Goshiki' AGM (syn. *O.* Tricolor)	Dense, slow-growing holly-like shrub	1–1.5m (3–5ft	1–1.5m (3–5ft)	Leaves marbled with cream, tinged pink. Variably shaped with spines	White, scented, produced in late summer	
O. heterophyllus 'Variegatus'	Dense, moderate rate of growth. Holly-like	2.5m (8ft)	2.5m (8ft)	Dark-green, edged with white. Variably shaped with spines	White, scented, produced in late summer	

Garden Design Considerations

This genus contains species and cultivars potentially very useful to the garden designer.

Osmanthus × burkwoodii is potentially a large, evergreen shrub or small tree. As a shrub, it can be employed to form a dense hedge or formal backdrop to a planting. The dark foliage contrasts well against, and displays, the white, scented flowers. Clipping after flowering will maintain a crisp shape but may reduce flowering in the following year. A more exciting possibility is to prune the shrub into a cloud-pruned, most interesting, multi-stemmed, small tree. Such pruning will expose a characteristic growth habit and the 'clouds' can be kept tight and well separated by clipping.

Osmanthus delavayi is also potentially a large, evergreen shrub. Its more open, arching habit of growth, whilst attractive to see, makes it less suitable for clipping into a restrained shape. *Osmanthus delavayi* 'Pearly Gates' is a very much more compressed cultivar and very useful in small or compact gardens.

As a species, *Osmanthus heterophyllus* is too large for small gardens, but its several variegated cultivars are much slower growing. *Osmanthus heterophyllus* 'Goshiki' may be clipped to produce a regular, rounded shape. This is best done carefully with secateurs to avoid cutting individual leaves. Tolerating light shade, it can be used to lighten and stand out in a shaded position. Its variously shaped leaves are fascinating upon close inspection and the scented flowers a surprise to those unfamiliar with the plant.

PACHYSANDRA

A small genus of low-growing, evergreen or semi-evergreen sub-shrubs, native to North America and eastern Asia. The flowers are monoecious, with male flowers at the top and female flowers at the bottom of the spike.

Growing Requirements

Hardiness: USDA 7b/8a. RHS H5, except *Pachysandra procumbens* which is USDA 6b/7a. RHS H6.
Aspect: Prefers north- or east-facing but will grow on west- or south-facing, providing there is adequate moisture in summer.
Shelter: Avoid cold winds.
Sun: Best in full shade or part shade but will tolerate full sun if there is adequate moisture in summer.
Soil: The soil can be of any texture provided it is well drained and retains adequate moisture.
pH: A wide range is accepted from alkaline to acid.

Garden Design Considerations

Pachysandra terminalis is an excellent, low, ground-covering plant, producing a dense canopy of leaves and growing well in a variety of conditions, including damp shade. If irrigated during establishment, it will also grow satisfactorily in dry shade. The low-growing sub-shrub forms a uniform carpet of green, which fills in at the base of small trees and shrubs enabling them to be well displayed. Newly planted plants are quite slow to spread and may need to be planted more thickly than other conventional low ground-cover subjects, such as *Vinca major* 'Variegata', if effective ground coverage is to be achieved in a short period.

The cultivars 'Green Carpet' and 'Green Sheen' are slightly more attractive, but slightly slower growing. As such, they are suitable for small-scale ground-cover in small or compact gardens.

Pachysandra terminalis 'Variegata' is the slowest growing of the cultivars, but where a lightening of a dark shaded area is required, it is excellent.

Pachysandra procumbens is more demanding in its requirements if it is to grow satisfactorily but may be chosen where something more unusual is required.

TABLE 4.79 SPECIES AND CULTIVARS OF *PACHYSANDRA*

Name	Habit	Height	Spread	Foliage	Flowers	Remarks
P. procumbens (Allegheny spurge)	Dwarf, creeping	15cm (5½in)	30–60cm (1–2ft)	Semi-evergreen. Leaves set on end of stems	5–10cm (2–4in) tall spike of pale-coloured flowers produced in spring	Does not like too much sun. Avoid alkaline soils
P. terminalis AGM (Japanese spurge)	Sub-shrub, forms a carpet of evergreen leaves	20cm (8in)	1–1.5m (3–4½ft)	Evergreen. Leaves diamond-shaped and coarsely toothed at distal end	10cm (4in) tall spike of white flowers produced in early summer	
P. terminalis 'Green Carpet' AGM	Sub-shrub, forms a carpet of evergreen leaves	15cm (5½in)	0.5–1m 20–40in)	Dark-green, glossy	10cm (4in) tall spike of white flowers produced in early summer	
P. terminalis 'Green Sheen' AGM	Sub-shrub, forms a carpet of evergreen leaves	15cm (5½in)	1m (40in)	Dark-green, very glossy	10cm (4in) tall spike of white flowers produced in early summer	
P. terminalis 'Variegata' AGM	Sub-shrub, forms a carpet of variegated evergreen leaves	20cm (8in)	0.1–0.5m (4–20in)	Leaves with a cream margin	10cm (4in) tall spike of white flowers produced in early summer	Spreads only slowly

Pachysandra procumbens. Producing a carpet of green leaves under dappled shade. The slightly yellowed growth is new leaves emerging.

Pachysandra procumbens. In flower in spring.

PEROVSKIA

Plants formerly of this genus have now all been transferred to the genus *Salvia* and will be found under this generic name.

PHILLYREA

A group of evergreen trees and shrubs native to the Mediterranean region and South-West Asia. They have opposite leaves and produce small flowers in the leaf axils of the previous year's growth.

Growing Requirements

Hardiness: USDA 7b/8a. RHS H5.
Aspect: The site should ideally be south-facing.
Shelter: Avoid cold, drying winds. They are well adapted to coastal gardens where strong, but not cold or drying winds are experienced.
Sun: Full sun or part shade.
Soil: A sandy or loamy soil is required. Avoid heavy clay or thin calcareous soils. The soil should be well drained and hold adequate moisture in the summer.
pH: An acid or neutral soil is required.
Pruning: As a group they stand and respond well to clipping.

Garden Design Considerations

Phillyrea angustifolia is used for hedging and topiary. It is well suited to this application and may make an alternative to *Buxus sempervirens* (box) where box blight is a problem. *Phillyrea angustifolia* f. *rosmarinifolia* is finer in texture and good for low and finely shaped hedging.

Phillyrea latifolia grows slowly into a well-shaped, leafy, small tree often similar in appearance to a small olive tree. It is suitable for use as a small tree in a garden where space is limited but may take time to reach a significant size. It can live for many years and old gnarled specimens found in old gardens should be treasured. The plant produces plentiful leaves, responds well to pruning and clipping, and as such is ideal as a project to form into a distinguished cloud-pruned specimen.

TABLE 4.80 *PHILLYREA* SPECIES

Name	Habit	Height	Spread	Foliage	Flowers
P. angustifolia	Compact, rounded bush	2.5–4m (8–13ft)	1.5–2.5m (5–8ft)	Dark-green evergreen linear leaves 6cm (2in) long	Produced in early summer. Small, white, fragrant
P. angustifolia f. *rosmarinifolia*	Compact, rounded bush	2.5–4m (8–13ft)	1.5–2.5m (5–8ft)	Leaves narrower and smaller than *P. angustifolia*	
P. latifolia	A small, olive-like tree	4–8m (13–26ft)	4–8m (13–26ft)	Dark-green, ovate leaves 8cm (3in) long	Produced in early summer. Small, dull white. Followed by small black fruit

Phillyrea angustifolia f. *rosmarinifolia* as a small, closely clipped hedge.

Phillyrea angustifolia f. *rosmarinifolia*, showing the fine texture of the hedge.

PHILADELPHUS

A genus of about eighty species widely spread over the northern hemisphere. It has opposite, simple leaves. The flowers have four petals and four sepals, are white and most are sweetly scented. The mock oranges or *Syringa*, as they are sometimes erroneously called, are extremely popular garden shrubs.

Growing Requirements

Hardiness: USDA 6b/7a. RHS H6. Except for *Philadelphus maculatus* cultivars that are USDA 7b/8a. RHS H5.
Aspect: A site facing any direction is acceptable, except for *Philadelphus maculatus* cultivars that do not do well in a north-facing site.
Shelter: No special provision is required, except for the *Philadelphus maculatus* cultivars that do require a sheltered site.
Sun: Full sun or part shade is required.
Soil: The soil can be of any texture, providing that it is well drained and moisture-retentive in summer.
pH: A very wide soil pH is satisfactory.

The mock oranges are extremely popular garden shrubs and there are a great many cultivars. Unfortunately, many get too tall or spread too widely for small or compact gardens. In Table 4.81, an attempt has been made to select only the best cultivars and to avoid those that grow too large.

Garden Design Considerations

Mock oranges and lilacs are the epitome of the English country garden, flowering around that point in late spring and early summer when such gardens are approaching their floral peak. As such both are essential in such gardens.

In any garden, flowers and scent are fundamental and *Philadelphus* supplies both. It is easy to grow but does benefit from correct regular feeding and pruning.

Positioning can be important. Although under warm, still conditions one philadelphus can perfume a whole garden, when placed near a path or entrance point, their scent and flowers can still be readily enjoyed under less favourable conditions. Those with downward-facing flowers benefit from where their flowers can be viewed from slightly below the horizontal, thus enabling the flowers to be better seen and appreciated. At the side of, and partway up, a flight of garden steps is a good example.

A woodland garden or a section of a garden in part shade is often a good situation to employ philadelphus. Here the flowers can appear to glow in the dappled light and the slowed air movement allows the scent to linger longer. Such a situation is excellent for *Philadelphus coronarius* 'Aureus'. The bright-yellow foliage will be seen to best advantage and will not be burnt by the sun. Imagine that coupled with the sweet fragrance!

Where space permits, some of the larger cultivars may be enjoyed to the full and some of the smallest, e.g. *Philadelphus* 'Monteau d'Hermine', are suitable subjects for the patio garden.

A constraint in the use of mock oranges is their comparatively short period of flower and rather dull appearance afterwards. When designing using these, ensure the planting is supplemented with other shrubs, possibly evergreen. Another good ploy is to use the mock orange to form a green background for later flowering herbaceous plants.

TABLE 4.81 *PHILADELPHUS* CULTIVARS

Name	Habit	Height	Spread	Flowers	Foliage	Remarks
P. 'Beauclerk' AGM	Bushy and arching	1.5–2.5m (5–8ft)	1.5–2.5m (5–8ft)	Short raceme of single, white flowers with a purple zone in the centre	Ovate	Flowers fragrant
P. 'Belle Etoile' AGM	Bushy	1.5–2.5m (5–8ft)	1.5–2.5m (5–8ft)	Single, white flower with a purple centre, 5cm (2in) across	Ovate	Flowers fragrant
P. coronarius 'Aureus' AGM	Bushy	1.5–2.5m (5–8ft)	1–1.5m (3–5ft)	Single, white, 2.5cm (1in) across	Ovate. Bright-yellow early summer, later yellow-green	Strongly scented. Best in part shade to prevent scorching of foliage
P. coronarius 'Variegatus' AGM	Bushy	1.5–2.5m (5–8ft)	1.5–2.5m (5–8ft)	Single, white, 2.5cm (1in) across	Ovate. Leaves with a broad white margin	Strongly scented
P. 'Innocence' AGM	Bushy	1.5–2.5m (5–8ft)	1.5–2.5m (5–8ft)	White, 3.5cm (1¼in) across	Green irregularly splashed with white	Scented
P. maculatus 'Mexican Jewel	Bushy, upright	1–1.5m (3–5ft)	60cm–1m (2–3ft)	White with a purple centre	Narrow	Flowers downward facing. Strongly scented. USDA7b/8a. RHS H5. Requires a sheltered site
P. maculatus 'Sweet Clare' AGM	Bushy, arching	1–2.5m (4–8ft)	1.5–2.5m (4–8ft)	White with a purple centre	Narrow	Flowers downward facing. Strongly scented. USDA7b/8a. RHS H5. Requires a sheltered site
P. 'Monteau d' Hermine' AGM	Bushy, small and compact	60–80cm (23–32in)	1–1.5m (3–5ft)	White double. 4cm (1½in) across	Small ovate	Scented
P. 'Silberregen' AGM (syn. *P.* 'Silver Showers')	Upright and compact	1–1.5m (3–5ft)	1–1.5m (3–5ft)	Single, white	Ovate	Scented

Philadelphus 'Belle Etoile'.

Philadelphus coronarius 'Aureus'. Scented flowers and coloured foliage.

Philadelphus maculatus 'Sweet Clare'.

Philadelphus coronarius 'Aureus'. The plant had been positioned where it received too much direct sunlight, resulting in severe scorching of the foliage.

Philadelphus maculatus 'Sweet Clare'. Well positioned where its arching habit and downward-facing, scented flowers can be fully appreciated.

POTENTILLA

A genus of over 300 species, nearly all of which are annuals, biennials or herbaceous perennials. There are only two shrubby species and only *Potentilla fruticosa* is commonly grown.

Growing Requirements

These are undemanding, hardy plants to grow.
Hardiness: USDA 6a-1. RHS H7.
Aspect: A site with any outlook will suffice.
Shelter: A sheltered or exposed site is suitable.
Sun: Full sun to part shade is required.
Soil: A soil of any texture is acceptable, provided it is well drained.
pH: A very wide range of soil pH is tolerated.

All cultivars have small, pinnate leaves and are deciduous. All cultivars have a long flowering season from early summer to early autumn.

TABLE 4.82 CULTIVARS OF *POTENTILLA FRUTICOSA*

Name	Habit	Height	Spread	Flower	Remarks
P. 'Abbotswood' AGM	Bushy	75cm (30in)	1.2m (4ft)	White	
P. BELLA SOL ('Hachdon')	Bushy	75cm (30in)	1m (3ft)	Rich-orange fading to golden-yellow	
P. 'Goldfinger'	Bushy, upright	1–1.5m (3–5ft)	90–120cm (3–4ft)	Large, yellow	
P. 'Goldstar'	Bushy, rounded	0.5–1m (2–3ft)	0.5–1m (2–3ft)	Golden-yellow	
P. 'Hopleys Orange' AGM	Bushy	0.5–1m (2–3ft)	0.5–1m (2–3ft)	Bright-yellow with brown-orange on reverse	
P. 'Pink Beauty'[PBR] AGM (syn. *P.* 'Lovely Pink')	Upright	1–1.5m (3–5ft)	0.5–1m (2–3ft)	Deep-pink	
P. PRINCESS ('Blink')	Low, bushy	60cm (2ft)	1m (3ft)	Pale-pink	
P. 'Limelight' AGM	Bushy	60–90cm (2–3ft)	90cm (3ft)	White with a yellow centre	
P. var. *pumila* 'Red Ace'	Low, bushy	60–90cm (2–3ft)	90–120cm (3–4ft)	Vermillion, yellow on reverse	Flower colour fades in sun
P. 'Tangerine'	Mounded	0.5–1m (2–3ft)	0.5–1m (2–3ft)	Yellow flushed with orange	

Potentilla fruticosa BELLA SOL ('Hachdon').

Potentilla fruticosa 'Goldfinger'.

Potentilla fruticosa 'Goldstar' showing the typical bushy habit of these shrubs.

Potentilla fruticosa 'Hopley's Orange'.

Potentilla fruticosa 'Goldstar'.

Garden Design Considerations

Cultivars of *Potentilla fruticosa* are the work horses of the gardening world, being very tough, undemanding and having a long flowering season. In the garden there are many different situations where these shrubs are suitable.

Their dense, bushy nature makes then very well adapted to form low, flowering hedges. They are useful for clothing banks and slopes where their low-maintenance requirement is a distinct advantage. Banks, slopes and gravel gardens can naturally have rather dry soil. These shrubby potentillas are drought-resistant and relish such situations.

Potentilla fruticosa cvs. are wind and salt air resistant, making them well suited for cultivating in cold windy gardens and windswept coastal gardens.

The long flowering season and range of flower colours available makes them valuable for inclusion in mixed borders, cottage-style and informal gardens.

Patio gardens and containers are other forms of gardening where these versatile plants can be gainfully employed, especially the compact cultivars with vivid flower colours.

RHODODENDRON

This is a very large genus containing about 1000 species and thousands of hybrids. The genus contains both deciduous and evergreen trees and shrubs, ranging from tiny alpine gems only 2–4cm (¾–1½in) tall to large trees. In the wild, they are widely distributed across the northern hemisphere, the main concentrations being found in China, Tibet and Myanmar.

Because of their tremendous potential as flowering shrubs, they have been the subject of much sustained attention by plant breeders, both professional and amateur, and an enormous number of hybrids have been produced.

What follows must be just a very brief selection of what is available to the garden designer and gardener. The main guide in the selection process has been the excellence of the plants included and the need for compact plants to suit today's small gardens.

Growing Requirements

Because of the tremendous range of environments that plants from this genus come from, a considerable range of requirements could be given.

Hardiness: All USDA 6b/7a. RHS H6, except where stated in Table 4.83.

Shelter: Most of the very large-leaved rhododendrons are large plants requiring considerable shelter, especially from cold winds, and a soil that remains constantly moist. Nearly all hybrid evergreen rhododendrons require some shelter from wind, especially cold, drying winds. A few of the hardy hybrid rhododendrons are resistant to wind damage and these may be used to provide shelter in wind-prone gardens. Some of the alpine species and dwarf cultivars require little or no shelter.

Sun: Most rhododendrons prefer light-dappled shade. Some will do well in full sun, but the flowers of some tend to fade or bleach.

Soil: This can be clay, loam or sand and in all cases should contain high organic matter content. If the soil does not naturally have much organic matter in it, then this should be incorporated prior to planting. High organic matter content should be maintained in subsequent years after planting and a mulch may be found to be required. Drainage must be good, but the soil must also retain adequate moisture throughout the summer. This is another reason why an annual mulch of well-rotten organic matter may be found to be highly beneficial.

pH: In the wild, rhododendrons grow in acid soils and they all prefer a pH of 4.0–6.0. This does limit where these plants can be successfully grown. The main reason for this requirement is that above pH6 iron becomes unavailable for the plants to take up from the soil. Iron is an essential plant nutrient and its deficiency in rhododendrons is first shown by a yellowing of the leaves, especially young growth, and subsequent very poor growth, accompanied by very poor flowering. Many different strategies have been developed by horticulturalists to try and overcome this limitation. The addition of organic matter to the soil prior to planting, especially if it is acidic in nature, such as chopped bracken or decayed pine needles, may help the situation but will do very little to reduce the soil pH. Dressing the soil with sulphur at the same time as adding organic matter may reduce the soil pH a little. There are several proprietary preparations of chelated iron that can be applied to the soil. In this chelated form the iron is protected from being 'locked up' in the soil and is available to the plant roots. Over time, this iron may be lost in soil drainage or become 'locked up' and re-application will be necessary.

It is claimed that some rhododendrons do show tolerance to a high soil pH. Amongst these are *Rhododendron ciliatum*, *Rhododendron* 'Praecox' AGM and *Rhododendron williamsianum*. Attempts have been made to try to develop lime-tolerant rootstocks using *Rhododendron* 'Cunningham's White' and this has led to the Inkarho rootstock that may be tolerant up to pH 7.5.

A method that definitely does work is to isolate the roots in some form of raised bed or container and grow the plant in a so-called ericaceous compost that contains no lime.

All the above strategies have some problems. In all cases, some extra cost is involved, either in materials or labour. In addition, with soil-based remedies there is the inevitable drift back to an alkaline soil. This may occur from soil water bringing lime with it or from lime in the irrigation water. Basically, unsuited alkaline soils also naturally have low organic matter content, due to its rapid breakdown by soil micro-organisms, and often have insufficient ability to retain adequate water during the summer.

RHODODENDRON SPECIES

Rhododendron ponticum (the common purple rhododendron) has become an invasive weed in many situations and its planting is no longer recommended. In the UK, this species is covered by the Wildlife and Countryside Act 1981 and it is an offence to plant or cause the species to grow in the wild.

TABLE 4.83 *RHODODENDRON* SPECIES SELECTED FOR GARDEN WORTHINESS AND COMPACTNESS

Name	Habit	Height	Spread	Foliage	Flowers	Remarks
R. augustinii	Fairly upright	2m (6ft)	2m (6ft)	Small evergreen leaves	Blue. Late spring, early summer	An excellent woodland shrub. USDA 8b/9a. RHS H4
R. dauricum	Bushy	1.5m (5ft)	1.5m (5ft)	Elliptic semi-evergreen leaves 2–3cm (¾–1¼in) long	Bright rose-purple. Late winter to early spring	Does well in an open position
R. forrestii	Prostrate shrub	30cm (1ft)	1.5m (5ft)	Rounded 2.5–3.5cm (1–1½in) evergreen	Large, bell-shaped, red. Late spring-early summer	Good for a slightly shaded rock garden
R. impeditum	Dwarf	40cm 16in)	60cm (24in)	Very small, 1–3cm (½–1¼in) evergreen	Light purplish-blue. Late spring early summer	
R. luteum (syn. *Azalea pontica*) (yellow azalea)	Bushy	3.5m (12ft)	3.5m (12ft)	Oblong 5–10cm (2–4in), deciduous	Yellow and very pleasantly fragrant	Leaves often develop bright autumn colour before falling. Good in sun or part shade
R. yakushimanum AGM	Dense, rounded shrub	1.3m (4ft)	1.5m (4ft)	Evergreen leaves leathery, rounded at their margins, green above and with a dense tomentum beneath	Flowers deep-pink in bud, opening to light-pink and finally white	A most desirable plant in its own right, this species has been used to produce a group of compact, free-flowering hybrids called collectively Yakushimanum hybrids

Rhododendron augustinii. Growing in a small garden.

Rhododendron augustinii. Showing the extraordinary colour of the flowers.

Rhododendron yakushimanum. The very compact nature of the species is clearly shown here.

Rhododendron luteum. A small, young plant. The leaves often develop good autumn colour before falling.

Rhododendron luteum. The flowers are very sweetly scented.

Rhododendron yakushimanum. In bud the flowers are pink; upon opening they gradually change to white, giving a beautiful overall effect.

Rhododendron yakushimanum. New foliage is covered with silvery hairs.

TABLE 4.84 EVERGREEN HYBRID RHODODENDRONS

Hardiness USDA7b/8a-6b/7a. RHS H5-6.

Name	Habit	Height	Spread	Foliage	Flowers	Remarks
R. 'Baden-Baden' AGM	Dwarf and compact	0.5–1m (20–40in)	1–1.5m (40–60in)	Small, dark-green, evergreen	Dark-red, early summer	Best in part shade
R. 'Blue Diamond'	Slow-growing and compact	1m (3ft)	1.5m (5ft)	Evergreen	Lavender-blue. Late spring. Early summer	Best in full sun
R. Blue Tit Group	Dense bush	1m (3ft)	1m (3ft)	Small green evergreen leaves	Lavender-blue. Mid-spring	Best in full sun
R. (Cilpinense Group) 'Cilpinense' AGM*	Rounded bush	1m (3ft)	1m (3ft)	Glossy green evergreen leaves	White, flushed pink. Late winter early spring	Flowers need protection from spring frosts
R. 'Curlew' AGM**	Low-growing and compact	10–50cm	10–50cm	Deep green, oval	Yellow. Early summer	
R. Elizabeth Group	Small, spreading shrub	1m (3ft)	1m (3ft)	Evergreen with small leaves	Dark-red. Spring	Does well in sun
R. 'May Day' AGM	Low-growing and wide-spreading	1.5m (5ft)	1.5m (5ft)	Evergreen with large leaves	Red or orange-red. Early summer	
R. 'Praecox' AGM*	Small and compact	1.3m (4ft)	1.3m (4ft)	Leaves partly deciduous	Rosy-purple. Late winter early spring	Does well in sun
R. 'Ptarmigan'**	Compact dwarf	0.9m (3ft)	0.9m (3ft)	Dark-green, small and elliptic	White. Early to mid-spring	
R. 'Razorbill' AGM**	Compact	1–1.5m (3–5ft)	1.15m (3¾ft)	Dark-green, oval	Rose-pink	
R. 'Snipe' AGM**	Dense-growing dwarf	10–50cm (4–19½in)	10–50cm (4–19½in)	Pale-green	Pale-pink. Late spring	
R. 'Sapphire'	Dwarf shrub with an open habit	0.5m (20in)	0.9m (35in)	Small, lance-shaped	Pale-blue clusters. Mid-spring	Best in part shade

* Those that flower in late winter or early spring need protection from late spring frosts and early morning sun. Avoid east-facing situations.

** These are members of the Glenoick Bird Series of dwarf hybrid rhododendrons bred by the father and son team Peter and Ken Cox. These rhododendrons are classified as dwarf Lepidote hybrids.

Rhododendron Blue Tit Group. Growing well in a raised bed filled with ericaceous compost.

Rhododendron Blue Tit Group.

Rhododendron 'May Day'. A cultivar flowering in early summer.

Rhododendron (Cilpinense Group) 'Cilpinense'. An early flowering cultivar requiring a sheltered site, avoiding early morning sun.

Rhododendron Elizabeth Group. A cultivar flowering in early summer.

Rhododendron 'Praecox'. A large, old plant that had been grown successfully in a half-tub for many years.

Rhododendron 'Praecox'. A cultivar that flowers in late winter/early spring being pollinated by a bumble bee.

Rhododendron 'Ptarmigan'. A compact, dwarf cultivar, well suited to growth in a container.

Rhododendron 'Razorbill'. A compact cultivar.

Rhododendron 'Razorbill'.

Rhododendron 'Sapphire'. Best grown in part shade.

'Hardy Hybrid' Rhododendrons

This group of rhododendrons, many of which had the very hardy species *Rhododendron catawbiense* involved in their parenthood, are hardy and suitable for planting in cold or exposed positions.

Hardiness: USDA 7b/8a-6b/7a. RHS H 5-6, except where stated.

Rhododendron 'Christmas Cheer'. An early flowering cultivar of this group.

Rhododendron 'Pink Pearl'. Here, a plant of this cultivar is developing into a typical rather tall overall shape.

TABLE 4.85 'HARDY HYBRID' RHODODENDRONS

Name	Habit	Height	Spread	Foliage	Flowers	Remarks
R. 'Britannia' AGM	Compact, rounded bush	1.5m (5ft)	2m (6½ft)	Dark-green, evergreen	Crimson-scarlet. Early summer	Good wind resister
R. 'Christmas Cheer' AGM	Dense and compact	1.2m (4ft)	1.8m (6ft)	Dark-green, evergreen	Pink fading to white. Late winter to early spring	
R. 'Cynthia' AGM	Dome-shaped	6m (20ft)	6m (20ft)	Evergreen with large leaves	Rose-crimson. Early summer	Very widely planted
R. 'Doncaster'	Small, dome-shaped	2m (6½ft)	2m (6½ft)	Very dark-green, glossy. Evergreen	Crimson scarlet. Black markings within funnel-shaped flowers. Early summer	
R. 'Pink Pearl' AGM	Ultimately a tall bush	4m (13ft)	4m (13ft)	Evergreen with large leaves	Large trusses of soft pink flowers. Early summer	USDA 8b/9a. RHS H4
R. 'Sappho'	Rounded bush	3m (10ft)	3m (10ft)	Evergreen with medium-sized leaves	Trusses of white flowers with a blotch of rich purple. Early summer	Does well in full sun
R. 'Vulcan' AGM	Compact, rounded dome	1.5–1.7m (5–6ft)	2m (6½ft)	Evergreen deep green	Strong red. Early summer	

Rhododendron 'Pink Pearl'. This is probably the most commonly planted member of the 'Hardy Hybrids'.

Rhododendron 'Vulcan'. A very good, strong red cultivar.

TABLE 4.86 *RHODODENDRON YAKUSHIMANUM* AND HYBRIDS

Hardiness: USDA 8b/9a. RHS H4.

Name	Habit	Height	Spread	Foliage	Flowers
R. 'Bashful' AGM	Low and spreading	60cm (2ft)	120cm (4ft)	Dark-green	Light-pink
R. 'Dopey' AGM	Compact and tight growing	100cm (3ft)	120cm (4ft)	Dark-green, lance shaped	Strong red
R. 'Dreamland' AGM	Spreading and low-growing	80cm (31in)	120cm (4ft)	Matt dark-green	Buds pink. Flowers apricot, flushed pink
R. 'Emanuela' AGM	Rounded mound	70cm (28in)	120cm (4ft)	Small matt green leaves	White with pink edges
R. 'Golden Torch' AGM	Rounded mound	70cm (28in)	120cm (4ft)		Yellow
R. 'Hatchman's Marlis' AGM	Spreading bush	90cm (3ft)	120cm (4ft)	Silvery young foliage	Mid pink with a white eye
R. 'Hatchman's Porzellan' AGM	Spreading and comparatively large	100cm (3ft)	200cm (6ft)	Matt leaves	White with green spotting
R. 'Percy Wiseman' AGM	Small and rounded	90cm (3ft)	120cm (4ft)	Glossy dark-green	Pink aging to cream
R. 'Surrey Heath'	Compact, dome-shaped	75–100cm (29– 36in)	120cm (4ft)	Deep green, narrow	Rose pink fading to cream

Rhododendron 'Dreamland'. This is a low-growing and spreading cultivar.

Rhododendron 'Golden Torch' This plant has grown happily in this container for many years.

Rhododendron 'Golden Torch'. Like many rhododendrons, the flower colour of this cultivar changes as the flower ages.

Rhododendron 'Percy Wiseman'. A very popular cultivar.

Rhododendron 'Surrey Heath'. This cultivar develops into a compact and dome-shaped shrub.

Rhododendron 'Surrey Heath'.

Azaleas

Azaleas are a group within the genus *Rhododendron*. Azaleas have five to seven stamens instead of the eleven to thirteen as found in rhododendrons. Azaleas can be further divided into evergreen and deciduous species and hybrids.

Deciduous Hybrid Azaleas

Hardiness: USDA 6b/7a/6a-1.RHS 6 or 7. These fall into several groups depending on where they were bred and the parent plants used.

- **Knap Hill-Exbury Hybrids.** Developed at Knap Hill Nursery and later at Exbury Gardens. The flowers are trumpet shaped, scentless and there is a wide range of colours available. Overall size: height 1.8m (6ft) and spread 2.5m (6½ft).
- **Mollis Azaleas.** These were developed by Belgian nurserymen to produce a group of hybrids with a diverse range of colouring. Overall size: height 1.2m (4ft) and spread 1.8m (6ft).
- **Occidental Hybrids**. Available in delicate pastel colours, this group of hybrids flower in early summer after the Mollis Azaleas. Overall size: height 1.8m (6ft) and spread 1.5m (5ft).
- **Rustica Hybrids.** This group has double flowers that are sweetly scented. Overall size: height 1.2m (4ft) and spread 1.5m (5ft).

In addition to flowers, most will produce good autumn leaf colour before leaf fall.

TABLE 4.87 DECIDUOUS HYBRID AZALEAS

Name	Habit	Height	Spread	Flowers	Group
R. 'Anneke'	Upright	1.5m (5ft)	1.5m (5ft)	Fragrant, yellow, late spring/early summer	Knap Hill-Exbury Hybrid
R. 'Baslisk'		1.2–3m (4–10ft)	1.2–3m (4–10ft)	Deep yellow, flushed with salmon-pink on outside.	Knap Hill-Exbury hybrid
R. 'Gibraltar' AGM	Open growing	1.5m (5ft)	1.5m (5ft)	Rich-orange. Early summer	Knap Hill-Exbury hybrid
R. 'Homebush' AGM	Compact and bushy	1.5m (5ft)	1.5m (5ft)	Semi-double, rose-pink flowers	Knap Hill-Exbury hybrid
R. 'Persil' AGM	Bushy	2m (6ft)	1–1.5m (3–5ft)	White flowers with a orange-yellow flare inside trumpet	Knap Hill-Exbury hybrid
R. 'Golden Eagle' AGM		1.5m (5ft)	1.7m (5½ft)	Orange flowers not scented. Early summer	Knap Hill-Exbury hybrid
R. 'Golden Lights'	Upright	1.5–1.8m (5–6ft)	1.5m (5ft)	Yellow fragrant. Early summer	Northern Lights hybrid
R. 'Koster's Brilliant Red'		1.8m (6ft)	1.8m (6ft)	Orange-yellow. Spring	Knap Hill-Exbury hybrid
R. 'Pucella'	Lax growing	1.5–2.5m (5–8ft)	1.5–2.5m (5–8ft)	Fragrant purplish-rose with an orange flair	Gent hybrid

Rhododendron 'Anneke'. This cultivar forms an upright-growing shrub.

Rhododendron 'Baslisk'. Well-shaped bush.

Rhododendron 'Gibraltar'. An open growing plant.

Rhododendron 'Gibraltar'.

Rhododendron 'Golden Eagle'.

Rhododendron 'Koster's Brilliant Red'. The flowers of this cultivar demonstrate what very strong and bright colours this group of plants can produce.

Rhododendron 'Pucella'.

Evergreen Hybrid Azaleas

Hardiness: USDA RHS H 4-5, except where stated in Table 4.88.

These evergreen and semi-evergreen rhododendrons come mainly from Japan and include the famous Kurume azaleas. Providing the roots are kept moist, they thrive in full sun, except for those cultivars whose flowers fade, and these are more satisfactory in partial shade.

There are currently seven groups of these evergreen azaleas and only a few examples can be given to give an idea of what is available. Most have small, evergreen leaves to match their small stature.

Rhododendron 'Blue Danube'. A compact cultivar.

Garden Design Considerations for *Rhododendron*

Apart from the sole constraint of their requirement for acid soil, there must be a rhododendron for every garden situation.

A major complaint made by some is their limited flowering period. Whilst it is true that by far the majority flower in late spring and early summer, by careful selection rhododendrons can be in flower outside from late winter to summer, with autumnal leaf colour if deciduous azaleas are included.

A second claim made by their detractors is that they have dull, evergreen foliage. A counter-claim is that this evergreen foliage is of immense value in furnishing and giving structure to the winter garden. It will also provide a degree of shelter for smaller plants and provide a uniform background for other plantings. *Rhododendron yakushimanum* and many of its cultivars have very attractive foliage with a dense indumentum on their undersides. All this is before you consider the autumn leaf colour of the deciduous azaleas.

Woodland and Dappled Shade

This is the situation perhaps most associated with rhododendrons. Here, indeed, rhododendrons do excel and there is an enormous number of cultivars to select from. Remembering that we are specifically considering the smaller and more compact gardens, care must be exercised in cultivar selection. The hardy hybrid rhodendrons may become too large, but they will respond well to being cut back hard in spring, providing that they are fed, watered and mulched afterwards. After such treatment, the gardener essentially starts again with new plants. If these are too large, then there are plenty of other evergreen hybrids and evergreen azaleas to choose from. These rhododendrons associate very well with, and help to create, the correct conditions for other choice woodland or shade loving plants.

Screening and General Planting

The dense, evergreen nature of rhododendrons makes them very effective for planting as a screen, using several plants to form an informal hedge to hide a poor view, or a single carefully positioned plant to prevent the prying eye.

Mixed Borders and General Planting

Where a mixed border is desired, provided that the soil is suitable, rhodendrons are admirable for inclusion, giving structure and seasonal colour. With the range of plant size and flower colour available, it is possible to create almost any plant and colour association desired.

Sunny Banks and Ground Cover

Rhododendrons are usually planted in moisture-retentive soils in part shade but, with careful selection, a sunny situation will result in even more flowering, providing adequate organic matter is added to the ground before planting and further mulching done annually to aid soil moisture-retention. With a sloping site, the extremely vivid colours of some of the deciduous or evergreen azaleas can be employed to produce a striking planting.

Rhododendrons with their dense canopy of evergreen leaves are very effective at preventing weed growth, providing that the ground is free of weeds before planting is done. They are also essentially very low in terms of maintenance requirement, needing an annual mulching to encourage good growth and removal of flower trusses after flowering to encourage new growth and profuse flowering the following year. On a small scale, the low-growing *Rhododendron* 'May Day' or the almost prostrate *Rhododendron* Elizabeth Group would be excellent for this purpose.

Inclusion in the Heather Garden

The attributes of this form of garden or garden feature have been discussed when considering the design of heather gardens. Rhododendrons associate beautifully in this type of garden or border, requiring the same soil conditions and adding extra height, textural leaf contrast and seasonal vivid colour to such a feature. A rhododendron of suitable size and colour can easily be selected.

Rock Gardens Raised Beds and Planters

On the rock garden, some of the smaller species and cultivars of evergreen rhododendrons, such as those in Glendoick Bird Series, fit in very well. They give a sense of permanence to the planting as so many other alpine and rock garden plants are spring and early summer flowering and then dormant or very low profile. Together with dwarf or truly slow-growing conifers, they can be employed to produce an attractive miniature landscape effect into which the other alpines can be introduced.

Small Formal Gardens, Patio Gardens, Raised Beds, Containers and Pots
Here smaller rhodendrons, of which there are a great many, fit in very well. The small, dense, regular dome shape of the smaller *Rhododendron yakushimanum* hybrids lend themselves to where space is restricted. If space is even more restricted, then dwarf hybrids, such as *Rhododendron* 'Curlew', would be appropriate. The need for formality is satisfied by their regular shape, and their compact growth fits them neatly into these small spaces.

Raised beds, containers and pots can be employed where the native soil is unsuitable or badly drained. With an annual top-dressing of ericaceous compost, coupled with feeding during the growing season and watering, when necessary, rhododendrons can be grown successfully in these situations for many years.

As a single plant in a large container, such as half a barrel, a rhododendron will make a very satisfactory specimen plant. A well-selected, slow-growing, evergreen rhododendron will be well furnished with leaves all the year round and will provide an intense display of colour during its flowering season.

ROSA

A genus of about 150 species found mainly in the northern hemisphere. It comprises deciduous or evergreen shrubs or climbers with the following characteristics: the flowers are either solitary or produced in clusters and vary greatly in size, colour and form, and have numerous stamens and carpels.

Flowering occurs in summer. The species usually flower only once per season. A few species and many hybrids are recurrent, also called remontant or repeat flowering, meaning that they produce several flushes of flowers per season. Non-recurrent plants produce a mass of flushes over one intense period, but recurrent plants produce a more extended display of perhaps less intense, but several flushes of flowers. In the tables R = recurrent and NR = non-recurrent.

The fruit is a hip that contains numerous dry seeds known as achenes. The fruits may be an important decorative feature, as in the masses of red, flagon-shaped hips produced in late summer and early autumn by *Rosa moyesii*, but are entirely absent in the many of the double-flowered hybrids. The leaves are alternate, pinnate with five to seven leaflets and usually have prominent stipules. In some cases, the foliage may be another very decorative feature, as in the case of *Rosa xanthina* with its fine almost fern-like pinnate leaves.

Thorns or prickles are usually present to a greater or lesser degree, although there are a few so-called thornless roses.

Grouping of Roses

Over centuries, roses have been the subject of tremendous attention by plant breeders and, in addition to the wild species, thousands of cultivars have been produced. To bring some order to this vast range of plants, they have been divided into groups based, where possible, on functional garden qualities. Several different grouping systems have been proposed, but the groups employed here are based upon the grouping system in the *RHS Plant Finder* (2020), which is currently the latest edition available. Only groupings with characteristics suitable for where space is limited and for smaller modern gardens have been considered. Where rose groups have been assigned other names in the past, these are shown in brackets after the current name.

Growing Requirements
Hardiness: Most roses are rated USDA 6b/7a. RHS H6, although some species or cultivars may be hardier and a few less so.
Aspect: All roses grow best in a sunny open position. For the great majority, a site of any aspect is accepted.
Shelter: No special provision for shelter is necessary.

TABLE 4.88 EVERGREEN HYBRID AZALEAS

Name	Habit	Height	Spread	Flowers	Remarks
R. 'Blue Danube' AGM (syn. 'Blaue Donau')	Compact	1.5–2.5m (5–8ft)	1–1.5m (3–5ft)	Violet-blue. Late spring	USDA RHS H3.
R. 'Azuma-kagami'	Upright	1.8m (5ft)	1.2m (4ft)	Phlox pink. Hose-in-hose. Early summer	Best in semi-shade
R. 'Bengal Fire'		1.8m (5ft)	1.2m (4ft)	Fiery-orange. Early summer	
R. 'Elsie Lee' AGM	Upright and compact	0.5–1m (2–3ft)	0.5–1m (2–3ft)	Pale-mauve. Early summer	
R. 'Florida' AGM	Upright	50–75cm (1ft 8in–2ft 4in)	60cm (2ft)	Red. hose-in-hose. Early summer	
R. 'Hino-crimson' AGM	Dwarf, dense-growing	60cm (2ft)	60cm (2ft)	Crimson-scarlet. Spring	
R. 'Irohayama' AGM	Dwarf and compact	60cm (2ft)	60cm (2ft)	White with a pale-lavender margin	
R. 'Kermesinum'	Low and compact	60cm (2ft)	60cm (2ft)	Purplish-red. Early summer	
R. 'Mother's Day' AGM	Compact dense mound	50–75cm (1ft 8in–2ft 4in)	60cm (2ft)	Deep-red. hose-in hose. Early summer	
R. 'Orange King' AGM		50–75cm (1ft 8in–2ft4in)	60cm (2ft)	Orange-red. Early summer	
R. 'Palestrina' AGM	Compact	1.2m (4ft)	1.2m (4ft)	White, late spring	
R. 'Squirrel' AGM	Dwarf and compact	0.5–1m (1ft 8in–3ft)	0.5–1m (1ft 8in–3ft)	Scarlet. Holds its colour well. Early summer	
R. Torchlight' AGM	Compact and spreading	0.6–1.2m	80cm	Purplish-red. Hose in hose. Early summer	

TABLE 4.89 *ROSA* SPECIES

Name	Habit	Height	Spread	Foliage	Flowers	R or NR*	Remarks
R. 'Helen Knight' (ecae hybrid)	Small, bushy, compact, many prickles	1.5–2.5m (5–8ft)	1–1.5m (3–5ft)	Small leaves with 5–9 pinnules	Small, bright-yellow, 2.5cm (1in) across. Late spring to early summer	NR	USDA 6b/7a. RHS H6
R. glauca AGM (syn. *R. rubrifolia*)	Arching and spreading	1.5–2.5m (5–8ft)	1.5–2.5m (5ft)–8ft)	Glaucous purple	Single, deep white in centre. Followed by rounded red hips	NR	USDA 6a-1. RHS H7. Excellent where coloured foliage is required
R. × *odorata* (Sanguinea Group) 'Bengal Crimson' AGM	Dense and fairly upright	2m (6½ft)	2–2.5m (6.5–8ft)	Wine-red when young, turning green	Single, bright-red, darkening with age	R	USDA 7b/8a. RHS H5. Flowers for a very long period
R. virginiana	Small, dense, suckering shrub	1–1.5m (3–5ft)	1–1.5m (3–5ft)	5–7 pinnules. Green during growing season. Orange-red autumn leaf colour	Single, deep-pink. Mid to late summer. Followed by small red hips	NR	USDA 6a-1. RHS H7. Good in sandy soils and in maritime situations

* R= remontant; NR = non-remontant.

Rosa glauca. In flower in early summer.

Rosa glauca. The single red and white flowers complemented by the foliage.

Rosa × *odorata* (Sanguinea Group) 'Bengal Crimson'. A medium-sized shrub.

Rosa × *odorata* (Sanguinea Group) 'Bengal Crimson'. Bright-red, single flowers are produced over a very long season.

Rosa glauca. In fruit. The hips are at their best in early autumn.

Sun: Full sun is preferred.
Soil: The soil must be well drained and can be of any texture, but it should be moisture-retentive in summer, which is why they are often said to prefer heavier soils. High humus content is recommended and, for most, high fertility is required.
pH: A wide range of soil pH is accepted, but extremes should be avoided.

Growing in Containers

Roses in containers need the same open, sunny position as those grown in border soil. Not all roses will grow well in containers. Those that will, need deep pots at least 23–35cm (9–14in) deep. A good mix to grow them in is John Innes Number 3 with the addition of 20 per cent of peat-free multipurpose compost or very well-rotted manure. The components extra to the J. I. No.3 are to increase moisture-holding capacity.

The roses will need the top 5cm (2in) of compost in the container removed and replaced with fresh compost each spring. This top-dressing is recommended to feed the roses, which are heavy feeders, and this may need to be supplemented by liquid feeding and regular watering during the growing season.

Old Garden Roses

Gallica

A group of roses that are compact, prickly and have distinctive, dull-green foliage. Flowering occurs in mid-summer and is profuse, very fragrant but non-recurrent.

TABLE 4.90 GALLICA ROSES

Name	Habit	Height	Spread	Foliage	Flowers	R or NR*	Remarks
R. gallica var. *officinalis* AGM (apothecary's rose, old red damask, red rose of Lancaster)	Bushy and upright	0.5–1m (1.75–3ft)	0.5–1m (1.75–3ft)	Distinctive dull green	Semi-double, rosy-pink, 8cm (3in) across, highly fragrant. Followed by small, red hips in autumn	NR	USDA 6a-1. RHS H7. Very suitable for short hedging
R. gallica 'Versicolor' AGM	Bushy and upright	0.5–1m (1.75–3ft)	0.5–1m (1.75–3ft)	Distinctive dull-green	Semi-double, rosy-pink, 7cm (1¾in) across with erratic white stripes. Some flowers only rosy pink. Followed by small red hips	NR	USDA 6a-1. RHS H7. Very suitable for short hedging

* R= remontant; NR = non-remontant.

Rosa gallica var. *officinalis*. The apothecary's rose.

Rosa gallica var. *officinalis*. A single bloom at its best. The flowers soon fade and then the petals drop.

Rosa gallica 'Versicolor'.

Rosa gallica 'Versicolor'. A single bloom showing the 'broken' colour in the petals.

Damask Portland

A group of roses that are upright, compact bushy shrubs with double or semi-double flowers that are often scented.

TABLE 4.91 DAMASK PORTLAND ROSES

Name	Habit	Height	Spread	Foliage	Flowers	R or NR*	Remarks
R. 'Comte de Chambord' (Madam Boll)	Fairly compact	1.2m (4ft)	0.9m (3ft)	Dark-green	Soft-pink, fully double, scented	R	
R. 'De Resht' AGM	Compact	0.5–1m (1.75–3ft)	0.5–1m (1.75–3ft)	Dark-green	Purple-crimson fully double, 6cm (7¾in) across, fragrant	R	USDA 6a-1. RHS H7
R. 'Duchess of Portland'	Fairly compact	1.25m (4ft)	0.9m (3ft)	Dark-green	Fuchsia-pink, semi-double flowers, scented and followed by hips	R	

* R= remontant; NR = non-remontant.

Spinosissima Hybrids Scots (or Scotch)

Selections from, or hybrids of, *Rosa spinosissima* (syn. *R. pimpinellifolia*) that form low-growing, suckering shrubs with prickly stems and finely divided almost fern-like pinnate foliage.

TABLE 4.92 SPINOSISSIMA HYBRIDS SCOTS (OR SCOTCH) ROSES

Name	Habit	Height	Spread	Foliage	Flowers	R or NR*	Remarks
R. spinosissima (syn. *R.pimpinellifolia* Scotch rose)	Small, freely suckering shrub	0.5–1m (1.75–3ft)	1–1.5m (3–5ft)	Delicate looking, fern-like	Small,single 4–5cm (1.5–2in) across, in early summer. Followed by small, black, rounded hips	NR	USDA 6a-1. RHS H7. Does well on sandy soil
R. spinosissima 'Dunwich Rose'	Forms a small shrub with elegant arching growth	0.5–1m (1.75–3ft)	1–1.5m (3–5ft)		Single, white flowers. Lightly scented	NR	
R. spinosissima 'Grandiflora'		2m (6½ft)	2.5m (8ft)	Delicate looking, fern-like	Single, creamy-white flowers 6cm (2in) across. Followed by rounded black hips	NR	USDA 6a-1. RHS H7
R. spinosissima 'double white'		1.5m (4ft)	1.25m (4ft)	Fern-like	Double, rounded, white, fragrant flowers	NR	
R. spinosissima 'William 111'	Forms a dense thicket of growth	0.9m (3ft)	0.9m (3ft)	Fern-like	Semi-double, purple-crimson, fragrant flowers. Followed by maroon hips	NR	H7

* R= remontant; NR = non-remontant.
A larger selection from the species

Modern Garden Roses

TABLE 4.93 CLUSTER-FLOWERED (FLORIBUNDA) ROSES

Name	Habit	Height	Spread	Foliage	Flowers	R or NR*	Remarks
R. 'Cinco de Mayo'[PBR] AGM	Bushy	0.6–0.9m (2–3ft)	0.9–1.2m (3–4ft)	Glossy dark-green	Semi-double, deep rose-red to pink. Scented	R	
R. HOT CHOCOLATE ('Wekpaltlez')	Bushy	0.9m (3ft)	0.6m (2ft)	Dark-green glossy	Double copper brown. Scented	R	Novelty Rose of the year 2006
R. MARGARET MERRIL ('Harkuly')	Bushy	1.5m (4ft)	1m (3ft)	Dark-green	Cup-shaped, pale-pink to white. Scented	R	
R. ORANGE MEILOVE ('Meipluria')	Bushy and compact	0.5–0.7m (1.75–2.25ft)	0.4–0.5m (1.5–1.75ft)	Medium to dark-green	Orange to light apricot, fragrant	R	ST
R. YELLOW MEILOVE ('Meitongos')	Bushy and compact	0.5–0.8m (1¾–2½ft)	0.4–0.6m (1.25–2ft)	Deep green	Yellow, moderately scented	R	
R. TRUMPETER ('Mactru') AGM	Bushy and compact	0.5–0.75m (1.75–2½ft)	0.5–0.75m (1¾–2½ft)	Green		R	
R. FOR YOUR EYES ONLY ('Cheweyesup'[PBR])	Bushy	0.9m (3ft)	0.9m (3ft)	Glossy dark-green	Single. Petals have a red eye with a prominent ring of stamens. Outer part of petals shades of pink	R	USDA 6b/7a. RHS H6

* R= remontant; NR = non-remontant.
ST = sold as a standard rose.

Rosa ORANGE MEILOVE ('Meipluria'). Shown here budded on to an upright briar stem to form a half-standard.

Rosa ORANGE MEILOVE ('Meipluria').

Rosa TRUMPETER ('Mactru').

Rosa FOR YOUR EYES ONLY ('Cheweyesup'[PBR]). This cultivar grows into a well-shaped shrub.

Rosa FOR YOUR EYES ONLY ('Cheweyesup'[PBR]). The single flower of this cultivar with its, so far, unique colour pattern.

Large-Flowered (Hybrid Tea)

Only relatively few of this very large group of roses can be considered compact enough to be listed here. Some that are suitable are listed in Table 4.94.

Rosa A WHITER SHADE OF PALE ('Peafanfare'[PBR]). Shown here as a young standard plant for sale on a plant nursery.

Rosa A WHITER SHADE OF PALE ('Peafanfare'[PBR]). Showing the large flower typical of this group and still loved by many.

TABLE 4.94 LARGE-FLOWERED (HYBRID TEA) ROSES

Name	Habit	Height	Spread	Foliage	Flowers	R or NR*	Remarks
R. A WHITER SHADE OF PALE ('Peaanfare'[PBR]) AGM	Compact	0.9m (3ft)	0.9m (3ft)	Small and glossy	Double, pale-pink, fragrant	R	ST USDA 6b/7a. RHS H6
R. DOUBLE DELIGHT ('Andeli')	Compact	0.9m (3ft)	0.6m (2ft)	Dark-green	Creamy-white, edged with pinkish-red. Very fragrant	R	ST
R. PAPA MEILLAND ('Meisar')	Compact	0.9m (3ft)	0.5m (1¾ft)	Glossy	Velvety rich-crimson. Very fragrant	R	ST
R. TROIKA 'Poumidor' AGM	Compact	0.5–1m (1.75–3ft)	0.5–1m (1.75–3ft)	Glossy, bronze tinted	Pointed, double, apricot flushed red. Scented	R	

* R= remontant; NR = non-remontant.
ST = sold as a standard rose.

Rosa TROIKA ('Poumidor'). Another large-flowered rose.

TABLE 4.95 PATIO ROSES (PATIO, MINIATURE FLORIBUNDA OR DWARF CLUSTER-FLOWERED)

Name	Habit	Height	Spread	Foliage	Flowers	R or NR*	Remarks
R. CAREFREE DAYS (Meirivoui'PBR) AGM	Compact	10–50cm (¼–1¾ft)	10–50cm (0.25–1.75ft)	Dark-green	Mid-pink, double, scented	R	ST
R. FLOWER POWER ('Frycassia'PBR) AGM	Compact	1m (3ft)	1m (3ft)	Dark-green	Double, copper-pink, scented	R	USDA 6b/7a. RHS H6
R. JOIE DE VIVRE ('Korfloci 01'PBR) AGM	Compact	50–100cm (1¼–3ft)	50–100cm (1¾–3ft)	Deep green	Rosette shaped, pale-pink in clusters	R	Rose of the Year 2011
R. LOVELY BRIDE ('Meiratcan')	Compact	60cm (2ft)	60cm (2ft)	Glossy	Double cupped light-pink fragrant	R	ST
R. QUEEN MOTHER ('Korquemu') AGM	Compact	10–50 cm (4–19½in)	10–50cm (4–19½in)	Dark-green	Semi-double, pale-pink	R	
R. RUBY ANNIVERSARY ('Habonny'PBR)	Compact	10–50cm (0.3–3ft)	50–100cm (0.3–3ft)	Dark-green	Semi-double, pale-pink	R	
R. SUNSEEKER ('Dicraker') AGM	Compact	10–50cm (4–19½in)	10–50cm (4–19½in)	Dark-green	Orange-red with a yellow centre	R	
R. SWEET DREAM ('Fryminicot') AGM	Compact	35cm (1ft)	35cm (1ft)	Dark-green	Cup-shaped, fully double, peach apricot	R	
R. THANK YOU (Chesdeep'PBR)	Compact	60cm (2ft)	45cm (1½ft)	Bronze tinted mid-green	Hot pink, slightly scented	R	

* R= remontant; NR = non-remontant.
ST = sold as a standard rose.

Rosa CAREFREE DAYS ('Meirivoui'PBR). A very free-flowering rose.

Rosa SUNSEEKER ('Dicraker').

Rosa CAREFREE DAYS ('Meirivoui'PBR). As well as being free-flowering, each flower is of good quality.

Rosa JOIE DE VIVRE ('Korfloci 01'). A good overall display from several bushes in a rose bed.

Rosa Joie de Vivre ('Korfloci 01').

TABLE 4.96 MINIATURE ROSES

Name	Habit	Height	Spread	Foliage	Flowers	R or NR*	Remarks
R. MY VALENTINE ('Mormyval')	Compact dwarf	10–50cm (4–19½in)	10–50cm (4–19½in)	Mid-green	Clusters of dark-red flowers	R	USDA 6b/7a. RHS H6
R. PEARL ANNIVERSARY ('Whitston'PBR)	Compact dwarf	60cm (2ft)	60cm (2ft)	Mid-green	Semi-double, light-pink	R	
R. SWEET MAGIC ('Dicmagic'PBR) AGM	Compact dwarf	10–50cm (4–19½in)	10–50cm (4–19½in)	Dark-green	Semi-double, orange-yellow	R	

* R= remontant; NR = non-remontant.

Climbing Miniature Roses

A new group of roses that is very well suited to the smaller garden.

TABLE 4.97 CLIMBING MINIATURE ROSES

Name	Habit	Height	Spread	Foliage	Flowers	R or NR*	Remarks
R. LAURA FORD ('Chewarvel') AGM	Upright	3m (10ft)	50–100cm (19½–39in)	Glossy and dark-green.	Yellow, semi-double	R	USDA 7b/8a. RHS H5
R. Little Bronze	Upright	2m (6½ft)	1.2m (4ft)	Dark-green	Red, double	R	

* R= remontant; NR = non-remontant.

Rosa LAURA FORD ('Chewarvel'). A miniature climber well suited to a small garden.

Rosa Little Bronze. Another miniature climber, useful where space is limited.

Rosa Nice Day ('Chewsea'PBR).

TABLE 4.98 POLYANTHA ROSES

Name	Habit	Height	Spread	Foliage	Flowers	R or NR*	Remarks
R. 'Bloomfield Abundance'	Compact	1.25–2.5m (4–8ft)	1.0–1.5m (3–5ft)	Glossy dark-green	Sprays of pale-pink. Light fragrance	R	USDA 7b/8a. RHS H6
R. 'Gruss an Aachen'	Compact	10–50cm (4–19½in)	10–50cm (4–19½in)	Dark-green	Sprays of full, double, blush-pink. Fragrant	R	
R. MARJORIE FAIR ('Harhero')	Compact	1–1.5m (3–5ft)	1.1.5m (3–5ft)	Glossy mid-green	Clusters of white, rosette-shaped flowers from red buds	R	
R. 'The Fairy'	Compact	50–100cm (19½–39in)	50–100cm (19½–39in)	Very glossy	Sprays of rosette-shaped, double, pink	R	
R. 'Yvonne Rabier'	Compact	10–50cm (0.3–1.75ft)	10–50cm (0.3–1.75ft)	Glossy light-green	Sprays of double creamy-white with a light scent	R	
R. 'White Pet' (syn. *R.* 'Little White Pet')	Compact	10–50cm (4–19½in)	10–50cm (4–19½in)	Glossy	Sprays of pompon, double, white flowers from red buds	R	

* R= remontant; NR = non-remontant.

Rosa 'The Fairy'.

Rosa 'The Fairy'. The picture shows the generous sprays of flowers produced by this cultivar.

TABLE 4.99 GROUND COVER

Name	Habit	Height	Spread	Foliage	Flowers	R or NR*	Remarks
R. × *jacksonii* 'Max Graf'	Procumbent	60cm (2ft)	2m (6½ft)		Deep-pink single	R	
R. Magic Carpet 'Jacklover' AGM	Low and spreading	45cm (1½ft)	2m (6½ft)	Dark and glossy	Lavender-pink	R	
R. 'Nozomii'	Low and spreading	30cm (1ft)	1.2m (4ft)	Dark and glossy	Light-pink, single		
R. 'Raubritter'	Arching and spreading	90cm (3ft)	2m (6½ft)		Pink, single		
R. ROTE MAX GRAF ('Kormax') (syn. *R.* 'Red Max Graf')	Arching and spreading	90cm (3ft)	1.8m (6ft)	Glossy	Single, red followed by orange hips		
County series of ground-cover roses							
R. BERKSHIRE ('Korpinka') AGM	Low and spreading	10–50cm (4–20in)	1–1.5m (3–5ft)		Semi-double, bright-pink	R	USDA 6b/7a. RHS H6
R. KENT ('Poulcov') AGM	Low and spreading	10–50cm (4–20in)	0.5–1m (1¾–3ft)	Dark-green	Semi-double, white	R	
R. LANCASHIRE ('Korstegli')PBR AGM	Low and spreading	10–50cm (4–10in)	0.5–1m (1¾–3ft)		Semi-double red	R	
R. SURREY ('Korlanum') AGM	Low and spreading	10–50cm (4–20in)	1–1.5m (3–5ft)		Double, soft-pink	R	
Flower carpet series of ground-cover roses**							
R. FLOWER CARPET AMBER 'Noa97400a'PBR) AGM	Procumbent	80cm (2.75ft)	80cm (2.75ft)	Dark-green, glossy	Orange buds opening to apricot	R	
R. 'Flower Carpet Coral'PBR AGM	Procumbent	60cm (2ft)	90cm (3ft)	Glossy	Single, coral-pink, slightly fragrant	R	
R. FLOWER CARPET RED VELVET (Noare'PBR) AGM	Procumbent	60cm (2ft)	90cm (3ft)	Glossy dark-green	Single, red	R	
R. FLOWER CARPET WHITE ('Noaschnne'PBR) AGM		50–100cm (1¾–3ft)	1–1.5m (3–5ft)		Cupped, double, white	R	ST

* R= remontant; NR = non-remontant.
** Just a few of this lengthy series are included here.
ST = sold as a standard rose.

Rosa 'Flower Carpet Coral'PBR.

Rosa FLOWER CARPET RED VELVET 'Noare'PBR. A generous spray of flowers is produced.

Rosa FLOWER CARPET RED VELVET 'Noare'PBR. Showing the arching and ground covering habit of this cultivar.

TABLE 4.100 ROSES GOOD FOR GROWING IN CONTAINERS

Name	Habit	Height	Spread	Foliage	Flowers	R or NR*	Remarks
R. 'Nice Day' AGM	Climbing	1.5–2.5m (5–8ft)	0.5–1m (1¾–3ft)	Dark-green	Sprays of double, salmon-pink, fragrant	R	A patio climber rose that needs some small supporting framework, e.g. a wire obelisk
R. Magic Carpet'	Procumbent	80cm (2½ft)	1.25m (4ft)	Glossy, dark-green	Semi-double, mauve and lilac, light fragrance	R	A ground-cover rose that will grow over and down the sides of the container
R. 'Robin Redbreast'	Compact	0.45m (1½ft)	0.6m (2ft)	Mid-green and glossy	Dense clusters of single, cupped, dark-red	R	A miniature rose
R. SWEET DREAM ('Fryminicot')	Compact	0.5m (1¾ft)	0.5m (1¾ft)		Cup-shaped, double, peach-apricot, fragrant	R	A patio rose

* R= remontant; NR = non-remontant.
ST = sold as a standard rose.

Standard, Half-Standard and Smaller Standard Roses

Many cultivars are sold which have been budded on to the vertical stem of a briar rose.

The briar should be a selected clone that is sufficiently vigorous to produce the vertical strong stem required and, ideally, not prone to suckering. When purchasing any form of standard rose, check for a lack of suckers that may be present on the vertical stem or emerging from below compost level. If suckers are appearing from below ground level, avoid the purchase. Suckers arising from the vertical stem are not so problematic but should be removed as soon as seen.

Check that at least three buds have been initially placed on to the vertical stem. This will help the young rose to develop a well-balanced head and be reasonably symmetrical.

The height at which standard roses are budded varies considerably. The height can be from 2m (6ft) to 50cm (1¾ft), smaller more compact cultivars often being budded at an even lower height.

The cultivars listed above that are known to be sold as a standard are indicated with ST in the remarks column.

Garden Design Considerations

Rosa is a very diverse genus and its cultivars have been made even more so by the activities of rose breeders. By virtue of their diversity, they have a wide range of applications in today's garden, which is often small and compact.

Rose Species

Many species roses are large shrubs with a loose, open habit and short, single period of flowering. As such, most are unsuitable for the gardens considered here.

A few rose species are compact or have particular characteristics that render them highly suitable. *Rosa* 'Helen Knight' is compact and provides a valuable, early splash of flower colour. *Rosa glauca*, whilst not particularly compact, does contribute a unique foliage colour, coupled with a good display of flowers and later hips. *Rosa* × *odorata* (Sanguinea Group) 'Bengal Crimson' AGM, whilst admittedly requiring space, does contribute a very long season of bright flower colour. *Rosa virginiana* differs from most other roses, flowering in late summer and then developing good autumn foliage colour. It may, therefore, associate better in a bed or planting intended to provide primarily autumn interest, rather than in a traditional rose border.

Gallica

Gallica roses form a distinct group with signature attractive, dull-green foliage and a stiff, upright habit. They are non-recurrent, but when in flower their flowering is generous and their fragrance legendary. Their habit makes them very suitable for semi-formal, low- to medium-height hedging.

Damask Portland

This group brings a sense of old garden rose character to any garden, whilst being fairly compact and recurrent flowering. They are well suited for inclusion in a mixed border, where a long season of interest is required.

Spinoisissima Hybrids, Scots (or Scotch)

These roses have their own semi-wild character. Their dense, prickly habit and delicate-looking fern-like foliage sets them apart and makes them very suitable for a less formal part of a garden. Their small, clustered flowers are added to by distinctive small rounded often black hips.

Modern Garden Roses

The breeding of these roses is considered to have begun after the cross-breeding of the recurrent hybrid perpetual roses with tea roses to create hybrid tea roses, later known as large-flowered roses.

Large-Flowered (Hybrid Tea)

These are the classic large-flowered roses loved by previous generations for their perfect, double-flower form and their often generous scent. Unfortunately, the bushes are rather large and open in growth and occupy too much space. A few, mainly more recent introductions, do, however, manage to combine perfect, large flowers in a recurrent manner on a compact plant.

Cluster-Flowered (Floribunda)

This group differs from large-flowered roses by having a greater number of smaller flowers clustered on one head of flowers. Their flower form is more diverse and includes single and semi-double flowers. In general, they are more recurrent and contribute more colour per given area than large-flowered roses. The more compact cultivars within this group are well suited to the smaller garden or where space is limited.

Patio (Patio, Miniature Floribunda or Dwarf Cluster-Flowered)
Modern rose-breeders have concentrated their efforts within this group. As a group they are compact, very floriferous and very well suited to the modern, small garden. Where summer colour and scent is required, it is difficult to fault their performance.

Miniature
A group of very small, compact roses, perhaps too small for inclusion in general border planting, but along with patio roses ideal for patio gardens, raised beds and planters.

Climbing Miniature
A new group of roses that combine the habit, but not the vigour, of traditional climbing roses. Their clusters of flowers are often small but produced in a recurrent manner. They are ideal to add the dimension of height in a small patio garden, trained on a structure, such as a small obelisk, or forming a backdrop on a low fence or wall.

Polyantha
A profuse flowering group compact and intermediate between cluster-flowered and patio roses.

Where the garden design calls for small or large-scale ground-cover, this group will provide a low-maintenance, colourful carpet. This could be considered to contrast with the somewhat unchanging shades of green provided by more traditional evergreen shrubs. Note that ground-cover roses do need sun to flower well and are not evergreen. Ground preparation prior to planting must be thorough, since the removal of weeds, especially perennial weeds, is not easy from amongst densely planted roses!

Take care in the selection of ground-cover roses. Some are very low-growing and spread slowly in a limited manner. Others cover the ground rapidly in a more irregular, arching type of growth. Not all cultivars are recurrent. Some provide just one flush of flowering, others flower over a longer period and some also provide autumn colour from hips.

Standard, Half-Standard and Smaller Standard Forms
The largest full-standard rose budded at a height of 1.8m (6ft) or more can be used to function as a tree in a small garden, giving height to the overall planting and determining a sense of scale. A spectacular focal point can be created by the inclusion of a not too vigorous climbing rose or ground-cover rose budded on to a standard stem. The resultant weeping rose when in full flower can be an arresting sight!

Standard or half-standard roses have been traditionally used to form a regular repeating pattern alongside paths, giving a pleasant, rhythmic feel to the visitor passing along the path. A very attractive feature, where space permits.

Used alone, a standard of appropriate height will make a focal point or accent plant at the end of a border.

In patio gardens and small gardens where space is very limited, patio roses, budded as short standards, or climbing miniature roses, on support structures, can be used to produce the same effects.

Containers
Some, but certainly not all roses are good subjects for container-growing. First, the size of container and compost selected must be correct, together with sunny situation.

Not all cultivars take well to growing in containers, but those that do will make a colourful, often scented, long season of display.

Styles of Garden
Roses, in general, can be used in both formal and informal garden designs. Where formality in a small garden is required, patio roses fit the bill ideally; where informality is wanted, somewhat larger, less compact cultivars can easily be selected.

ROSMARINUS

All plants previously known as *Rosmarinus* have been transferred to the genus *Salvia*.

RUBUS

A large genus of over 250 species that are spread widely over the globe. In addition, in Europe and North America, there are several apomictic species that produce many microspecies.

The most distinct common characteristic of the genus is the fruit. This consists of many seeds, each of which is embedded in a juicy drupe. The drupes are united together to form a rounded, cap shape, which fits on to a cone-shaped receptacle. This fruit is exemplified in the bramble (*Rubus fruticosus*) and raspberry (*Rubus idaeus*). Other common characteristics within the genus are spiney stems and alternate leaves.

Many of the species are not garden-worthy and of those that are, most are too vigorous and spreading for inclusion in small or compact gardens. Only those of suitable size and vigour are included here.

Growing Requirements

Rubus is of easy cultivation, but like all plants rewards good husbandry.
Hardiness: UDSA. 7b/8a or 6b/7a.
RHS H5 or 6.
Aspect: Most species will grow well on a site of any orientation.
Shelter: They do not require special shelter.
Sun: Tolerate full sun or part shade
Soil: The soil can be of any texture but must be well drained.
pH: A very wide range of soil pH is accepted.
Pruning: Those grown for winter-stem effect are best cut down to ground level every spring and then fed and mulched. This will encourage the production of strong, new canes every year, which exhibit the best stem colour.

Garden Design Considerations

Rubus cockburnianus 'Goldenvale' and *Rubus thibetanus* are outstanding for winter-stem colour. They are well worthy of any border or garden specifically planted for winter colour and interest, and associate with the coloured stem *Cornus*, also grown for winter-stem colour. Both *Rubus* also contribute to the summer garden. *Rubus cockburnianus* 'Goldenvale' produces leaves that are bright-yellow in spring and gradually expand and fade to an attractive lime-green by summer. *Rubus thibetanus* produces fern-like foliage, which is silvery-green on the top surface and white beneath. This has strong decorative value and may explain why it is often erroneously sold as *Rubus thibetanus* 'Silver Fern' or *Rubus* 'Silver Fern'. Both these *Rubus* are of only moderate vigour and can be easily controlled in the garden situation.

Rubus spectabilis 'Olympic Double' is unusual in the brambles in that it produces an excellent display of brightly coloured, large flowers. Where there is concern about its spread, it can be constrained by hard landscaping, such as a concrete path, or it may be successfully grown in a large, isolated container.

TABLE 4.101 *RUBUS* SPECIES AND CULTIVARS

Name	Habit	Height	Spread	Leaves	Flowers	Remarks
R. 'Betty Ashburner'	Prostrate, spreading, forming a mat	15cm (6in)	1.5 2.5m (5–8ft)	Evergreen, glossy and rounded	White, concealed by foliage. Produces a few orange fruit	Very good low maintenance ground cover. Less vigorous spreader than *R. tricolor*
R. cockburnianus 'Goldenvale' AGM	Arching stems	1.5–2.5m (5–8ft)	1.5–2.5m (5–8ft)	Deciduous. Bright-yellow in spring, fading to lemon-green. Very attractive	Rose-purple. Not significant	Striking chalky white stems in winter. Less vigorous and spreading than *Rubus cockburnianus*
R. spectabilis 'Olympic Double'	Suckering. Forms a thicket	1.2–1.8m (4–6ft)	1.5–2.5m (5–8ft)	Deciduous, trifoliate, mid-green	Large 3cm (1in) across, double, rose-magenta fragrant. Spring	Needs a site that is not north-facing and receives full sun*
R. thibetanus AGM (ghost bramble) (syn. *R. thibetanus* 'Silver Fern')	Upright stems	1.8–2m (6–7ft)	1.5–2.5m (5–8ft)	Deciduous. Pinnate, fern-like, greyish above, white beneath	White, produces a few red edible fruit	Stems purple with bristles and a white bloom
R. tricolor (Chinese bramble)	Prostrate and spreading	10–50cm (4in–20in)	1.5–2.5m (5–8ft)	Dark glossy green	White, produces a few red, edible fruit	Produces long, trailing growth. Good for large-scale ground-cover. Where space is more restricted, *R.* 'Betty Ashburner' is a better choice

* This species is listed on Schedule 9 of The Wildlife (Northern Ireland) Order (1985) as amended, and on Schedule 3 of the EU (Birds and Natural Habitats) Regulations 2011 in the Republic of Ireland as an invasive species. This does not prevent it from being sold or grown in gardens in Ireland, but care should be taken with managing it and disposal of unwanted plant material.

Rubus cockburnianus 'Goldenvale'. The white stems of the bramble add to this winter border scene.

Rubus cockburnianus 'Goldenvale'. The lime-yellow foliage in summer.

Rubus cockburnianus 'Goldenvale'. Detail of the fine summer foliage.

Rubus spectabilis 'Olympic Double'. The double, pink flowers produced in the early summer by this bramble.

Rubus thibetanus. The greyish-white winter-stem effect.

Rubus thibetanus. The growth produced in the summer.

Both *Rubus* 'Betty Ashburner' and *Rubus tricolor* are useful to form evergreen, effective, low-maintenance ground-cover and are well suited to covering banks and slopes where other plantings may be difficult to maintain. *Rubus tricolor* is a vigorous, spreading plant and where the area to be covered is small, *Rubus* 'Betty Ashburner' would be a more manageable choice.

Rubus thibetanus. A close up of this plant's foliage, explaining why this bramble is sometimes incorrectly called *Rubus thibetanus* 'Silver Fern'.

SALVIA

This is a large entry since this genus has recently had several previously separate genera incorporated into it, including *Rosmarinus* (rosemary) and *Perovskia* (Russian sage).

Before the addition of the two genera just mentioned, *Salvia* was already a large genus containing between 800 and 900 species, widely distributed through tropical, subtropical and temperate regions. A 'hot spot' is Mexico, which contains 275 species. The genus contains annuals, biennials, herbaceous, deciduous and evergreen shrubs. Only shrubs with a hardiness rated at USDA8b/9a. RHS H4, i.e. hardy in an average UK winter or hardier, are considered here. This means that some plants that are very popular with gardeners, such as *Salvia* 'Amistad' AGM, *Salvia elegans* (pineapple sage), *Salvia fulgens*, *Salvia guarantica*, *Salvia guarantica* 'Blue and Black' and *Salvia involucrata*, plus its cultivars, are not included here because they are USDA 9b/10a. RHS H3 or even more tender, and not hardy enough to warrant inclusion.

For ease of the reader and those thinking of selecting small shrubs for their garden or planting scheme, the plants are divided into tables:

- Table 4.102: *Salvia greggii* and its cultivars
- Table 4.103: *Salvia* × *jamensis* and its cultivars
- Table 4.104: *Salvia microphylla,* its varieties and cultivars
- Table 4.105: other hardy salvias
- Table 4.106: *Salvia officinalis*, culinary sage and cultivars
- Table 4.107: plants previously classified as *Rosmarinus* but now classified as *Salvia*
- Table 4.108: plants previously classified as *Perovskia* but now classified as *Salvia*

Growing Requirements

Hardiness: Plants have been selected to be as hardy as USDA 8b/9a. RHS H4 or hardier. A hardiness rating for each plant is included in Tables 4.102–4.108.
Aspect: A site of any orientation, other than north-facing, is acceptable.
Shelter: A sheltered site is required, especially from cold, drying, easterly winds.
Sun: A sunny, open site is recommended with, preferably, full sun or just light shade. Sun is required to encourage flowering, and to ripen and harden wood before the challenge of winter.
Soil: The soil can be of any texture but must be well drained. This is of vital importance since wet, cold conditions in the winter will kill these plants. Heavy soils can be lightened, and their drainage improved by the generous incorporation of grit. Very light soils may benefit from a light application of organic matter to improve water-retention in the height of summer.
pH: A wide range of soil pH is acceptable. Drought-tolerant.

New cultivars are constantly increasing the already very wide colour range.

Salvia × *jamensis*

This is a bushy shrub that is evergreen in mild localities.
Height: 100cm (39in).
Spread: 50cm (20n).
Leaves: Aromatic and mid-green.
Flowers: Typical of other salvias in being two-lipped, but the flower colour is very varied and can be red, rose-pink, salmon-pink, orange or creamy-yellow. The flower is usually only one colour in contrast to most of the *Salvia greggii* cultivars that tend to be bicoloured.

The plant naming between *Salvia greggii* cultivars and *Salvia* × *jamensis* cultivars is confused in nursery listings. Wherever possible the names applied by *The RHS Plant Finder 2022* are used. However, some cultivars

TABLE 4.102 *SALVIA GREGGII* (AUTUMN SAGE, TEXAS SAGE) AND CULTIVARS

Name *	Height	Spread	Leaves	Flowers	Remarks
S. greggii	30–50cm (12–20in)	30–50cm (12–20in)	2–3cm (1¼–1¾in) long. Deep-green, fragrant. All the same	Can be red, purple, pink, yellow or violet	Flowers late summer into autumn. Drought tolerant
S. 'Alba'	90cm (37in)	60cm (24in)		White	
S. 'Amethyst Lips'	90cm (37in)	75cm (29in)		Top lip tipped purple-blue, rest white. Lower lip purple-blue	
S. 'Desert Blaze' (syn. 'Sparkler')	40cm (17in)	30cm (12in)	Leaves edged with white	Red	
S. 'Dyson's Joy'	90cm (35in)	90cm (35in)		Top lip deep-pink. Bottom lip light-pink	
S. 'Emperor'	60cm (24in)	50cm (20in)		Deep red-purple	
S. 'Icing Sugar'[PBR]	65cm (26in)	65cm (26in)		Top lip red-purple. Bottom lip light-purple	
S. 'Lipstick'	60cm (24in)	60cm (24in)		Reddish-pink	
S. 'Mirage Deep-Purple'	60cm (24in)	60cm (24in)		Purple	
S. 'Peach'	60–75cm (24–30in)	60cm (24in)		Top lip yellowish-pink. Bottom lip deep-red	Profuse in flowering
S. 'Strawberries and Cream'	75cm (30in)	75cm (30in)		Top lip dark-pink. Bottom lip pale-pink with cream edge	
S. 'Sungold' (syn. 'Devon Cream')	50cm (20in)	50cm (20in)		Lemon yellow	Less profuse in flowering

Hardiness: All USDA8b/9a. RHS H4
Habit: The species and its cultivars all have basically the same habit being dwarf bushy evergreen shrubs.

are only listed in the UK and others only listed in the USA. Hardiness of *Salvia* × *jamensis* varies from cultivar to cultivar and, to an extent, as to which nursery is consulted. Where available, hardiness ratings are given in terms of RHS and USDA ratings. These may not correspond directly, and this may be due to climatic factors other than temperature.

Growing Requirements

In general, similar *to Salvia greggii*.

Aspect: A site that is south- or west-facing is required.

Shelter: Shelter is required. To overwinter, full sun and good drainage is required. *Salvia* × *jamensis* cultivars tend to less hardy than those of *Salvia greggii*, but there are exceptions.

Sun: The site must be in full sun.

Leaves: All aromatic mid-green.

TABLE 4.103 *SALVIA* × *JAMENSIS* CULTIVARS

Name	Height	Spread	Flowers	Remarks
S. × *jamensis**	100cm (39in)	50cm (20in)	Many different colours	RHS H5. USDA 7b/8a
'Californian Sunset'	90cm (35in)	100cm (39in)	Orange	RHS H5-H4. USDA 7b-9
'Cherry Lips'	80cm (32in)	70cm (28in)	Pink lower lip. Top lip white with a pink tip	RHS H3. USDA 9b/10a
'Dark Dancer'	91cm (36in)	122cm (48in)	Purple-red	RHS H4. USDA 8b/9a
'James Compton'	60cm (24in)	100cm (39in)	Deep-crimson	RHS H4. USDA 8b/9a
'Javier' AGM	100cm (39in)	100cm (39in)	Violet	RHS H4. USDA 8b/9a
'La Luna'	90cm (35in)	90cm (35in)	Yellow	RHS H5. USDA 7b/8a
'La Siesta'	90cm (35in)	50cm (20in)	Pink	RHS H5. USDA 7b/8a
'Maraschino'	90cm (35in)	50–100cm (20–39in)	Red	RHS H5. USDA 7b/8a
'Moonlight Serenade'	75cm (30in)	50cm (20in)	White	RHS H3. USDA 9b/10a
'Pat Vlasto'	91cm (36in)	91cm (36in)	Orange pink	RHS H4-H3. USDA 8b/9a-9b/10a
'Peter Vidgeon' AGM	90cm (36in)	50cm (20in)	Pink lilac	RHS H5. USDA 7b/8a
'Pluenn'PBR	80cm (32in)	80cm (32in)	Bright-pink	RHS H4. USDA8b/9a. Claimed to be an improvement on 'La Siesta'
'Raspberry Royale'	60–91cm (24–36in)	60–91cm (24–36in)	Carmine red	RHS H5. USDA 7b/8a
'Red Velvet'	90cm (35in)	90cm (35in)	Red	RHS H4. USDA 8b/9a
'Sierra San Antonio'	80cm (32in)	90cm (35in)	Light peach	RHS H4. USDA 8b-10

* New cultivars are constantly being introduced.

Salvia × *jamensis* 'Maraschino'.

Salvia × *jamensis* 'Maraschino'.

Salvia × *jamensis* 'Peter Vidgeon'.

Salvia × *jamensis* 'Flammenn'PBR.

Salvia × *jamensis* 'Red Velvet'.

Salvia microphylla. **(syn. *S. neurepia, S. macellaria*) (Baby Sage)**

A small bushy evergreen shrub.
Height: 1–1.5m (3–5ft).
Spread: 0.5–1m (1.6–3ft).
Leaves: Bright-green, aromatic and 5cm (2in) long.
Flowers: Bright-red flowers fading to a bluish red are borne in terminal clusters from summer into autumn.

Growing Requirements

Aspect: The site should be south- or west-facing.
Shelter: Shelter is required.
Sun: Full sun.
Soil: The soil can be of any texture but must be well drained.
pH: A wide soil pH is tolerated.

TABLE 4.104 *SALVIA MICROPHYLLA* CULTIVARS

Name	Height	Spread	Leaves	Flowers	Remarks
S. microphylla	1–1.5m (39–60in)	0.5–1m (20–39in)		Bright-red	RHS H4. USDA 8b/9a
S. 'Aphrodite'	60–90cm (24–35in)	40–50cm (16–20in)		Peach	RHS H3. USDA 9b/10a
S. 'Cerro Potosi' AGM	70–80cm (27–32in)	70–80cm (27–32in)		Pink	RHS H4. USDA 8b/9a
S. 'Delice Fiona'	70–80cm (27–32in)	70–80cm (27–32in)	Bright-green, aromatic 5cm long	Pink/white	RHS H4. USDA 8b/9a
S. 'Glitter'	78–80cm (30–32in)	50–60cm (20–23.5in)		Lilac	RHS H3. USDA 9b/10a
S. 'Kew Red'	90cm (35in)	90cm (35in)		Red	RHS H4. USDA 8b/9a
S. microphylla var. *microphylla*	120cm (47in)	90cm (35in)		Pink	RHS H4. USDA 8b/9a. Blackcurrant sage
S. 'Newby Hall'	50–100cm (20–39in)	50–100cm (20–39in)		Blood-red	RHS H4. USDA 8b/9a
S. ' Blush Pink'	90cm (35in)	45cm (18in)	Yellowish green	Upper lip purple-rose. Lower lip red purple	RHS H4. USDA 8b/9a
S. 'Wild Watermelon'	75cm (30in)	75cm (30in)		Magenta-pink	RHS H4. USDA 8b/9a
S. 'Wine and Roses'	60–80cm (24–32in)	60–80cm (24–32in)		Red and purple	RHS H4. USDA 8b/9a

Salvia microphylla 'Kew Red'. An established plant of this cultivar can make a small shrub 1.5m (5ft) tall and 1.5m (5ft) wide by the end of a growing season.

Salvia 'Hot Lips'. An extremely popular cultivar.

Salvia 'Nachtvlinder'.

Salvia 'Hot Lips'. Seen here flowering profusely.

Salvia 'Royal Bumble'.

Other Hardy, Shrubby Salvias

These are similar to those already listed but whose parentage may be unknown or disputed.

Growing Requirements

The same as for *Saliva greggii*, *Salvia × jamensis*, *Salvia microphylla* and their cultivars.

Salvia officinalis (Common Sage, Culinary Sage)

An evergreen, slow-spreading sub-shrub.
Height: 75cm (27in).
Spread: 0.5–1m (18–39in).
Leaves: Oblong-ovate entire, very aromatic, grey-green, felty 8cm (3in) long.
Flowers: Short spikes of pale-blue flowers in early summer.

Growing Requirements

Hardiness: USDA 7b/8a. RHS H5.
Aspect: Any aspect will be tolerated.
Shelter: Shelter is required.
Sun: The site should be sunny or only part-shaded.
Soil: The soil can be of any texture but should be well drained.
pH: A wide pH range is accepted.

Salvia officinalis 'Purpurascens'.

TABLE 4.105 OTHER HARDY SHRUBBY SALVIAS

Name	Height	Spread	Leaves	Flower	Hardiness
Salvia AMETHYST LIPS ('Dyspurp')	0.5–1m (20–39in)	0.5–1m (22–39in)	Evergreen, mid-green, aromatic	Bicoloured. Top lip white. Bottom lip purple	RHS H4. USDA 8b/9a
Salvia 'Hot Lips' AGM	100cm (39in)	100cm (39in)	Evergreen, ovate, mid-green, aromatic	Bicoloured. Early summer tends to be red all over. Mid-summer. Top lip white. Bottom lip red. Autumn may become all-white	RHS H5. USDA 7b/8a
Salvia lavandulifolia (lavender-leaved sage)	60cm (24in)	60cm (24in)	Evergreen, narrow, blue-green, downy leaves	Racemes of blue flowers	RHS H5. USDA 7b/8a
Salvia 'Nachtvlinder' AGM (syn. *S. × jamensis* 'Nachtvlinder')	75cm (30in)	50cm (20in)	Evergreen, aromatic small, toothed, mid-green	Dark maroon	RHS H5. USDA 7b/8a
Salvia 'Royal Bumble' AGM	60cm (24in)	40cm (16in)	Evergreen, aromatic, glossy, deep green	Red flowers with purple black calyx	RHS H4. USDA 8b/9a
Salvia 'Silas Dyson'	90cm (35in)	40cm (16in)	Evergreen, aromatic, mid-green	Pinkish red and velvety	RHD H4. USDA 8b/9a

TABLE 4.106 *SALVIA OFFICINALIS* (CULINARY SAGE) AND CULTIVARS

Hardiness and remarks: All USDA 7b/8a, RHS H5

Name	Height	Spread	Leaves	Flowers
S. officinalis (common sage, culinary sage)	75cm (30in)	50–100cm (20–39in)	Evergreen, oblong-ovate, very aromatic. Grey-green, felty, 8cm (3in) long	Short spikes of pale-blue flowers
S. officinalis 'Berggarten' AGM	60cm (24in)	100cm (39in)	Leaves a little larger than *S. officinalis*	Short spikes of pale-blue flowers
S. officinalis 'Icterina' AGM	60cm (24in)	100cm (39in)	Variegated with yellow and pale-green	Short spikes, of pale-blue flowers
S. officinalis 'Purpurascens' AGM	60cm (24in)	100cm (39in)	Purple-leaved sage. Leaves 3–6cm (1–2in) long	Short spikes, of purple-blue flowers
S. officinalis 'Tricolor'	60cm (24in)	100cm (39in)	Variegated with cream. Youngest leaves also a little purple	Short spikes of pale-blue flowers

Salvia officinalis 'Purpurascens'. Flowering in early summer.

Salvia rosmarinus (Syn. *Rosmarinus officinalis*) (Common Rosemary)

An evergreen shrub commonly associated with remembrance and widely grown and loved. Native to southern Europe and Asia Minor in rocky, hot, dry conditions and in cultivation preferring poor soil. A dense-growing plant.
Height: 2m (6ft 7in).
Spread: 2m (6ft 7in).
Leaves: The stems are clothed with opposite, paired, linear, green or greyish-green leaves, which are white on their undersides. Heavily aromatic when rubbed and widely used as a culinary herb.
Flowers: The flowers are small, blue, two-lipped and produced in axillary clusters on the previous year's growth in early summer or sometimes earlier, and again sometimes in late summer.

Growing Requirements

Hardiness: All cultivars USDA 8b/9a, RHS H4, but all, especially *Salvia rosmarinus* Prostrata Group, may be susceptible to light damage due to late spring frost, which can be easily pruned off and will not usually cause long-term damage.
Aspect: The site can be of any orientation, except north-facing.
Shelter: Shelter is not necessary.
Sun: Full sun is preferred.
Soil: A soil of any texture will suffice, but the drainage must be good, especially in the winter. A soil low in nutrients is beneficial; rich soil encourages soft growth, which will be liable to winter damage.
pH: A wide pH range is accepted.

TABLE 4.107 PLANTS PREVIOUSLY CLASSIFIED AS *ROSMARINUS* BUT NOW CLASSIFIED AS *SALVIA*

Name	Habit	Height	Spread	Leaves	Flowers	Remarks
S. rosmarinus	Dense but variable	2m (78in)	2m (78in)	Green or greyish green	Blue. Late spring or early summer and often again later	Common rosemary, widely used for culinary purposes
S. rosmarinus 'Benenden Blue'	Dense but variable	0.5–1m (20–39in)	0.5–1m (20–39in)	Dark-green leaves	Bright blue	More compact and less variable than straight species
S. rosmarinus 'Green Ginger' AGM	Dense but variable	1m (39in)	1m (39in)	Greyish green	Light blue	Leaves scented and flavoured with a ginger like overlay
S. rosmarinus 'Lady in White'	Arching and spreading	0.6m (24in)	1–1.5m (39–59in)	Dark-green	White	
S. rosmarinus 'Miss Jessopp's Upright' AGM	Upright	2m (79in)	2m (79in)	Greyish-green	Blue	Strong growing
S. rosmarinus Prostrata Group	Forms a prostrate mat	0.5m (20in)	1.5m (59in.)	Greyish-green	Blue	Low-growing
S. rosmarinus 'Severn Sea' AGM	Has arching stems	0.5–1m (20–39in)	0.5–1m (20–39in)	Greyish-green	Brilliant blue	Growing to a medium height

Salvia yangii AGM (Previously Classified as *Perovskia atriplicifolia*)

The former genus of *Perovskia* contained seven species of late-flowering, aromatic sub-shrubs native to central Asia and the Himalayas, all with deeply toothed or finely cut leaves. Of these, *Salvia yangii* (syn. *Perovskia atriplicifolia*), commonly known as Russian sage, is horticulturally the most important and is described here.

A woody-based, deciduous perennial with an upright habit.
Height: 0.5–1.2m (20–47in).
Spread: 0.6–1.2m (24–47in).
Stems and leaves: The stems have a square cross-section and are covered in an indumentum giving them a silvery appearance. The grey-green leaves are arranged in opposite pairs. Each pinnate leaf is deeply incised, 3–5cm (1.5–2in) long and 0.8–2cm (¾–1in) long.
Flowers: The inflorescence is in the form of a much-branched panicle each, panicle bearing a raceme of small purple flowers, the overall effect sometimes being described as a haze of flowers.
Pruning: The plant naturally dies back during the winter and in cultivation is best pruned back to where it is still alive at the end of winter/beginning of spring.

Numerous cultivars have been developed with the aims of reducing the height of the plant and increasing the depth of the leaf incisions, this making the plant more desirable as a compact, attractive plant. Some are derived directly from *Salvia yangii*, others may be hybrids between *S. yangii* and *Salvia arbrotanoides*, which itself is like *S. yangii* but is distinguished by having bipinnate leaves. They are brought together here in Table 4.108 as having similar overall appearance and use in the garden.

Salvia rosmarinus.

Growing Requirements

Hardiness: USDA 7b/8a. RHS H5.
Aspect: A site with a southerly or western aspect is needed.
Shelter: A little shelter is beneficial.
Sun: Full sun.
Soil: The soil can be of any texture but must be well drained in winter.
pH: A wide pH range is tolerated. Once established the plant is resistant to drought and maritime conditions.

Garden Design Considerations

Tables 4.102–4.105 contain similar plants that are small, evergreen sub-shrubs with the same role in the garden. They all provide exceptionally vivid summer and autumn flower colour. A very wide range of colours is available, together with several bicoloured cultivars. They all require a sunny situation to flower profusely and good drainage, especially in winter. Provided these requirements are met, all should prove hardy in an average UK winter, but those rated H5 should be significantly hardier than those only graded H4. After winter, all these plants will require pruning back to where they are still alive but should regrow rapidly from this. In terms of garden design, these plants are ideal to provide colour in summer and early autumn at the front of sunny borders, in raised beds, planters and patios.

Salvia officinalis and its cultivars in Table 4.106 have an important culinary role and should feature in any herb garden or border. As with all sages, a sunny, well-drained site is required. In the decorative garden, they can be used to furnish the edge of borders, making a good colour and textural contrast with *Bergenia* cultivars used for the same purpose. *Salvia officinalis* and its cultivars also produce strong leaf colour plant associations with other dwarf shrubs with vivid leaf colour and different texture, such as *Nandina* and *Choisya*.

Table 4.107 contains common rosemary and its cultivars. Together with culinary sage, it is important in the herb garden, but its role extends to also being important in the decorative garden. *Salvia rosmarinus* 'Miss Jessopp's Upright' can be used as a striking feature plant with its distinctive erect habit being used to good effect. In contrast, *Salvia rosmarinus* Prostrata Group looks very attractive planted on top of, and allowed to drape over, stonework, especially dry-stone walls with their natural very sharp drainage. In full sun it makes a novel ground-cover and provides another growth form in the extensive rock garden.

TABLE 4.108 PLANTS PREVIOUSLY CLASSIFIED AS *PEROVSKIA*, BUT NOW CLASSIFIED AS *SALVIA*

Name	Habit	Height	Spread	Leaves	Flowers	Remarks
S. yangii	Upright	0.5–1.2m (20–47in)	0.6–1.2m (24–47in)	Pinnate, deeply incised, 3–5cm long 0.8–2cm wide. Grey-green	Small, pale-purple, appearing as a haze. From mid-summer to early autumn	Stems have a silvery-grey appearance
S. yangii 'Blue Jean Baby'	Upright, branched	0.8m (32in)	0.9m (35in)	Grey-green, deeply cut		
S. yangii 'Blue Shadow'		60 cm (24in)	60cm (24in)	Grey-green, deeply cut		
S. yangii 'Blue Steel'		0.5–06m (20–24in)	0.4–0.7m (16–28in)	Grey-green, deeply cut	Very dark blue	
S. yangii SILVERY BLUE ('Lissvery'PBR)		60cm (24in)	60cm (24in)	Heavily dissected, silvery foliage	Very dark blue	Very compact. Smaller than 'Little Spire'
S. 'Blue Spire' AGM		1.2m (47in)	0.6–1.2m (24–47in)	Grey-green, deeply cut foliage	Larger panicles of darker blue flowers than *S. yangii*	
S. 'Filigran'		1.2–1.3m (47–51in)	0.6–1.2m (24–47in)	Fern-like foliage	Larger panicles of darker blue flowers than *S. yangii*	
S. LACEY BLUE ('Lisslitt'PBR)		50cm (20in)	60cm (24in)	Grey-green deeply cut	Large purple blue	Very compact. Smaller than 'Little Spire'
S. 'Little Spire'PBR		0.6m (24in)	60cm (24in)	Less deeply cut, but greyer than 'Blue Spire'	Purple-blue	

Russian sage (*Salvia yangii*) and its cultivars are listed and described in Table 4.108. They are a distinctive group, many new compact cultivars having been released and designed to be very appropriate in today's garden where space is at a premium. In summer and autumn, they provide leaf colour and texture, plus flowers in a light and airy form. In winter, they have yet another function. The

Salvia yangii 'Blue Spire'. In flower in late summer.

Salvia yangii 'Blue spire' AGM. The white winter-stems contribute to colour and interest in the garden during winter.

Salvia yangii 'Blue Spire'.

light-coloured stems can be made a feature in a winter-plant association, their stem colour adding to the stem colours provided by the likes of *Cornus alba* cultivars. As with all salvias, a sunny, well-drained site is required.

This common requirement for sun and good drainage, coupled with drought-resistance in summer, makes the use of salvias in Mediterranean-style and some maritime or coastal gardens very appropriate. Mediterranean gardens are famed for their spring and early summer colour, but the salvias can extend this into late summer and autumn.

In all a very useful range of plants, with many applications in both the culinary and decorative garden.

SANTOLINA

A genus of about ten species from the Mediterranean region. Collectively they are small, evergreen, mounded sub-shrubs.
Height: 10–60cm (4–24in).
Leaves: May be entire and minute, or pinnately divided, usually silvery and aromatic. They are composites.
Flowers: Form on thin stems 10–25cm (4–10in) long, are gathered into a dense, rounded head called a capitulum 1–2cm (½–¾in) in diameter. There are no ray florets, and the flower heads are frequently referred to as buttons.

Growing Requirements

Hardiness: USDA 7b/8a. RHS H5.
Aspect: A south- or east-facing site is strongly preferred.
Shelter: Shelter from cold winds is necessary.
Sun: Full sun is required.
Soil: Chalk, loam or sand soils, even poor thin soils are acceptable, but avoid heavy clay. Drainage must be good.
pH: A wide range of soil pH from alkaline to acid is accepted.

Garden Design Considerations

In *Santolina* we have a range of small or very small, fine-textured, evergreen shrubs with many possible applications in the garden.

Traditionally, they have been used as low or dwarf hedging or edging of borders. The cultivar *Santolina chamaecyparissus* 'Lambrook Silver' is especially suited to this purpose. This does make a neat, aromatic, dwarf hedge, but the need for a well-drained, sheltered, sunny situation must be remembered. Plants will need to be clipped in

TABLE 4.109 *SANTOLINA* SPECIES AND CULTIVARS

Name	Habit	Height	Spread	Foliage	Flowers	Remarks
S. chamaecyparissus (syn. *S. incana*) (cotton lavender)	Dwarf, evergreen shrub	50cm (20in)	1m (39in)	Narrow, silvery, dissected, 4cm (1½in) long	Yellow buttons in summer	Aromatic leaves
S. chamaecyparissus 'Lambrook Silver'	Slightly taller cultivar	75cm (30in)	1m (39in)	Very bright, silver-grey leaves	Lemon yellow	A taller cultivar, favoured for hedging
S. chamaecyparissus 'Lemon Queen'	A compact cultivar	60cm (24in)	60cm (24in)	Green foliage	Pale-yellow	Aromatic leaves
S. chamaecyparissus 'Nana' AGM	A very compact cultivar	25cm (10in)	50cm (20in)	Narrow, silvery, dissected	Yellow buttons	Aromatic leaves
S. chamaecyparissus 'Pretty Carroll' AGM	A compact cultivar	40cm (16in)	40cm (16in)	Silvery leaves	Yellow	Aromatic leaves
S. pinnata subsp. *neapolitana* AGM (rosemary-leaved cotton lavender)	Dwarf, evergreen shrub	75cm (30in)	1m (39in)	Finely divided, grey-green leaves, 4cm (1½in) long	Bright lemon-yellow flowers	Aromatic leaves
S. pinnata subsp. *neapolitana* 'Edward Bowles'	Dwarf, evergreen shrub	70cm (28in)	0.5–1m (20–39in)	Grey-green leaves	Pale primrose	Aromatic leaves
S. rosmarinifolia (syn. *S. virens* and *S. viridis*)	Dense, rounded bush	60cm (24in)	1m (39in)	Very finely divided, bright-green, 5cm (2in) long	Yellow	Very aromatic leaves
S. rosmarinifolia 'Lemon Fizz' AGM	Dense, rounded bush	50cm (20in)	1m (39in)	Thread-like lemon green	Lemon yellow	
S. rosmarinifolia subsp. *rosmarinifolia*	Dense, rounded bush	60cm (24in)	1m (39in)	Finely dissected green	Bright-yellow	
S. rosmarinifolia subsp. *rosmarinifolia* 'Primrose Gem' AGM	Dense rounded bush	60cm (24in)	1m (39in)	Finely dissected, bright-green. 5cm (2in) long	Primrose yellow	

Santolina pinnata 'Sulphurea'. Photographed in the winter when the silvered, finely divided foliage is still very attractive.

Santolina rosmarinifolia 'Lemon Fizz' and Geranium ROZANNE ('Gerwat'[PBR]) AGM make a vivid colour contrast.

spring and after flowering to keep the hedge neat and the plants vigorous. Even with this care, such hedges do tend to be short-lived.

They can be used as small-scale, evergreen ground-cover in a sunny, dry situation, such as a south-facing slope, but they will need clipping after flowering to prevent gaps appearing. Liking hot, dry, sunny situations, they are well suited to gravel gardens and the dwarfer selections, such as *Santolina chamaecyparissus* 'Nana', are in scale for the sunny rock garden. They add fine foliage texture and colour to the cottage and informal garden.

They are ideal for compact, patio, container and raised bed gardens, where their small size and evergreen nature makes them especially valuable.

The vivid foliage colour of some cultivars, such as *Santolina rosmarinifolia* 'Lemon Fizz', is exciting when used to produce colour contrasts with other small, coloured plants, such as *Ajuga reptans* 'Catlin's Giant', with its purple crinkled foliage and, later in the season, with *Salvia patens* with deep-blue flowers.

Finally, the strongly aromatic foliage of many members of the genus makes them eminently suitable for inclusion in a sensory garden where such foliage is an important constituent.

SARCOCOCCA

A genus of small or dwarf shrubs known collectively as Christmas box, due to their habit of flowering in mid to late winter but often just after Christmas. Their main attributes for the gardener are their attractive, evergreen, glossy foliage and scented winter flowers, both being worthy of cutting for indoor decoration.

The plants are monoecious, having separate male and female flowers produced on the same plant in the leaf axils. The tiny apetalous flowers appear in short racemes, the males above the female. The fruit are small black or red berries.

Spread is often by suckering from the base, resulting in a dense mass of stems.

Growing Requirements

Hardiness: USDA 7b/8a. RHS H5.
Aspect: A site of any orientation is tolerated.
Shelter: Some shelter from cold winds is required.
Sun: Full shade and partial shade are enjoyed. Full sun is accepted but may result in chlorosis and leaf drop, if adequate soil moisture is not present.
Soil: A moderately fertile soil of any texture is suitable, but it must be well drained. The group is exceptional for growing well in dry shade once established and then tolerating a degree of neglect.
pH: The group is again unusual for evergreens in accepting a soil of any pH, including that which is chalky with a high pH.

Garden Design Considerations

Sarcococca is a genus of exceptionally useful, if not very visually exciting, small evergreen shrubs. Where they are employed most extensively is perhaps as low-growing, dense, evergreen ground-cover in part or heavy shade. This role is extended by their acceptance of high soil pH and dry soil, once established. They are almost unique in this range of tolerance; they will also put up with a degree of atmospheric pollution!

In late winter, one can watch a group of non-gardeners passing a border edged with *Sarcococca*. Suddenly one will turn and bend towards the plant, others will follow; they have detected the sweet aroma of the tiny flowers. *Sarcococca* is an essential ingredient for a winter border or winter garden. As well as clouds of fragrance, dark-green foliage provides a background for early spring bulbs, especially snowdrops, *Galanthus*, *Iris reticulata* cultivars or *Crocus*.

The white flowers, although very small, can be produced in profusion, which, together with shiny black fruits, are readily noticeable in winter when the eye is searching for extra interest.

Positioning is very important. Planting by a door leading outside, by a frequently used path or in a prominently positioned container planted for winter interest, makes certain that their glorious scent will be sensed and fully appreciated. *Sarcococca confusa* makes a good evergreen, low-growing hedge, giving a dark-green semi-formality, with winter flowers and scent in many situations.

Sarcococca confusa. Despite its common name of Christmas box, this plant always flowers later, generally early in the New Year.

Sarcococca confusa. Showing the sweetly scented flowers and the black, shiny fruits. The fruits are at their best one year after the flowers that produced them.

Sarcococca hookeriana var. *digyna*. This plant has a more upright habit than *S. confusa*.

Sarcococca hookeriana var. *digyna*. This plant has similar flowers but narrower and more linear leaves than *S. confusa*.

TABLE 4.110 *SARCOCOCCA* SPECIES AND CULTIVARS

Name	Habit	Height	Spread	Leaves	Flowers	Fruit	Remarks
S. confusa AGM (sweet box)	Suckering, dense growth	2m (6ft)	1–1.5m (3–5ft)	Elliptic, wavy edged, glossy dark-green, 5cm (2in) long, 2cm (1.25in) wide	White, each with two to three stigmas, scented	At first green, becoming, shiny black, 5mm (¼in) diameter	Origin unknown, but probably China
S. hookeriana (Himalayan sweet box)	Erect growing and rhizomatous	0.3–0.6m (1–2ft)	1m (3ft)	Lanceolate, deep-green, each 9cm (3.5in) long	White, each with three stigmas, scented	Black, 5mm (¼in) diameter	
S. hookeriana var. *digyna*	Suckering, erect, but slenderer than above	1–1.5m (3–5ft)	1.5–2.5m (3–8ft)	Lanceolate but narrower than above. Deep-green, 8cm (3in) long, 1cm (¾in) wide	White, each with two stigmas, leading to only two seeds in each fruit. Scented	Black 5mm (¼in) diameter	Stems green or purple
S. hookeriana var. *humilis*	Dwarf, suckering form	0.5–1m (1.6–3ft)	0.5–1m (1.6–3ft)	Elliptic, dark-green, glossy, 8cm (3in) long, 1cm (¾in) wide	White, scented	Black 5mm (¼in) diameter	
S. hookeriana 'Purple Stem' AGM	Upright, new shoots flushed purple	1.2m (4ft)	2m (6ft)	Leaf petioles and midrib flushed purple	White, tinged pink. Scented	Black 5mm (¼in) diameter	
S. hookeriana WINTER GEM ('Pmoore03'[PBR])	Upright	0.5m (1½ft)	0.5m (1½ft)	Leaves very glossy	White, profuse, highly scented	First red becoming black	
S. ruscifolia (fragrant sweet box)	Stiff, but arching stems	0.5–1m (1½–3ft)	0.5–1m (1––3ft)	Ovate, tapered, dark-green, 6cm (2.5in) long	White, highly scented	Dark-red 5mm (0.25in) diameter.	
S. ruscifolia var. *chinensis* (Chinese fragrant sweet box)		0.5–1m (1½–3ft)	0.5–1m (1½–3ft)	Narrowly ovate, glossy	White, highly scented	Dark-red	
S. ruscifolia var. *chinensis* 'Dragon Gate' AGM	Upright	0.5–1m (1–3ft)	0.5–1m (1½–3ft)	Small, glossy, dark-green	White, profuse, highly scented	Red berries, in summer	

SKIMMIA

A small genus of low-growing, evergreen shrubs, with alternate leaves and terminal flower panicles. The plants are grown for their attractive leaves, flowers, which in some cases may be scented, and fruits.

The plants may be:

- Monoecious, where male and female flowers are born on the same plant, but separately.
- Dioecious, when male and female flowers are born on separate plants.
- Hermaphrodite, where male and female organs are found in the same flower.

Where male and female flowers are only born on separate plants, plants of both sexes must be grown together if fruits are to be produced.

Growing Requirements

Hardiness: USDA 7b/8a. RHS H5.
Aspect: A site that is not south-facing is generally best.
Shelter: The site should be sheltered.
Sun: Part shade is preferred. An excess of sun will often cause leaf chlorosis (leaf yellowing) and poor growth. *Skimmia* × *confusa* is an exception, tolerating a site of any orientation and even full sun is accepted, provided that the plant is given otherwise good growing conditions.
Soil: The soil can be of any texture, providing that it contains plenty of organic material, is moisture-retentive in summer and well drained in winter.
pH: Soil pH can be acid or alkaline and *Skimmia* is not calcifuge, although growth is often best on acid or neutral soils. Poor soil, drought, excessive wet in winter or other stress may cause the plants to become chlorotic. Organic matter should be added to soil before planting, especially where the soil pH is high.

Garden Design Considerations

A comparatively small, but useful group of widely planted evergreen shrubs. The predictability of known sex cultivars means that these are the most commonly planted.

As long as the correct favourable growing conditions are provided, as a genus, they are exceptional in supplying winter fruit colour, interest and structure for the winter garden, plus scent in the spring. They associate well with shrubs such as *Sarcococca confusa* and coppiced *Cornus alba* cultivars when placed in a border designed to perform in the winter.

The low-growing cultivars are particularly suitable for use as low-maintenance, evergreen ground-cover in part-shaded situations.

As young plants they are particularly appropriate to furnish planters, large pots, window boxes and hanging baskets intended to provide a green backdrop and splash of red winter berry colour.

Skimmia × confusa 'Kew Green'. Showing a plant in full flower in spring. The scent carries through the garden.

Skimmia × confusa 'Kew Green'. A complete head of the very fragrant flowers.

Skimmia japonica. A head of female flowers. The operative stigma can be seen in the centre of each flower, as can the ring of inoperative anthers that fail to develop pollen.

Skimmia japonica 'Fragrans'. This plant is slightly chlorotic due to stressful growing conditions, and this has stimulated the plant to flower profusely.

Skimmia japonica 'Fragrans'. A close-up of a head of the male flowers that this cultivar produces. The flowers are very fragrant and they have operative anthers but inoperative stigmas.

Skimmia japonica subsp. *reevesiana*. A close-up of the very colourful fruits this cultivar produces. This cultivar also produces male flowers.

Skimmia japonica 'Rubella'. A male cultivar that produces large heads of flowers that are dark-pink in bud over winter and then open to give white, heavily scented flowers in spring.

A planted container designed to give a long season of interest. The planting includes a plant of *Skimmia japonica* 'Rubella'.

TABLE 4.111 *SKIMMIA* SPECIES AND CULTIVARS

Name	Habit	Height	Spread	Leaves	Flowers	Remarks
S. × *confusa* 'Kew Green' AGM (m)	Bushy and dense	0.5–1m (20–39in)	1–1.5m (3–5ft)	Dark-green, lance shaped, 12cm (5in) long	Large clusters of creamy-white heavily fragrant flowers in spring	
*S. japonica** (plants may be male or female)**	Can be creeping, low-growing or erect	1.5–2.5m (4–8ft)	1.0–1.5m (3–5ft)	Dark-green	***If pollinated followed by red fruits in autumn 8cm (3in) across	As a species the plant can be very variable in habit
S. japonica 'Fragrans' AGM (m) (syn. *S. japonica* 'Fragrant Cloud')	Dome-shaped	0.5–1m (20–39in)	0.5–1m (20–39in)	Dark-green, obovate, 10cm (4in) long	White buds opening to white fragrant flowers in spring	
S. japonica 'Kew White' (f)		1.5m (5ft)	1.5m (5ft)	Dark-green	Scented white flowers in spring followed by white fruits	
S. japonica 'Magic Marlow'[PBR] (m)	Compact dwarf	0.5m (20in)	0.5m (20in)	Variegated, cream and green marbled	Buds green opening in spring to scented creamy-white flowers	
S. japonica 'Nyman's' AGM (f)	Bushy and spreading	0.5–1m (20–39in)	1.0–1.5m (3–5ft)	Dark-green, 12cm (5in) long, lance shaped	White flowers in spring, followed by red berries in autumn	
S. japonica OBSESSION 'Obsbolwi'[PBR] (m/f)		0.5–1m (20–39in)	0.5–1m (20–39in)	Dark-green, obovate and acute tipped	Small panicles of white flowers in spring followed by masses of red berries in autumn	
S. japonica subsp. *reevesiana* (m and f)	Compact and bushy	0.5–1.m (20–39in)	0.5–1.m (20–39in)	Dark-green, obovate, and acute tipped	Creamy-white flowers in spring followed by red fruits in autumn	Hermaphrodite and self-fertile
S. japonica 'Rubella' AGM (m)	Bushy	1–1.5m (3–5ft)	1–1.5m (3–5ft)	Dark-green, obovate, 10cm (4in) long	Red buds in winter followed by fragrant white flowers in spring	

* Plants labelled just *Skimmia japonica* are usually available only as unsexed plants. Selected cultivars are clones of known sex.
** Male plants have large, conspicuous flower-heads of sweetly scented flowers in spring.
*** Female plants have smaller flower-heads, which, if pollinated, have red fruits in autumn and winter.
f = female.
m = male.

SORBARIA

A genus of deciduous shrubs with pinnate leaves, white flowers with five petals and numerous stamens. As a genus they are separated from *Spiraea* by having pinnate leaves, whereas *Spiraea* has simple leaves.

Sorbaria is, in general, a group of large, suckering shrubs and only the cultivars of one species are considered suitable for compact spaces or small gardens.

Growing Requirements

Hardines: USDA7b/8a. RHS H5.
Aspect: A site of any orientation is accepted.
Shelter: The planting position does not need to be especially sheltered.
Sun: Full sun or part shade is satisfactory, although a site in full sun will result in more flowering.
Soil: The soil can be of any texture, providing that it is well drained and retains adequate moisture in the summer.
pH: Soil pH can be alkaline or neutral.

Maintenance

A couple of years after planting, annual maintenance should consist of removing any old shoots down to ground level, cutting off old flower stems and removing any excess suckers. This should be done in the dormant season.

Spread by suckers is quite slow and may not be a problem. If necessary, such removed suckers are usually rooted and can be regarded as new material for planting.

Garden Design Considerations

Both cultivars are very desirable, compact shrubs for the small garden. Their foliage is distinctively fine-textured and unusually coloured during most of the growing season. As such, they associate well with other small shrubs. They contrast in leaf texture against coarser textured shrubs, such as *Viburnum davidii*, and contrast in leaf colour with many other small, coloured-leaved shrubs, such as the yellow-leaved *Berberis thunbergii* 'Aurea' or *Choisya* × *dewitteana* GOLDFINGERS ('Limo'[PBR]). Only where space is very restricted is the moderate suckering habit likely to be a problem.

TABLE 4.112 *SORBARIA SORBIFOLIA* CULTIVARS

Name	Habit	Height	Spread	Foliage	Flowers
S. sorbifolia PINK HOPI ('Cousorb05')	Upright, suckering, spreading. Said to produce fewer suckers and to spread less than *S.* 'Sem'PBR AGM	1–1.5m (3–5ft)	1m (3ft)	Finely pinnate. In spring red, pink and cream. Later leaves are similarly coloured, but with white veins. By summer leaves are green. Colouring of leaves said to last longer than *S.* 'Sem'PBR AGM	Stiffly erect panicles of white flowers in summer
S. sorbifolia 'Sem'PBR AGM	Upright, suckering, spreading	1–1.5m (3–5ft)	1.1.5m (3ft)	Finely pinnate. In spring pink and bronze, later yellow-green, later green, developing red tones in autumn before leaf fall	Panicles of white flowers in summer

Sorbaria sorbifolia 'Sem'PBR. A plant in spring, showing the overall effect of the foliage.

Sorbaria sorbifolia 'Sem'PBR. The shrub flowering in summer. Note how the older foliage has become green.

Sorbaria sorbifolia 'Sem'PBR. The colourful foliage in the spring.

Sorbaria sorbifolia 'Sem'PBR. Detail of a flowerhead showing how it is composed of many small white flowers.

SPIRAEA

A genus of about eighty species whose origin is spread over Europe, North Asia and North America. No species is native to the UK, but *Spiraea salicifolia* has become naturalised. The majority are of easy cultivation. Only those that are compact and suited to today's small garden are considered here.

Growing Requirements

Essentially very easy.

Hardiness: All species and cultivars mentioned here. USDA Zone 6b/7a. RHS rating 6.

Shelter: Shelter is not required, except for the most exposed of sites.

Sun: They will grow readily in full sun or part shade.

Soil and pH: Any soil of any pH will be tolerated, provided that it is moist but well drained in winter.

Pruning: Prune immediately after flowering. First remove all dead, damaged or diseased wood. Then remove at, or near, ground level some, usually about one-third, of the oldest shoots. Feed and apply a mulch. This treatment will encourage strong, new growth that will flower next year.

Garden Design Considerations

Excellent for low-maintenance, colourful ground-cover and low hedges. Good for seaside gardens and all except the most exposed situations.

Spiraea japonica (Japanese Meadowsweet)

A freely branching deciduous shrub with erect stems, small, serrated leaves and rosy-pink flowers in late summer. In horticulture, it is mainly the many coloured-leaved and more compact forms that are cultivated. Many have been introduced, and tried and tested or new and different cultivars are mentioned in Table 4.114.

Pruning: All *Spiraea japonica* cultivars are best cut back hard during the dormant season to encourage vigorous new growth the following spring when the vivid foliage colour will be well displayed.

TABLE 4.113 *SPIRAEA* 'ARGUTA' (BRIDAL WREATH, FOAM OF MAY)

Name	Habit	Height	Spread	Foliage	Flowers	Remarks
S. 'Arguta'	Slender, arching branches	1.5–2.5m (5–8ft)	1.5–2.5m (5–8ft)	Deciduous. Leaves small, green, entire, oval	Clusters of small, white flowers produced all along the branches in late spring	Produces a bright white display

Spiraea 'Arguta'. In spring this shrub provides a welcome splash of white.

Spiraea 'Arguta'. Close-up of the arching sprays of pure white flowers.

Spiraea japonica DOUBLE PLAY Artisan ('Galen').

Spiraea japonica DOUBLE PLAY GOLD ('Yan').

Spiraea japonica 'Goldflame'. Showing the size and habit of this shrub.

Spiraea japonica 'Goldflame'. Photographed as the foliage was changing from bronze-red to bright-yellow.

Spiraea japonica DOUBLE PLAY BIG BANG.

Spiraea nipponica 'Snowmound'

Pruning: After flowering, prune back flowered shoots to strong buds and, with well-established plants, cut down to ground level about one-quarter of the oldest shoots.

Garden Design Considerations

Excellent for low-maintenance, colourful ground-cover and low hedges. Good for seaside gardens and all except the most exposed situations. Very good for foliage colour contrasts, e.g. against the more compact, dark-foliaged cultivars of *Physocarpus opulifolius*. Useful for creating an accent point using strong colour.

Because of the range of cultivars with differing size, spread and leaf colour available, a selection can be made that will fit precisely into the space available and the colour combination required.

All cultivars are rated as excellent for plant pollinators.

STYRAX

Plants in the genus *Styrax* may be shrubs or trees. All *Styrax* are considered in Chapter 3.

SYRINGA

A genus of about thirty species found in South-East Europe, China and Japan. They are all deciduous and have the following characteristics in common: all have opposite leaves, flowers in panicles, a corolla that is tubular with four lobes, a bell-shaped calyx and two stamens. In addition, most have beautifully scented flowers!

Growing Requirements

Hardiness: USDA 7b/8a, 6b/7a, or 6a-1. RHS H5, H6 or, H7. Consult tables for specific information.
Aspect: A site with south-west- or east-facing orientation is satisfactory, but avoid a north-facing site.
Shelter: There is no need to select a particularly sheltered position.
Sun: Full or part shade. For the most profuse flowering, try to select a site offering full sun.
Soil: Provided the soil is well drained, the soil can be of any texture.
pH: A wide pH range is accepted. *Syringa* will grow well in a soil with a high pH but avoid a very acidic soil.

TABLE 4.114 *SPIRAEA JAPONICA* AND CULTIVARS

Name	Habit	Height	Spread	Foliage	Flowers
S. japonica	Erect stems	1.2–2m (4–6½ft)	1.2–2m (4–6½ft)	Deciduous. Small and serrated	Rosy-pink flowers in late summer
S. japonica 'Gold Mound'	Dense and bushy	90cm (36in)	1.2m (4ft)	Deciduous. New foliage golden, changing to lime-green as the seasons progress	Pink
S. japonica 'Goldflame'	Bushy	75cm (30in)	75cm (30in)	Deciduous. New foliage burnished bronze-red, changing to bright-yellow and finally mid-green	Pink
S. japonica GOLDEN PRINCESS ('Lisp')	Bushy	0.5–1m (1¾–3¼ft)	0.5–1m (1¾–3¼ft)	Deciduous. New foliage red-bronze later changing to mid-green	Pink
S. japonica DOUBLE PLAY ARTISAN ('Galen')	Bushy	90cm (36in)	90cm (36in)	Deciduous. New foliage purple-red	Deep-pink
S. japonica DOUBLE PLAY BIG BANG	Bushy	90cm (36in)	90cm (36in)	Deciduous. New foliage burnt-orange gradually changing to golden-yellow, finally orange before leaf fall	Pink

TABLE 4.115 *SPIRAEA NIPPONICA* 'SNOWMOUND'

Name	Habit	Height	Spread	Foliage	Flowers	Remarks
S. nipponica 'Snowmound'	Dense mound of growth	2.5m (8ft)	2.5m (8ft)	Deciduous. Small, rounded, green leaves	Clusters of white flowers in early to mid-summer	Equally tough as the other spiraeas, but useful where something slightly larger, but still compact, is required

Syringa meyeri cultivars

All form small, dense bushes and are thus ideal for the compact garden. All are also available as top-worked at 80cm (31in) or a little more in height. Sometimes referred to as 'standards' or 'lollipop trees'. As such, they can also function as small or miniature trees.

TABLE 4.116 *SYRINGA MEYERI* CULTIVARS

Name	Habit	Height	Spread	Leaves	Flowers	Remarks	Hardiness
S. meyeri (Flowerfesta Series) FLOWERFESTA PINK ('Anny 200817'PBR)	Dense and bushy	1.25m (4ft)	1.25m (4ft)	Small, broadly ovate	Small, pink, scented	Flowers late spring, early summer and intermittently through summer into autumn	USDA 6a-1. RHS H7
S. meyeri (Flowerfesta Series) FLOWERFESTA PURPLE ('Anny 200809'PBR)	Dense and bushy	1.25m (4ft)	1.25m (4ft)	Small, broadly ovate	Small, purple, scented	Flowers late spring, early summer and intermittently through summer into autumn	USDA 6a-1. RHS H7
S. meyeri (Flowerfesta Series) FLOWERFESTA WHITE ('Anny200810'PBR)	Dense and bushy	1–1.25m (3–4ft)	1–1.5m (3–5ft)	Small, broadly ovate	Small, white, scented	Flowers late spring, early summer and intermittently through summer into autumn	USDA 6a-1. RHS H7
S. meyeri 'Palibin' AGM	Dense and bushy	1–1.5m (3–5ft)	1–1.5m (3–5ft)	Small, broadly ovate	Small, lilac pink, scented	Flowers late spring and early summer	USDA 7b/8a. RHS H5

Syringa meyeri 'Palibin'. A very free-flowering, small shrub.

Syringa meyeri 'Palibin'. The flowers have the typical lilac fragrance.

Syringa × persica

TABLE 4.117 *SYRINGA × PERSICA* AND CULTIVAR

Name	Habit	Hardiness	Height	Spread	Foliage	Flowers
S. × persica (Persian lilac)	Slender branches. Bushy overall	USDA 6b/7a. RHS 6	1.2–2.4m (4–7ft)	1.5–3m (5–10ft)	Lanceolate, dark-green	Panicles 5–7cm (2–2¾in) long. Pale-pink, richly fragrant flowers in early summer
S. × persica 'Alba'	Slender branches. Bushy overall	USDA 6b/7a. RHS 6	1.2–2.4m (4–7ft)	1.5–3m (5–10ft)	Lanceolate, dark-green	Panicles 5–7cm (2–2¾in) long. White, richly fragrant flowers in early summer

Syringa pubescens microphylla 'Superba' AGM

TABLE 4.118 *SYRINGA PUBESCENS MICROPHYLLA* 'SUPERBA' AGM

Name	Habit	Hardiness	Height	Spread	Foliage	Flowers
S. pubescens microphylla 'Superba' AGM	Slender, branched but bushy overall	USDA 6b/7a. RHS 6	1.5–2.5m (5–8ft)	1.5–2.5m (5–8ft)	Leaves small, mid-green	Produced in loose panicles of pink, fragrant blooms. Mid to late spring

Syringa vulgaris Cultivars

Hardiness: All USDA 6b/7a. RHS H6.

All *Syringa vulgaris* cultivars grown as shrubs should be grown on their own roots. In the past, cultivars were budded or grafted on to *Syringa vulgaris* (common lilac) or *Ligustrum* (privet). In the garden this often led to the problem of many suckers arising.

As can be seen from Table 4.119, *Syringa vulgaris* cultivars eventually make large shrubs and occupy a lot of space. This does not stop them being very popular shrubs because of their beautiful flowers and scent. The problem of excessive size can be prevented to some extent by carrying out hard, rejuvenative pruning, once the plant has become too large. This involves cutting the plant hard down, to ground level if necessary, during the dormant season. In the following growing season strong regrowth from the base will occur, followed in the following growing season by a heavy crop of flowers.

Syringa vulgaris cultivars are also produced as plants top-worked on to an upright stem of *Syringa vulgaris* at 80cm (31in) or more above ground. This has the effect of converting the shrub into what is effectively a small tree. To avoid the problem of suckers arising from this rootstock, avoid damaging the roots at any time. This often occurs when forking through a border to improve its appearance or weeding but cultivating too deeply and causing root damage.

Whether grown as a shrub or small tree, they benefit from having old flower heads removed once they are finished. This prevents the plant diverting energy into seed production and encourages the formation of flower buds for the following season.

Garden Design Considerations

Despite being rather large for today's smaller gardens, cultivars of *Syringa vulgaris* remain very popular, just as they were in the recent past. This must be because of the very attractive and highly scented flowers. As well as being sold as traditional shrubs, they are now produced as 'standard' or so-called 'lollipops', enabling them to function as small trees. This has several advantages. As they are closer to head height, the flowers with their strong scent can be better appreciated. They can be used to give a dimension of height to what otherwise might be a rather undistinguished overall planting and leave space beneath them for underplanting.

TABLE 4.119 *SYRINGA VULGARIS* CULTIVARS

Name	Habit	Height	Spread	Foliage	Flowers
S. vulgaris 'Anderken an Ludwig Späth' AGM	Bushy	2.5–4m (9–13ft)	2.5–4m (9–13ft)	Heart-shaped. Mid-green	Single, deep-purple
S. vulgaris 'Katherine Havemeyer' AGM	Upright and bushy	2.5–4m (9–13ft)	2.5–4m (9–13ft)	Heart-shaped. Mid-green	Double, lavender-blue
S. vulgaris 'Charles Joly' AGM	Erect growing and bushy	2.5–4m (9–13ft)	2.5–4m (9–13ft)	Heart-shaped. Mid-green	Magenta, double
S. vulgaris 'Madame Lemoine' AGM	Bushy	2.5–4m (9–13ft)	2.5–4m (9–13ft)	Heart-shaped. Mid-green	Yellow buds opening to white double flowers
S. vulgaris 'Primrose' AGM	Bushy	2.5–4m (9–13ft)	2.5–4m (9–13ft)	Heart-shaped. Mid-green	Creamy-yellow
S. vulgaris 'Sensation' AGM	Rather loose growing but bushy	2.5–4m (9–13ft)	2.5–4m (9–13ft)	Heart-shaped. Mid-green	Purple/red, edged white

Another criticism heard about lilacs generally is that they only have a brief period in which they flower, but after this are rather dull to look at during the rest of the growing season. This is rather too true, since little or no autumn leaf colour is produced. Help is at hand!

The small overall size of *Syringa meyeri* cultivars makes them very well suited to small or compact gardens. As with common lilac, a limitation has been their short duration of flower in late spring. This may have been overcome by the recently introduced Flowerfesta Series of *S. meyeri*. This series flower in late spring/early summer, but also intermittently through the summer and into the autumn. Just how profuse this later flowering is still has to be judged and currently it is only *Syringa meyeri* 'Palibin' that holds the AGM.

The range of *Syringa meyeri* cultivars is now being produced as standards and dwarf standards. This makes both shrubs, and now the standards, very suitable for container and patio gardening where not only their compact size, but their extended flowering season and scent will be much appreciated.

Syringa × persica cultivars and *Syringa pubescens microphylla* 'Superba' are intermediate in size between the two previously mentioned groups and fit in equally well where this size of shrub is required.

VERONICA

All plants previously placed in the genus Hebe have now been transferred to the genus Veronica.

The leaves are arranged in four perpendicular rows in decussate pairs. The flowers are arranged in a spike. Lowland and coastal species tend to be large-leaved, those coming from higher altitudes have smaller leaves and alpine species have leaves reduced to scales and are said to have a whipcord appearance.

Growing Requirements

Hardiness: Generally, USDA RHS H4, a few are USDA RHS H5 and some are USDA RHS H3.
Aspect: Most will grow well in a site facing any direction, except north.
Shelter: They vary a little in the need for shelter. Most need a degree of shelter, especially from cold, drying winds. They will tolerate the milder, but still occasionally strong, coastal winds. It is low temperatures and desiccating winds they dislike. Although reliable for seaside planting, many are vulnerable to low winter temperatures experienced inland.
Sun: All like full sun or partial shade.
Soil: A soil of almost any texture is suitable, providing that it is well drained.
pH: A soil pH ranging from alkaline to neutral is accepted.

Veronica shrubs do not stay in a neat condition for longer than five to ten years, after which they tend to become rather untidy with gaps in their normally neat appearance. Rejuvenative pruning, cutting the shrubs hard back to where new young shoots can be seen at the base of the plant, can be done in late spring. The pruning should be followed by an application of fertiliser and mulching. This leaves a rather unattractive plant, but regrowth back to an attractive plant form is rapid. Alternatively, the plants can be regarded as disposable and just replaced, as vegetative propagation is very rapid and easy.

Garden Design Considerations

Veronicas form a large group of garden-worthy, evergreen shrubs. From the gardener's and designer's viewpoints they can be considered to fall roughly into two main groups: those grown mainly for their foliage and distinctive habit; and those grown mainly for their flowers, although a few combine both attributes. This is indicated in Table 4.120.

Hardiness is another factor that needs to be carefully considered when selecting hebes for inclusion in a garden or design scheme.

Those grown for their foliage and distinct habit tend to be hardier, although notable exceptions are the heavily variegated cultivars. The hardier hebes have a range of applications, from a single, isolated specimen grown for its distinctive, dense, regular shape, e.g. *Veronica rakaiensis*, which can substitute for clipped specimens of box *Buxus sempervirens*, to larger planting where their regular, distinctive habit can be used to form a pleasing pattern in a landscape. This group's degree of tolerance to pollution can be useful in industrial landscape plantings.

Those grown for their summer flowers or very pronounced variegation are most often less hardy. Their use is better limited to sheltered, warmer gardens, where low winter temperatures or cold, desiccating winds are not likely to be expected. The smaller, very floriferous cultivars or those with very bright variegation also find good uses in seasonal displays in patio plantings, raised beds, planters and window boxes. It should also be mentioned that all these summer-flowering shrubs are beneficial to insect pollinators.

All hebes can be freely used where low winter temperatures and cold desiccating winds are not expected. They excel in seaside plantings and the favoured gardens of south-west England and similarly favoured areas of western Scotland. Here, their resistance to strong, but not cold, winds and a degree of drought-resistance can be used with considerable advantage.

Veronica 'Blondie'. Very attractive coloured foliage.

Veronica 'Goldrush'[PBR]. Grows to produce a compact bushy plant that is grown for both its flowers and foliage.

Veronica odora.

Veronica odora. Produces a neat, rounded shrub.

Veronica 'Pink Elephant'. Used here to produce a contrast against the surrounding ericas.

Veronica 'Pink Elephant'. Showing its unusual habit, leaf texture and colour.

Veronica pinguifolia 'Pagei'. Useful for small-scale ground-cover, where its grey foliage is distinctive.

Veronica pinguifolia 'Sutherlandii'. Similar in colour to *V. pinguifolia* 'Pagei' but making a more rounded, compact shrub.

Veronica pinguifolia 'Sutherlandii'.

Veronica pimeleoides 'Quicksilver'.

Veronica pimeleoides 'Quicksilver'. Blue-grey leaves make an interesting combination with blue flowers.

Veronica rakaiensis. When young this species may be clipped to make a very neat ball of foliage.

Veronica albicans Recurva Group.. A little less regular in shape than other finer textured cultivars.

Veronica vernicosa. Makes a regular, mounded shrub.

Veronica 'Red Edge'.

Veronica vernicosa. The fine, regular, overall texture is shown.

TABLE 4.120 *VERONICA* SPECIES AND CULTIVARS

Name	Habit	Height	Spread	Leaves	Flowers	Remarks*
V. albicans AGM	Dense-growing dwarf rounded shrub	60cm (2ft)	90cm (3ft)	Thick glaucous green 1.5–3cm long leaves	White in summer	L. Prominent glaucous grey foliage. USDA 8b/9a. RHS H4
V. × *andersonii* 'Variegata'	Loose open habit	1.8m (6ft)	1.5–2.5m (5–8ft)	Leaves 10cm (4in) long. Leaves margined and splashed creamy-white	Lavender-blue fading to white in late summer	L. A very boldly variegated cultivar
H. 'Autumn Glory'	Loose, open habit	50–100cm (2–3ft)	50–100cm (2–3ft)		Many dense racemes of purple-blue flowers in late summer and autumn	F
V. 'Calcedonica' AGM (syn. *Hebe.* 'E.B. Anderson', *H.* 'Knightshayes')	Rounded bush	50–100cm (2–3ft)	50–100cm (2–3ft)	Dark-green	Violet. Late spring-early autumn	L
V. 'Emerald Gem' AGM (syn. *Hebe.* 'Green Globe', *H.* 'Emerald Green')	Rounded bush	10–50cm (4–20in)	10–50cm (4–20in)	Small, pointed	Small, white flowers. Summer	L. USDA 8b/9a RHS H4
V. 'Great Orme' AGM	Rounded and compact	1.2m (4ft)	1–1.5m (3–5ft)	Leaves long and pointed, 5–7.5cm (2–4in) long	Pink fading to white	F. USDA 8b/9a RHS H4
V. 'Goldrush'[PBR]	Compact and bushy	60cm (2ft)	60cm (2ft)	Each leaf 1.5cm (½in) long. Grey-green and yellow, edged in red	Pink in early summer	L/F
V. hulkeana	Small shrub with rather an open habit	60–100cm (2–3ft)	1m (3ft)	Glossy, green, toothed leaves	Large panicles of lavender-blue flowers in early summer	F. The most beautiful species that is mainly grown for its flowers
V. macrantha AGM	Upright, rather stiff and open	10–50cm (4–20in)	50–100cm (20– 40in)	Oval and bright-green	Each flower 2cm (1in) across, white, in early summer	L/F. USDA 8b/9a RHS H4
V. 'Margaret' AGM	Bushy, compact and rounded	10–50cm (4–20in)	10–50cm (4–20in)	Bright-green	Light-blue flowers in early summer	L/F. USDA 8b/9a RHS H4
V. 'Magic Summer'[PBR]	Compact and bushy	50–100cm (20–36in)	50–100cm (20–340n)	Small, narrow, white margined, turning red-purple in spring and early summer	Purple-blue flowers in summer	L/F. Note lack of winter hardiness. USDA 9b/10a. RHS H3
V. 'Midsummer Beauty' AGM	Rounded and upright	2m (6½ft)	1–1.5m (3–5ft)	Small with a narrow, white margin turning red-purple in spring and summer	Long, purple-blue racemes in summer	L/F. USDA 8b/9a RHS H4
V. 'Mrs Winder' AGM (syn. *Hebe. vernicosa* 'Waikiki')	Rather spreading	50–100cm 20–36in)	1–1.5m (3–5ft)	Narrow, dark-green, tinged with reddish-brown	Small violet-blue flowers in a raceme 7cm (3in) long	L/F. USDA 8b/9a RHS H4
V. 'Nicola's Blush' AGM	Bushy and rounded	50–100cm	1–1.5m	Lance-shaped 4cm (1.5in) long	Magenta-pink fading to white	F. USDA 8b/9a RHS H4
V. 'Neil's Choice' AGM (syn. *Hebe* 'Purple Emperor')	Bushy and rather lax	1–1.5m (3–5ft)	1–1.5m (3–5ft)	Narrow leaves	Dense racemes of violet flowers. Summer/early autumn	F. USDA 8b/9a RHS H4

Name	Habit	Height	Spread	Leaves	Flowers	Remarks*
V. ochracea 'James Stirling' AGM	Dense, low-growing	10–50cm (4–20in)	10–50cm (4–20in)	Reddish-yellow, scale-like leaves appressed to stem giving a 'whipcord' appearance	Small, white flowers in late spring	L. Very distinctive appearance, unlike other hebes. USDA 8b/9a RHS H4
V. odora 'Nana' (syn. *Hebe buxifolia* 'Nana')	Compact and rounded	30cm (1ft)	45cm (18in)	Green, crowded leaves	White in early summer	L. USDA7b/8a. RHS H5
V. 'Pascal' AGM	Dwarf shrub	10–50cm (4–20cm)	10–50cm (4–20cm)	Narrow, lance-like leaves. Leaves green, but turn purplish in winter	Violet-blue, early summer	L/F. USDA 8b/9a RHS H4
V. 'Pewter Dome' AGM	Compact and small	10–50cm (4–20cm)	10–50cm (20–40in)	Grey-green 'pewter-coloured' 2cm (1in) long	White flowers in dense spikes each 3cm (1in) long. Early summer	L. Neat, mounded shape. USDA 8b/9a RHS H4
V. pinguifolia 'Pagei' AGM	Very low-growing, dwarf shrub	30cm (1ft)	90cm (3ft)	Small and silvery-green	White in a dense spike. Late spring/early summer	L. Good for low, small scale ground cover. USDA 8b/9a RHS H5
V. pinguifolia 'Sutherlandii'	Small, compact, rounded shrub	45cm (18in)	45cm (18in)	Grey-green, small, 1.5cm (1in) long	Short spike of white flowers in early summer	L.
V. pimeleoides 'Quicksilver' AGM	Dwarf, spreading shrub	10–50cm (4–20in)	50–100cm (20–40in)	Bright blue-grey leaves, contrast against dark-coloured young shoots	Pale-lilac in spikes, 4cm (1.5in) long	L. USDA 8b/9a RHS H4
V. rakaiensis AGM	Compact and rounded	90cm (3ft)	1.5m (5ft)	Small, bright-green leaves	Small, white spike in early summer	L. Forms a very neat mound when young. USDA 8b/9a RHS H4
H. 'Red Edge' AGM	Compact and rounded	10–50cm (4–20in)	10–50cm (4–20in)	Oblong, blue-grey leaves, tinged red, especially in winter	Mauve-white	L. USDA 8b/9a RHS H4
V. 'Silver Queen' AGM (syn. *Hebe × franciscana* 'Variegata', *H. elliptica* 'Variegata')	Forms a dense mound	50–100cm (2–3ft)	50–100cm (2–3ft)	Broadly elliptic green with creamy-white variegation	Purple in short racemes. Mid-summer to autumn	L/F. Probably the most strikingly variegated cultivar but note lack of hardiness. USDA 9b/10a RHS H3
V. *topiara* AGM	Very compact	10–50cm (4–20in)	10–50cm (4–20in)	Grey-green leaves	Short, dense white racemes of flowers. Early to mid-summer	L. Dense neat growth. USDA 8b/9a RHS H4
V. verniculosa AGM	Bushy and compact	10–50cm (4–20in)	10–50cm (4–20cm)	Small, glossy, dark-green	White racemes, early summer	L. USDA 8b/9a RHS H4
V. 'Wiri Cloud' AGM	Bushy	10–50cm (4–20in)	10–50cm (4–20in)	Pale-green 1.5cm (1in) long leaves and yellow-green shoots	White flowers from pink buds in summer	L/F. USDA 8b/9a RHS H4
V. 'Youngii' AGM (syn. *Hebe*'Carl Teschner')	A spreading, dwarf shrub	20cm (8in)	50cm (20in)		Short racemes of violet-blue flowers in early summer	L/F. USDA 8b/9a RHS H4

* L = a plant is principally grown for its leaves or overall habit; F = a plant is grown principally for its flowers; L/F = where both flowers and foliage are important.

VIBURNUM

A large genus of approximately 200 species that are found mainly in the northern temperate areas of the world. They comprise a few trees and many shrubs, which may be evergreen or deciduous. The leaves are simple and produced in opposite pairs.

The white, or more seldomly pink, flowers are produced in panicles or cymes and are often followed by colourful fruits. In species or cultivars mainly grown for their fruits, it is often advisable to plant two or more plants together to enable cross-pollination to occur, which will result in greater fruit set.

Several species produce two distinct types of flowers. One is sterile and showy, whose function is to attract potential pollinators. The second is fertile, smaller and less showy. In three species plant breeders have managed to produce cultivars whose inflorescences consist entirely of sterile flowers and are more attractive from a horticultural viewpoint.

Only species and cultivars small enough to be suitable for small or compact gardens will be considered here.

Growing Requirements

Most viburnums are easy shrubs to grow, being undemanding in their requirements.
Hardiness: This ranges from USDA 6a-1 RHS H7 to USDA8b/9a, RHS H4 and is indicated for each plant in Table 4.121.
Aspect: A site of any orientation is acceptable.
Shelter: There is no special need for a sheltered position, except for the winter-flowering *Viburnum farreri* and *Viburnum* × *bodnantense*, where the plants are perfectly hardy, but the winter flowers will benefit from some shelter. *Viburnum davidii* is also hardy, but for its large evergreen foliage to over-winter in pristine condition, some shelter from physically damaging winds is advisable.
Sun: Full sun or part shade.
Soil: Soil of any texture is acceptable, but it must be well drained and moisture-retentive in summer.
pH: A very wide range of soil pH is accepted.

Garden Design Considerations

This one diverse genus offers a range of opportunities for the gardener and garden designer.

Winter Colour and Scent

Although rather large to be considered ideally suited to the small or compact garden, *Viburnum* × *bodnantense* cultivars and *V. farreri* and its cultivars are of tremendous value when planning for winter. Very few, if any, plants can offer the duration of winter colour and scent as these shrubs, flowering as they do right through winter. When frosted, those flowers currently out are killed and the display ends, only to start again in the next mild spell when more buds open! This can be repeated several times in a winter. Perhaps *V. farreri* 'Nanum' should be chosen to save space?

Late Spring and Early Summer

This is the peak season for the flowering of many viburnums. You can select from the list, depending on the size of shrub required. Some will provide vivid white flowers, some a hint of pink and others, such as *V.* × *juddii*, wonderful scent in a compact, space-saving shrub.

Autumn and Winter

Again, the viburnums have a lot to offer. The various selected cultivars of *V. plicatum* start with flowers in late spring and early summer, then provide colour in autumn from colourful leaves, plus fruits and later in winter the distinct tiered habit provides winter interest. A powerful incentive for the gardener to find some space for these multi-seasonal plants.

Evergreens

The two compact evergreen cultivars of *V. tinus* have several uses in the garden. In winter, they provide the vital structure needed, plus flowers. Although compact, they can be used to form effective dividers of garden spaces and clipped to form dense hedges. Such compact dark-green plants can also function as backdrops to small, light-coloured, winter and early spring beauties, such as *Galanthus* (snowdrops) and *Iris reticulata* cultivars.

Evergreen *Viburnum davidii* is much employed by garden designers because of its range of attributes. As an isolated specimen, its dark, broad, coarse-textured foliage can be contrasted against fine-textured shrubs, such as a clipped ball of box (*Buxus sempervirens*) or the regular fine texture of *Veronica rakaiensis*. Its low-growing, evergreen nature also lends itself to employment as ground cover on a small or larger undulating scale.

Viburnum × *bodnantense* 'Charles Lamont'. This cultivar has flowers that are slightly paler pink than *V.* × *bodnantense* 'Dawn' AGM.

Viburnum × *bodnantense* 'Deben'. A large, old specimen plant.

Viburnum × *bodnantense* 'Deben'. Showing the sheer profuseness of flowering of these *V.* × *bodnantense* cultivars.

Viburnum × *bodnantense* 'Dawn'. A close-up of the deep-pink, sweetly scented flowers.

Viburnum × *bodnantense* 'Deben'. The flowers become almost pure-white once opened.

Viburnum davidii. The evergreen, broad, deeply veined, deep-green leaves of this species. The blue fruit are only produced by a female plant once it has been pollinated and fertilised by the pollen from a male plant.

Viburnum davidii. The flowers of a male plant. If fruiting is required, then a planting of female mixed with male plants must be ensured.

Viburnum farreri. This plant is beginning to flower in autumn before all the leaves have fallen. Winter-flowering viburnums flower on and off during the winter, flowering corresponding with mild spells.

Viburnum farreri 'Candidissimum'. The beautiful white-flowered cultivar.

Viburnum farreri 'Nanum'. More compact than the species, this cultivar is useful where space is limited.

Viburnum × juddii. A medium-sized, deciduous, spring-flowering shrub.

Viburnum × juddii. The numerous flowers are deliciously and heavily scented.

Viburnum lantana. The large heads of slightly off-white flowers are produced in early summer.

Viburnum opulus. Heads of white flowers are produced in late spring.

Viburnum opulus. Bright-red fruits and red leaf colour are developed in autumn. The cultivar *Viburnum opulus* 'Compactum' produces the flowers, fruit and autumn leaf colour in just the same manner, but on a much more compact plant. Very useful in the small garden.

Viburnum plicatum f. *tomentosum* 'Mariesii'. A single plant in full flower in early summer.

Viburnum plicatum f. *tomentosum* 'Mariesii'. Close-up of the flowers.

Viburnum plicatum f. *tomentosum* 'Mariesii'. A plant displaying autumn leaf colour.

Viburnum plicatum f. *tomentosum* 'Mariesii'. Detail of the autumn leaf colour.

Viburnum plicatum f. *tomentosum* 'Mariesii'. In winter, especially when the plant is encased in frost, the semi-tabulate habit of the plant is well displayed.

Viburnum plicatum f. *tomentosum* 'Pink Beauty'. The plant in full flower. The habit of this cultivar is more upright than *V.* 'Mariesii'.

Viburnum plicatum f. *tomentosum* 'Pink Beauty'. Close-up of the flowers. Flower colour is variable from year to year, not always pink, sometimes white.

Viburnum plicatum f. *tomentosum* 'Pink Beauty'. A plant displaying autumn leaf and fruit colour.

Viburnum plicatum f. *tomentosum* 'Pink Beauty'. The fruit is produced on upright, short stalks displaying it to advantage.

Viburnum tinus 'Gwenllian'. This is an evergreen, compact and free-flowering cultivar of the species *V. tinus*. Useful where space is limited or where a low, flowering hedge is required.

TABLE 4.121 SPECIES AND CULTIVARS OF *VIBURNUM*

Name	Habit	Height	Spread	Leaves	Flowers	Remarks
V. × *bodnantense* 'Charles Lamont' AGM	Initially upright, broadening with age	1.5–2.5m (5–8ft)	1–1.5m (3–5ft)	Deciduous. Green, ovate with red petioles	Buds dark-pink opening to bright-pink. Sweetly scented from autumn to spring	USDA 6b/7a RHS H6
V. × *bodnantense* 'Dawn' AGM	Initially upright, broadening with age	1.5–2.5m (5–8ft)	1–1.5m (3–5ft)	Deciduous. Green, ovate with red petioles	Buds dark-pink opening to light-pink. Sweetly scented, from autumn to spring	USDA 6b/7a RHS H6
V. × *bodnantense* 'Deben' AGM	Initially upright, broadening with age	1.5–2.5m (5–8ft)	1–1.5m (3–5ft)	Deciduous. Green, ovate with red petioles	Buds pink opening white. Sweetly scented from autumn to spring	USDA 6b/7a RHS H6
V. davidii AGM	Low, rounded mound	1.5m (5ft)	1.5m (5ft)	Evergreen, large, 15cm (6in) long, shiny oval leaves each deeply three-veined	Flattened clusters of small, white flowers in late spring followed by blue fruits	USDA 7b/8a RHS H5. Both male and female plants are needed to obtain fruits. Only some suppliers will offer sexed plants
V. 'Eskimo' AGM	Rounded bush	1–1.3m (3–4ft)		Semi-evergreen, dark-green leathery leaves	Rounded heads of white, unscented flowers each 7cm (2.5in) across in early summer	USDA 7b/8a RHS H5
V. farreri AGM (syn. *V. fragrans*)	Initially upright, broadening with age. Becoming very dense if unpruned	1.5–2.5m (5–8ft)	1.5–2.5m (5–8ft)	Deciduous, oval leaves	Clusters of pink buds opening to light-pink, scented flowers. From autumn to spring	USDA 6b/7a RHS H6 Flowers less profuse than V. × bodnantense cvs
V. farreri 'Candidissimum'	Initially upright, broadening with age. Becoming very dense if unpruned	1.5–2.5m (5–8ft)	1.5–2.5m (5–8ft)	Deciduous, oval leaves	Clusters of green buds opening to white, scented flowers. From autumn to spring	USDA 6b/7a RHS H6
V. farreri 'Nanum'	Upright and dense	90cm–1.20m (3–4ft)	1–1.6m (3–5ft)	Deciduous, oval leaves	Clusters of pink buds opening to light-pink scented flowers. From autumn to spring	USDA 6b/7a RHS H6. A useful compact form. Slightly less profuse in flowering
V × *juddii*	Bushy and compact	1–1.5m (3–5ft)	1–1.5m (3–5ft)	Deciduous, dark-green, broad, ovate leaves	From mid to late spring produces, from deep-pink buds, sweetly scented, white flowers in domed inflorescences	USDA 6b/7a RHS H6. Smaller and more compact than the larger and less compact, but also spring flowering, sweetly scented *V.* × *burkwoodii* and *V. carlesii*
V. lantana (wayfaring tree)	Upright	2.5–4m (8–13ft)	2.5–4m (8–13ft)	Deciduous, broadly ovate with a dense tomentum (tiny hairs) on underside	Dense head of white flowers in early summer followed by red fruits, which mature to black	USDA 6a-1 RHS H7. Too large to be considered compact and only included for growth as a hedge since its height can be maintained at 1m (3ft) by clipping

Name	Habit	Height	Spread	Leaves	Flowers	Remarks
V. opulus 'Compactum' AGM	Initially upright	1–1.5m (3–5ft)	1–1.5m (3–5ft)	Deciduous. Leaves green and palmately lobed. Autumn colour pink or red	Flat heads of fertile flowers surrounded by larger sterile flowers. Followed by red translucent berries	UDA 6b/7a RHS H6. A compact cultivar of the Guelder rose. Native to Europe
V. plicatum f. *tomentosum* KILIMANJARO ('Jww1'[PBR])	Pyramidal with layered branches	2.5–4m (8–13ft)	2.5–4m (8–13ft)	Deciduous. Green, deeply veined leaves that turn purple in autumn	White 'lacecap' inflorescences. Followed by red berries that mature to black. Scattered flower heads may appear later in summer	USDA 6b/7a RHS H6
V. plicatum f. *tomentosum* KILIMANJARO SUNRISE ('Jww5'[PBR])	Tiered branches	2.5–4m (8–13ft)	2.5 –4m (8–13ft)	Deciduous. Green, deeply veined leaves that turn purple in autumn	White tinged pink 'lacecap' inflorescences. Followed by red berries that mature to black. Scattered flower heads may occur later in summer	USDA 6b/7a RHS H6. Chelsea Plant of the Year 2015 and People's Choice Plant of the Decade
V. plicatum f. *tomentosum* SUNRISE	Bushy and tiered	2.5–3.5m (8–11½ft)	2.5–4m (8–13ft)	Deciduous. Dark-green, pleated, producing purple autumn colour	White inflorescences produced in a double row at each leaf node along each branch. 'Lacecap' flowers with fertile flowers in the centre and larger sterile flowers outside. Produced in late spring, early summer	USDA 6b/7a RHS H6. Shrub slightly wider and less tall than *V.* 'Pink Beauty'
V. plicatum f. *tomentosum* 'Pink Beauty'	Bushy and tiered	2.5–4.5m (8–15ft)	2.5–3.5m (8–12ft)	Deciduous. Dark-green, pleated producing purple autumn colour	White, aging to pink inflorescences produced in a double row at each leaf node along each branch. 'Lacecap' flowers with fertile flowers in the centre and larger sterile flowers outside. Produced in late spring, early summer	USDA 6b/7a RHS H6. Attractive red fruit produced in same manner as previous inflorescences
V. sargentii 'Onodaga' AGM	Upright bush	1.5–2.5m (5–8ft)	1–1.5m (3–5ft)	Deciduous. Palmate 3-lobed leaves each 12cm (5in) long. Bronze-purple when young, developing to green, finally developing purple autumn colour	In early summer produces 'lacecap' inflorescences with the red buds of the fertile central flowers surrounded by larger sterile white flowers. Red fruit in autumn	USDA 6b/7a RHS H6
V. tinus 'Eve Price' AGM	Dense bush	1.5–2.5m (5–8ft)	1.5–2.5m (5–8ft)	Evergreen. Oval, dark-green leathery leaves	Flattened heads of deep-pink buds surrounded by white flowers form early winter to spring. May be followed by blue fruits	USDA 8b/9a RHS H4. A compact form of *Viburnum tinus* laurustinus
V. tinus 'Gwenllian' AGM	Dense bush	1.5–2.5m (5–8ft)	1.5–2.5m (5–8ft)	Evergreen. Oval dark-green leathery leaves	Flattened heads of deep-pink buds surrounded by white flowers form early winter to spring. May be followed by blue fruits	USDA 8b/9a RHS H4. A compact form of *Viburnum tinus* laurustinus

WEIGELA

A genus of only approximately ten species of hardy deciduous shrubs from temperate Eastern Asia. They have simple, ovate leaves and on the previous year's growth produce tubular flowers 2–4 cm (¾–1½in) long, with a five-lobed corolla in early summer. As a group they are easily grown.

Growing Requirements

Hardiness: USDA 7b/8a or 6b/7a. RHS H5 or 6.
Aspect: A site of any orientation is acceptable.
Sun: Full sun or part shade.
Soil: Soil of any texture is acceptable, providing that it is well drained.
Shelter: The growing position can be either exposed or sheltered.
pH: Within a very wide range, the plants are not sensitive to soil pH.
Pruning: They are essentially low-maintenance shrubs. To maintain youthfulness and good flower performance, after flowering prune about 20 per cent of the oldest stems down to ground level.

Garden Design Considerations

The older, slightly larger cultivars, such as *Weigela* 'Bristol Ruby', are excellent plants for providing a splash of early summer flower colour and, in some cases, more extended variegated or coloured foliage interest. They are tolerant of non-ideal conditions and are a good choice for exposed gardens or new gardens in the process of being established.

More compact cultivars, especially those with an extended flowering period, such as the All Summer Series, are well suited to where space is limited and where colour and interest are required for a more extended period. Such smaller, compact, colourful cultivars are ideal for the patio garden and in raised beds and containers.

Weigela middendorffiana and its cultivar 'Mango', whilst hardy, are somewhat more demanding in their growing requirements. Provided these can be satisfied, they will produce an attractive, if rather short-lived, display of an unusual flower colour.

Weigela ALL SUMMER PEACH ('Slingpeach'PBR). Showing how compact this cultivar is.

Weigela ALL SUMMER PEACH ('Slingpeach'PBR). The flowers are produced over a long period throughout the summer.

Weigela ALL SUMMER RED ('Slingcol'PBR). Equally as long flowering and as compact as *W.* All Summer Peach.

Weigela BLACK AND WHITE ('Courtacad 1'PBR). A very compact cultivar with flowers that contrast very strongly against the dark foliage.

Weigela BRIANT RUBIDOR ('Olympiade'). There is a strong colour contrast between the yellow-green leaves and the deep red flowers.

Weigela florida 'Folies Purpureis'. An established plant of this cultivar.

TABLE 4.122 *WEIGELA* SPECIES AND CULTIVARS

Name	Habit	Height	Spread	Leaves	Flowers	Remarks
Weigela ALL SUMMER PEACH 'Slingpink'PBR	Bushy	75cm (30in)	75cm (30in)	Mid-green	Pink. Late spring to early autumn	H5. Compact with a very long flowering period
Weigela ALL SUMMER RED 'Slingcol'PBR	Bushy	75cm (30in)	75cm (30in)	Mid-green	Reddish pink. Late spring to early autumn	H5. Compact with a very long flowering period
Weigela BLACK AND WHITE 'Courtacad1'PBR	Bushy and compact	40–50cm (16–20in)	60–80cm (24–32in)	Dark-green	White. Spring to early summer	H6
Weigela Briant Rubidor 'Olympiade'	Bushy	1.5m (5ft)	1.4m (4½ft)	Yellow-green	Deep red. Early summer	H5. Strong colour contrast between foliage and flowers
Weigela 'Bristol Ruby'	Bushy and upright	1.5–2.5m (5–8ft)	1.5–2.5m (5–8ft)	Mid-green	Dark crimson. Early summer	H6
Weigela CARNAVAL 'Courtalor' AGM	Bushy	1–1.25m (3–4ft)	1–1.25m (3–4ft)	Mid-green	Red, pink, and white. Early summer and again later	H6
Weigela coraeenis AGM	Bushy, tall	2.5–4m (8–13ft)	2.5–4m (8–13ft)	Bright-green	Cream aging to pink	H6 A large shrub.
Weigela florida 'Alexandra'PBR AGM	Bushy	1.8m (6ft)	2.0m (6½ft)	Purple tinted becoming darker later in season	Reddish pink. Early summer	H6
Weigela florida 'Folies Purpureis'	Compact	0.5–1m (20–40in)	1–1.5m (3–4ft)	Dull purple green	Outside pink, paler inside trumpet. Early summer	H6
Weigela florida MONET 'Verweig'PBR	Compact	60cm (2ft)	60cm (2ft)	Variegated. Green centre to leaf with a cream margin and flushed pink	Pink. Early summer	H6
Weigela 'Florida Variegata'	Bushy	1.5–2.5m (5–8ft)	1.5–2.5m (5–8ft)	Light-green leaves edged with creamy-white	Pink. Early summer	H6
W. middendorffiana	Bushy, upright	1–1.5m (3–5ft)	1–1.5m (3–5ft)	Bright-green	Bell-shaped, pale-yellow with throat marking, varying from yellow to red. Early summer	H6. Not north-facing. A sheltered and part shaded site is necessary
W. middendorffiana 'Mango'	Bushy	1–1.5m 3–5ft)	1–1.5m (3–5ft)	Bright-green	Bell-shaped, mango-orange coloured. Early summer	H6. Not north-facing. A sheltered and part shaded site is necessary

Weigela florida 'Folies Purpureis'. The distinct, purple-green foliage of this compact cultivar is clearly seen.

Weigela 'Florida Variegata'. This older cultivar produces a larger, more open-growing shrub than newer cultivars, but is still very popular.

CHAPTER 5

DIRECTORY OF SUGGESTED CONIFERS

In some cases, garden design considerations are given for individual genera. In addition, all genera are all considered under general design considerations at the end of the chapter.

INITIAL OBSERVATIONS ON CONIFERS

Conifers are a very large group of plants consisting of mainly evergreen trees distributed predominately in the temperate and subtropical regions.

Their leaves vary considerably, for example:

- Bundles of long needles as in pines (*Pinus*).
- Short, pointed leaves of the firs (*Abies*), spruces (*Picea*) and junipers (*Juniperus*).
- Small, scale-like leaves of cupressus (*Cupressus*), false cypresses (*Chamaecyparis*), *Xanthocyparis* and arborvitae (*Thuja*).

To add to this complexity, when grown from seed conifers show a sequential development on their foliage, often having juvenile, semi-juvenile and adult foliage forms. Occasionally, seedlings will become 'stuck' at a certain stage and continue to grow whilst exhibiting a juvenile or semi-juvenile foliage form and not moving on to producing the normal next stage of foliage. These so-called 'juvenile fixations' can be attractive in their own right and have been propagated by horticulturalists to produce new cultivars.

In addition, there are several genera of deciduous conifers, of which *Ginkgo* is included here.

The flowers of conifers are considered primitive and usually take the form of catkin-like structures called strobili. The separate male and female strobili are borne on the same tree or different trees. Pollination is by wind. The fruit containing the seeds is usually a woody cone but can be a fleshy fruit as in yew (*Taxus*) and juniper (*Juniperus*).

Cones can be an extremely decorative feature of a conifer tree. Cones are only produced when a conifer tree reaches sexual maturity and, in the case of many, this does not occur until the tree is several years old and large. In a few species, cone production will start with a small, young tree; some dwarf or slow-growing cultivars will cone whilst small.

DWARF AND SLOW-GROWING CONIFERS

There are relatively few true dwarf conifers. They may occur as mountain forms of normally tall species and a few species are naturally low-growing, e.g. the creeping forms of *Juniperus horizontalis*.

Slow-growing forms have arisen over time, occurring in a number of ways, including:

- Seedling variation, e.g. juvenile fixations.
- Sports or mutations, which appear sporadically and naturally. These abnormal growths being noticed and propagated by horticulturalists.
- Witches' brooms. Congested abnormal growths that occur and again are propagated to produce new cultivars.
- Culti-variants. These may occur when cuttings are taken from different parts of a conifer tree. A cutting taken from a horizontal branch from a normally upright tree may grow on to produce a cultivar with an abnormal horizontal and perhaps slow rate of growth.

ABIES

The silver firs consist of about fifty species of evergreen trees distributed over the northern hemisphere. The species all have an erect, symmetrical habit with branches coming off the main stem in whorls or tiers, one being produced each year. The leaves are linear and have two white bands of stomata on their undersides. They are always spirally arranged, but this may appear to be flattened into two rows. The female cones are erect and disintegrate before falling.

Growing Requirements

Hardiness: USDA 6b/7a or 6a-1. RHS 6 or 7.
Aspect: A site facing any orientation is acceptable.
Shelter: A sheltered site is required for those listed in Table 5.1. They prefer a moist climate.
Sun: Full sun is required.
Soil: The soil should be well drained, retain moisture in the summer and can be of any texture.
pH: They dislike lime and a soil with a high pH should be avoided.

Pinus contorta 'Chief Joseph'. A slow-growing cultivar showing the bright orange-yellow needle colour that intensifies as the cold strengthens during winter.

TABLE 5.1 *ABIES* SPECIES AND CULTIVARS

Name	Habit	Height	Spread	Foliage	Cones	Remarks
A. balsamea Hudsonia Group	Very dense, dwarf shrub with a rounded top	0.5–1.5m (20in–5ft)	1–1.5m (3–5ft)	Dark-green, linear	No cones are produced	A true dwarf and very slow-growing
A. koreana AGM	Pyramidal	12m (40ft)	4–8m (12–26ft)	Dark-green, linear, 1–2cm (½–¾in) long	Upright violet or purple cones 5–7 cm (2–2¾in) long	Cones are produced even on very young trees. The tree is slow-growing but may eventually become too large for the space available
A. koreana 'Silberlocke' AGM	Small, bushy tree	2.5–4m (8–15ft)	1–1.5m (3–5ft)	Dark-green, but leaves twisted so the silvery undersides are revealed	Upright violet or purple cones 5–7cm long	Cones are produced on very young trees
A. nordmanniana (Caucasian or Nordmann fir)	Large tree forming a columnar crown	12m+ (40ft)+	4–8m+ (16–28ft)	Dark-green	Small, reddish when young	Often sold as 'Christmas trees', but not suitable for small gardens
A. nordmanniana subsp. *nordmanniana* 'Golden Spreader' AGM	Flattened shrub, wider than tall	0.5–1m (20–40in)	1–1.5m (3–5ft)	Bright, golden-yellow, more so in winter		Slow-growing, spreading

Abies balsamea Hudsonia Group. A very small tree that is already several years old.

Abies koreana 'Icebreaker'. The leaves grow bent upwards, displaying the very prominent silvering on their undersides.

Abies Koreana 'Silberlocke' showing the very beautiful, young female cones.

Abies koreana 'Silberlocke' with old and new cones. The old cones have disintegrated, as they do naturally, leaving just the central stem to which the ovuliferous (seed-bearing) scales were attached.

Abies koreana 'Silver Shadow', with silvering prominent on the underside of the leaves, plus grey-blue cones.

Abies nordmanniana 'Golden Spreader'. Showing the cultivar's layered habit and bright, new, yellow growth.

Cedrus atlantica 'Glauca Pendular'. This picture shows what fantastic shapes this cultivar can be grown into. It is excellent for training up and over a tall gateway.

Abies nordmaniana. A young tree prepared for sale as a 'Christmas tree'.

CEDRUS

The cedars are a small genus of four species. As young trees they have a conical shape, but over time become large trees with massive trunks and characteristic horizontal branching. The leaves are narrow and needle-like. The branches are of two kinds. Leading shoots, which can extend considerably and bear single needles in extended spirals and short shoots. Short shoots only extend a little, less than 2.5cm (1in) per year and the needles are crowded into rosettes.

Cones are produced in the autumn. The male cones are upright and bright-yellow and produce masses of buoyant pollen. The female cones are erect, stout, barrel-shaped and take two years to develop. The female cones finally disintegrate on the tree.

As a genus, it has produced relatively few dwarf or slow-growing forms suitable for use where space is limited.

Growing Requirements

Hardiness: USDA RHS H6.
Aspect: A site with any orientation is acceptable.
Shelter: The site may or may not be sheltered, but top-worked standard specimens are better in a sheltered situation.
Sun: Full sun is required.
Soil: The soil can be of any texture providing that it is well drained.
pH: A wide range of soil pH is tolerated.

Cedrus atlantica 'Glauca Pendular' AGM

Such a remarkable blue, needled, weeping cultivar of the Atlantic cedar that it deserves this space on its own. Left to its own devices, it just sprawls over the ground; grown over a garden bank, it can look very attractive. A main stem can be trained upward to a desired height and then allowed to adopt its pendular habit, whereupon it becomes a feature weeping tree. Another technique is to train it up and over an archway and a cascading veil of growth is formed. It grows 20–40cm (8–16in) per year and can be pruned as required. A truly very adaptable, versatile and useful plant.

Cedrus deodara

TABLE 5.2 CULTIVARS OF *CEDRUS DEODARA*

Name	Habit	Height	Spread	Leaves	Remarks
C. deodara 'Feelin' Blue' AGM	Almost prostrate, forms a mounded shape	0.1–0.5m (4–20in)	1.5–2.5m (5–8ft)	Blue-green needles	Is available as a grafted standard tree. Slow-growing
C. deodara 'Golden Horizon'	Semi-prostrate with weeping branches	0.4–0.6m (16in–24in)	1.5–1.8m (5–6ft)	Green in summer, yellow in winter	May also produce vertical shoots, which if not removed will cause the plant to grow into a pyramidal tree. Height and spread are difficult to predict because of variable growth habit

CHAMAECYPARIS

The false cypresses are a genus of evergreen trees consisting of five species: three from North America and two from Japan and Taiwan. The adult foliage has small, opposite leaves, densely arranged and scale-like. The ends of branches are flattened and frond-like. Male and female strobili are produced on the same tree. The male is very small and coloured red or yellow. The female cones are small, rounded and composed of six to twelve scales.

Chamaecyparis lawsoniana (Lawson Cypress)

A large, cone-shaped tree with drooping branches and broad bands of foliage in flattened sprays. This species has proved useful for hedging and screening. A large number of cultivars have arisen through seedling variation and, to a lesser extent, form branch sports. The cultivars are here divided into dwarf, slow-growing trees and larger cultivars principally used for hedging and screening.

Growing Requirements

Hardiness: USAD 6a-1. RHS H7.
Aspect: A site of any orientation is acceptable.
Shelter: Shelter is not required.
Sun: Full sun.
Soil: A wide range of soil types are accepted, but the drainage must be good.
pH: Slightly acid soils are preferred, but some degree of alkalinity is tolerated.

Chamaecyparis obtusa

The hinoki cypress from Japan. The species is a large, evergreen conifer with a broadly conical habit. The branches spread out horizontally from the main stem. The small, scale-like leaves are produced in unequal pairs. The species is seldom grown as an ornamental conifer, but it has given rise to a number of miniature and slow-growing forms.

Growing Requirements

Hardiness: USDA RHS H7.
Aspect: A site of any orientation, except north-facing, is acceptable.
Shelter: The site should be sheltered, especially from cold winds in winter.
Sun: Full sun or partial shade is required.
Soil: The soil should be well drained, but moisture-retentive.
pH: A soil with an acid or neutral pH is preferred.

Chamaecyparis lawsoniana 'Ellwoodii'. An example of just how large this 'slow-growing' conifer can get.

Chamaecyparis lawsoniana 'Ellwoodii'. The feathery, semi-juvenile foliage of this cultivar.

Chamaecyparis lawsoniana 'Little Spire'. The distinctive, upright habit of this slow-growing cultivar.

Chamaecyparis lawsoniana 'Little Spire'. This cultivar produces, every spring, a very large number of conspicuous, red, male cones, which give the plant a very unusual overall appearance.

TABLE 5.3 DWARF AND SLOW-GROWING CULTIVARS OF *CHAMAECYPARIS LAWSONIANA*

Name	Habit	Height	Spread	Foliage	Remarks
C. lawsoniana 'Ellwoodii'	Columnar	4–8m (13–26ft)	1.5–2.5m (5–8ft)	Semi-juvenile, fine. Grey-green	Fairly slow-growing, but if care not taken will outgrow space
C. lawsoniana 'Ellwood's Gold'	Small and conical	2.5–4m (8–13ft)	0.5–1m (1.6–3ft)	Semi-juvenile, fine. Grey-green with yellow tips	Slow-growing
C. lawsoniana 'Gnome'	Bun-shaped dwarf	30cm (1ft)	35cm (1.2ft)	Semi-juvenile. Deep green	Grows less than 2.5cm (1in) per year
C. lawsoniana 'Green Globe' AGM	Rounded dwarf	0.1–0.5m (4–20in)	0.1–0.5m (4–20in)	Dense flattened sprays	Remove coarser growth if this should suddenly appear
C. lawsoniana 'Minima Aurea' AGM	Dense, compact cone	1m (3ft)	50cm (20in)	Adult foliage. Fan shaped sprays of yellow-green, becoming stronger yellow in winter	
C. lawsoniana 'Snow White' AGM	Dwarf, compact	1.2m (4ft)	70cm (1.25ft)	Semi-juvenile. Tips white when young later blue green	
C. lawsoniana 'Tamariscifolia'	Wide-spreading, flat-topped	60cm (2ft)	1.2m (4ft)	Blue green and largely horizontal	A slow-growing, but ultimately large shrub

TABLE 5.4 DWARF AND SLOW-GROWING CULTIVARS OF *CHAMAECYPARIS OBTUSA*

Name	Habit	Height	Spread	Foliage	Remarks
C. obtusa 'Fernspray Gold' AGM	Rather irregular	1.5m (5ft)	1–1.5m (3–5ft)	Fern-like foliage gold-yellow tipped with gold in autumn	
C. obtusa 'Kosteri' AGM	Flat-topped when young developing into an irregular shape	40–60cm (1.25–2ft)	30–45cm (1–1½ft)	Glossy, dark-green, bronzed in winter	
C. obtusa 'Nana Gracilis' AGM	Forms an irregular, dense cone	70–90cm (2.25–3ft)	60–70cm (2–2.25ft)	Shell-like sprays of green foliage with silvery markings beneath	Slow-growing
C. obtusa 'Nana Lutea' AGM		45cm (1½ft)	20–40cm (10–20in)	Bright-yellow sprays of foliage	Very slow-growing

Chamaecyparis obtusa 'Fernspray Gold'. Showing the overall habit of the plant.

Chamaecyparis obtusa 'Fernspray Gold'. Showing the sprays of coloured foliage.

Chamaecyparis obtusa 'Nana Gracilis'. An extremely popular cultivar.

Chamaecyparis obtusa 'Nana Gracilis'. Here, the distinct shell-like sprays of foliage can be clearly seen.

Chamaecyparis pisifera

The sawara cypress from Japan. An evergreen conifer, which in Japan grows to 21–30m (70–100ft). The tree is broadly conical in habit, bearing branches of horizontally flattened sprays of dark-green foliage. The leaves are small, scale like and pointed. Female cones are brown, pea-sized and about 6mm (¼in) in diameter. It is seldom seen in cultivation.

Growing Conditions

Hardiness: USDA 6a-1. RHS H7.
Aspect: A site of any orientation is acceptable.
Shelter: The site can be sheltered or exposed.
Sun: Full sun.
Soil: A wide range of soil textures are accepted, but the soil must be well drained and retain adequate moisture in the summer.
pH: Acid or neutral soils are preferred, but slightly alkaline soils will be tolerated.

Chamaecyparis pisifera 'Boulevard'. A good-sized mature plant.

Chamaecyparis pisifera 'Filifera Aurea'. This slow-growing cultivar will grow into a medium-sized but shapely tree.

Chamaecyparis pisifera 'Boulevard'. The characteristic silvery foliage is shown, as is this cultivar's rather unfortunate feature of the older foliage tending to die and then remain. This can lead to the plant looking rather poor as it matures.

Chamaecyparis pisifera 'Filifera Aurea'. The branches produce light-yellow, slightly pendulous, rather thin growth, which is developed in overlapping layers and looks most attractive.

TABLE 5.5 DWARF AND SLOW-GROWING FORMS OF *CHAMAECYPARIS PISIFERA*

Name	Habit	Height	Spread	Foliage	Remarks
C. pisifera 'Boulevard' AGM	Pyramidal when young, becoming conical when older	1.5–1.8m (5–6ft)	60–90cm (24–36in)	In the Squarrosa Group and has semi-juvenile finely divided foliage, which is silver-grey	May carry unattractive old dead foliage when older, but this can be largely removed by careful pruning
C. pisifera 'Compacta'	Bun-shaped	20cm (8in)	50cm (20in)	Normal adult foliage produced in light-green sprays	Stays very compact
C. pisifera 'Filifera Aurea' AGM	Starts low and spreading but develops into a broad pyramid	90–1.20cm (3–4ft)	90–1.20cm (3–4ft)	In the Filifera Group with golden-yellow thread-like foliage	Ultimately will get larger but can be pruned to limit size, but retain attractive shape and habit
C. pisifera 'Filifera Nana'	Low-spreading mound	40–60cm (18–24in)	60–90cm (24–36in)	Green and thread-like	

Cryptomeria japonica

A monotypic genus of which *Cryptomeria japonica* is the only species. It is native to Japan and China where it forms a large, evergreen, pyramidal tree growing up to 45m (150ft) tall. Male and female strobili are produced on the same tree. The female cones are globular 1–3cm (½–1¼in) in diameter and brown at maturity. The adult foliage consists of awl-shaped, narrow leaves, 0.5–1.5cm (¼–¾in) long, spirally arranged and appressed to the branches.

The species is a very important timber tree, especially in Japan, but elsewhere seldom grown as an ornamental specimen tree.

Unusually for a conifer, this species will coppice successfully and such trees, allowed to only get a few years old after coppicing and then cut down again, are a feature of some Japanese gardens where they make very striking vertical features.

Growing Requirements

Hardiness: USDA RHS H6. Note that the forms mentioned in Table 5.6 tend to be less hardy and require shelter from cold winds in winter.

Shelter: A site of any orientation, except north-facing, is acceptable. A sheltered site is required, especially from cold, desiccating winds in winter.

Sun: Full sun or partial shade is required.

Soil: The soil can be of any texture but should be well drained and retain moisture in the summer.

pH: A wide pH range is accepted, but acid to neutral is preferred.

Note that whilst all the selections in Table 5.6 are smaller than the original species, some, given time, can attain moderate size, whilst others remain very small or dwarf.

Cryptomeria japonica 'Globosa Nana'. A young plant forming the typical, slow-growing, irregular dome shape.

Cryptomeria japonica 'Globosa Nana'. Detail of branches and foliage.

Cryptomeria japonica 'Globosa Nana'. A fine old specimen of this cultivar.

Cryptomeria japonica 'Cristata' (syn. 'Sekkan-sugi'). A young tree showing the slightly pendulous habit.

Cryptomeria japonica 'Cristata' (syn. 'Sekkan-sugi'). Showing the yellowish-green of the young foliage.

Cupressus macrocarpa 'Goldcrest'. Note how this cultivar can grow into a medium-sized tree quite quickly.

TABLE 5.6 CULTIVARS OF *CRYPTOMERIA JAPONICA*

Name	Habit	Height	Spread	Foliage	Remarks
C. japonica 'Bandai-sugi' AGM	Irregular with a rounded outline	1.5–2.5m (5–8ft)	1–1.5m (3–5ft)	Very dense foliage that is green in summer and bronzed in winter	Slow-growing. Often used as a subject for bonsai
C. japonica Elegans Group 'Elegans'	Upright, but irregular	2–3m (6–10ft)	1–1.5m (3–5ft)	Soft, semi-juvenile foliage, which is light-green in the summer and bronze, purple in winter	May need pruning to maintain a regular shape
C. japonica 'Elegans Compacta'	Neater and more rounded than *C, japonica* Elegans Group 'Elegans'	1.5–2.5m (2.5–9ft)	1–1.5m (3–5ft)	Juvenile foliage in which the leaves are longer and thinner than adult foliage. Green in summer, soft purple in winter	Needs a sheltered site
C. japonica 'Elegans Viridis' AGM	Loosely conical	8–12m (26–40ft)	2.5–4m (8–13ft)	Juvenile foliage that remains pale-green throughout the year	
C. japonica 'Globosa Nana' AGM	A low-growing dome shape	0.5–1m (20in–3ft)	0.5–1m (20in–3ft)	Dense foliage, green in summer, purple in winter	
C. japonica 'Cristata' (syn. 'Sekkan-sugi') AGM	Small, conical tree with semi pendulous branches	8–12m (28–40ft)	2.5–4m (8–13ft)	New leaves a pale creamy-yellow that darken to green as they mature	Needs protection from cold winds. Produces small brown cones
Cryptomeria japonica 'Spiralis' AGM (granny's ringlets)	Given time it can make a most interesting architectural shape	4–8m (13–28ft)	2.5–4m (8–13ft)	Awl-like leaves are twisted and spiraled around the stem giving an almost rope-like appearance	Slow-growing semi-dwarf. May need formative pruning when young

CUPRESSUS

The true cypresses are now considered to be a genus of about fifteen species of evergreen conical or columnar trees coming from western North America and North Africa across to China.

The leaves are small, scale-like and adpressed to the stem. The branches produce side branches radially outwards from the main stem and not in flat fans of growth, as in *Chamaecyparis*. Male and female strobili are produced on the same tree. The female strobili develop into woody cones, larger than those of *Chamaecyparis*, and remain on the trees for several years.

As a genus, it is less hardy than *Chamaecyparis* and this has limited its use in Europe.

Growing Requirements

Hardiness: Please see the hardiness rating for individual species.
Aspect and shelter: A site oriented in any direction is acceptable, but shelter from cold, drying winds is required.
Sun: Full sun.
Soil: The soil can be of any texture but must be well drained. Dry sites are well tolerated.
pH: Can be within a very wide range and high alkalinity is not a problem. This combination makes cupressus well suited to thin, dry calcareous soils.

Cupressus arizonica var. *glabra* 'Blue Ice'

Hardiness: USDA RHS H5.
Height: 12m+ (40ft+).
Spread: 2.5–4m (8–13ft).
Habit: Forms a narrow cone.
Foliage: Striking silvery blue-grey.
Pruning: This species is unusual in that it lends itself well to clipping and topiary. Consequently, it can be used for hedging but needs regular clipping and is available commercially as spiral, cloud-pruned and pom-pom shapes. Providing that its use is regarded as temporary, or the tree is regularly pruned, it can be used in a mixed planting of medium-sized conifers, where it contributes a distinctive, vivid foliage colour and can be topiarised to give a contrasting shape.

Cupressus macrocarpa (Monterey Cypress)

Hardiness: USDA 8b/9a. RHS H4.

Cupressus sempervirens (Italian Cypress, Mediterranean Cypress)

Hardiness: USDA RHS H5.
The classic dark-green, narrow, vertical tree, so typical of Mediterranean gardens, where it enjoys growing in hot, dry summers and mild, wet winters. Such a vertical exclamation mark is the dream of many gardeners and garden designers, but its lack of hardiness and intolerance of cold winds must be remembered. Although its shape makes it appear ideal for where ground space is limited but vertical space is not restricted, remember the tree's need for full sun and the possibility of root damage to any adjacent buildings.

Both cultivars in Table 5.8 are more manageable where space is limited, but no hardier than the species. They may need tying in and/or clipping to maintain the desired pencil shape, and are liable to damage by snow fall. The trees are best planted when small, since larger trees are very liable to be blown over or wind damaged.

TABLE 5.7 *CUPRESSUS MACROCARPA* AND CULTIVARS

Name	Habit	Height	Spread	Foliage	Remarks
C. macrocarpa	Large, broad-headed tree	12m+ (40ft+)	4m+ (13ft+)	Dark-green	Frequently seen forming a shelter belt in coastal areas. Not suitable for use where space is limited. Can be used for hedging, where cold winds are not experienced but will require clipping
C. macrocarpa 'Goldcrest'		12m+ (40ft+)	4–8m (13–26ft)	Juvenile, feathery, bright golden-yellow	Height 3–5m (10–16ft) in ten years. May be used in a planting of mixed, medium-sized conifers to give vivid colour, but ultimate size must be remembered. Frequently sold as a small tree intended for use as a house plant. Here its use will be of only short duration. It is frequently planted as a young tree in large pots and containers but, again, its ultimate size must be considered
C. macrocarpa 'Wilma' AGM	Upright, slow-growing tree	1.8m (6ft)	1m (3ft)	Juvenile golden green	Well suited to planting in a group of slow-growing conifers to give contrasting shape and foliage colour. Good as a specimen in a pot of for hedging

Cupressus macrocarpa 'Goldcrest'. Detail of the coloured foliage.

Cupressus macrocarpa 'Wilma'. The foliage is very similar, both in colour and texture, to *C. macrocarpa* 'Goldcrest'.

TABLE 5.8 *CUPRESSUS SEMPERVIRENS* AND CULTIVARS

Name	Habit	Height	Spread	Foliage	Remarks
C. sempervirens	Very narrow and upright	12m+ (40ft+)	2.5–4m (8–13ft)	Dark-green, scale-like	Note this tree's large ultimate size
C. sempervirens 'Swane's Gold'	Very narrow and upright with a slightly irregular outline	4–8m (13–26ft)	1–1.5m (3–5ft)	Greenish yellow all year round	
C. sempervirens 'Totem Pole'	Very narrow and upright	8–12m (26–40ft)	0.5–1m (1½–3ft)	Dark-green, scale-like	

Cupressus sempervirens 'Swane's Gold'. Forming a very narrow, slightly irregular, small tree.

Cupressus sempervirens 'Swane's Gold'. The yellowish-green foliage stays the same colour throughout the year.

Cupressus sempervirens 'Totem Pole'. Displaying the very narrow, upright habit of this striking cultivar.

Cupressus sempervirens 'Totem Pole'. Detail of the grey-green foliage.

GINKGO BILOBA L. (SYN. *SALISBURIA ADIANTIFOLIA*) (MAIDENHAIR TREE)

Belonging to a monotypic genus, this species is a remarkable survivor of bygone ages whose ancestors occurred in many parts of the world and are found in the fossil record of about 190–200 million years ago, in widespread locations such as Robin Hood's Bay in Yorkshire, UK and Yunnan in China.

Long thought to be extinct in the wild, the species has survived in China where the impact of glaciation was minimal. Yun-Peng Zhao, Guangyi Fan and Song Ge (2019) have discovered three refugia where this species has survived in China.

A slow-growing deciduous tree with an irregular, spreading crown. In China, very large specimens can be found, some claimed to be in excess of 2000 years old. In general cultivation, the species makes a medium-sized tree whose eventual shape is difficult to predetermine.

Growth of branches takes two forms: short shoots with leaves in whorls and long shoots with leaves arranged in a spiral around the branch. This pattern of growth gives this species a distinctive winter habit.

The leaves are unique in appearance being fan-shaped with two lobes, hence the specific epithet of *biloba*. This shape is like that of the subdivisions of the frond of maidenhair fern (*Adiantum*), giving the tree its common name.

Trees are either male or female. Female trees produce yellow, plum-like fruits, which fall in the autumn. The soft tissue of the fruit, when rotting on the ground, smells foul and, therefore, only male trees are generally planted. The hard seeds can be eaten but are toxic in excess and have been widely used in Chinese herbal medicine. Indeed, the author when ill with altitude sickness in Lhasa was prescribed a mixture of ground gingko, plus Western medicine and survived!

When grown from seed, the resulting trees are extremely variable in habit. Fastigiate, pendular, dwarf, variegated, 'normal' and other mutant forms can be produced. Over the years, many selections have been made and named by nurserymen. The upright, male clone *Gingko biloba* 'Princeton Sentinel' has been widely planted as a street tree, for which purpose it is admirably suited.

Other dwarf selections have been made and are the main subject for inclusion here.

Selected clones can be propagated by semi-ripe cuttings or by grafting.

Growing Requirements

Hardiness: USDA 6b/7a. RHS H6.
Aspect: A site facing any direction is acceptable.
Shelter: Some shelter is required from cold, drying winds.
Sun: Best in full sun but will tolerate light shade.
Soil: The soil can be of any texture and should be well drained.
pH: This species is very tolerant of a wide range of both acidity and alkalinity.
Tolerance: Tolerates atmospheric pollution well and some degree of soil compaction.

Garden or Landscape Design Considerations

A selected male, upright-growing tree, such as *Ginkgo biloba* 'Princeton Sentinel', could be employed where floor space, but ultimate height is not restricted.

The dwarf and slow-growing cultivars, mentioned above, are eminently suited for use as specimens in: compact city gardens, courtyard gardens, patios, container-planting and bonsai. In these situations, they will provide fascination from their leaf shape and growth pattern in winter. In autumn, trees provide beautiful leaf colour changing from dark-green to bright-yellow and, finally, bronze before fall.

TABLE 5.9 RECOMMENDED COMPACT CULTIVARS OF *GINGKO BILOBA*

Name	Habit	Height	Spread	Foliage	Remarks
'Fastigiata Blagon'	Narrowly conical tree	15m (48ft)	2.5m (8ft)		The narrowness of the cone can be increased by seasonal light clipping
'Troll'	A compact, very slow-growing dwarf	1–1.5m (3–5ft)	0.5–1m (1¾–3ft)		
'Saratoga'	Upright, normally with a central leader	4–8m (13–26ft)	1.5–4m (5–13ft)	Leaf blade narrow and fans out irregularly	Lends itself well to cloud pruning. A male clone

Ginkgo biloba. Here, used as a street tree. This species is excellent for this purpose, but much too large for small or compact gardens.

Ginkgo biloba 'Fastigiata Blagon'. The foliage of this cultivar.

Ginkgo biloba 'Fastigiata Blagon'. This cultivar naturally has an upright habit, but here this has been encouraged and the growth thickened by clipping.

Ginkgo biloba 'Troll'. A compact, dwarf cultivar.

Ginkgo biloba 'Troll'. The foliage of this is slightly abnormal, but still very attractive.

Ginkgo biloba 'Saratoga'. The foliage is narrower and more divided than that of the species.

Ginkgo biloba. Golden autumn foliage colour is reliably produced before leaf fall.

Ginkgo biloba 'Saratoga'. A slow-growing cultivar, here trained and pruned into a cloud-pruned, small tree.

JUNIPERUS

A coniferous genus of about fifty species widely distributed throughout the northern hemisphere. As plants they range from large, upright trees through shrubs to creeping or prostrate plants. The genus contains many small trees and prostrate plants, very well suited for use in small gardens or where space is severely restricted.

The juvenile leaves are needle-like, pointed and prickly to the touch. The adult leaves are scale-like and crowded together on the branches. Depending on the species or cultivar selected, a juniper plant may have just juvenile or adult foliage or a mixture of both. Male and female strobili (flowers) in some species are produced on the same plant, but in other species are produced on separate plants. The cones are unlike other conifers and are rounded and berry-like once mature.

Growing Requirements

This group's tolerance of full sun, sharp drainage and high pH, makes them very well suited to covering sunny, excessively well-drained banks and slopes, where the soil may be very thin and limey, e.g. on chalk or soft limestone-derived soils.

Hardiness: USDA 6b/7a or 6a-1. RHS 6 or 7.
Aspect: A site of any orientation is satisfactory.
Shelter: For most, the site can be exposed or sheltered, but some cultivars do require some shelter. This is mentioned in the text.
Sun. Full sun or only partial shade is best.
Soil: A soil of virtually any texture is accepted, but good drainage is required.
pH: A very wide soil pH is tolerated including those with a high pH.
Protective clothing: Plants of this genus may cause skin irritation. The wearing of gloves and suitable clothing is recommended when handling these plants.

Juniperus chinensis

Hardiness: USDA 6b/7a. RHS H6.
This is a widely distributed and very variable species, typically appearing as a tall, columnar tree bearing both adult and juvenile foliage. In China, several very beautiful clones have been selected, each clone having a distinctive habit of growth that can often be exaggerated and enhanced by judicious pruning.

Juniperus communis (Common Juniper)

Hardiness: USDA 6a-1. RHS H7.

This species is one of the only three British native conifers. It is unusual in being extremely variable in habit and the author remembers a demonstration bed once present in Cambridge University Botanic Gardens where a considerable number of selected forms had been deliberately planted together to demonstrate this species' remarkable, natural diversity of habit.

Several of these variants have been selected by horticulturists and propagated vegetatively to form cultivars that are very well suited for use where space is limited.

Juniperus horizontalis (Creeping Juniper)

Hardiness: USDA 6a-1.RHS H7.

A low-growing shrub from North America, where it grows on sea cliffs and gravelly slopes forming a mat of ground-covering growth. It should come as no surprise that this species has produced several low-growing cultivars whose main horticultural function is low-maintenance, evergreen ground-cover.

A plant bed consisting of *Juniperus communis* intended to display the enormous variation between individual plants of this species.

Juniperus communis 'Compressa' a trio of the Noah's ark juniper. Very desirable but notoriously susceptible to weather, and pest and disease damage.

TABLE 5.10 CULTIVARS OF *JUNIPERUS CHINENSIS*

Name	Habit	Height	Spread	Foliage	Remarks
J. chinensis 'Aurea' AGM	Slender, conical tree	8–12m (26–40ft)	2.5–4m (8–13ft)	A male clone showing bright-yellow juvenile and adult foliage	Young's golden juniper. Can give a bright architectural effect
J. chinensis 'Kaizuka' AGM	An erect growing shrub, eventually a small tree with long, spreading branches giving an irregular outline	4–8m (13–26ft)	2.5–4m (8–13ft)	Bright-green	Hollywood juniper. Where space allows, an effective isolated specimen

TABLE 5.11 CULTIVARS OF *JUNIPERUS COMMUNIS*

Name	Habit	Height	Spread	Foliage	Remarks
J. communis 'Compressa' AGM	A tight, regular column	0.5–1m (20–39in)	0.1–0.5m (4–20in)	Grey-green fine juvenile foliage	The Noah's ark juniper. Susceptible to damage from cold winds and frost, needs a sheltered site. Susceptible to attack by red spider mite. Slow-growing. Despite susceptibilities, very popular because of 'classic' shape
J. communis 'Green Carpet' AGM	Forms a low-spreading mat	10–15cm (4–6in)	1–1.5m (3–5ft)	Prickly, bright-green in summer, darker in winter	Very useful for ground cover in sunny, dry sites. Will tolerate very light shade, but not heavy or full shade
J. communis 'Hibernica'	Narrow, upright form	1.8–2.5m (5–8ft)	25–30cm (10–14in)	Upright branches bearing needle-like grey-green foliage	Irish Juniper
J. communis 'Repanda' AGM	Forms a slightly irregular spreading mat	10–50cm (4–20in)	1.5–2.5m (5–8ft)	Needle-like foliage green in summer but bronzed in the winter	Useful for evergreen ground-cover in sunny or partially shaded sites

TABLE 5.12 CULTIVARS OF *JUNIPERUS HORIZONTALIS*

Name	Habit	Height	Spread	Foliage
J. horizontalis 'Bar Harbor'	Prostrate forming a low-growing mat	10cm (4in)	3m (10ft)	Glaucous grey-green linear growth that turns reddish-purple in winter
J. horizontalis 'Blue Chip'	Prostrate forming a low-growing mat	10–50cm (4–20in)	2–2.5m (6–8ft)	Silvery blue
J. horizontalis 'Golden Carpet' AGM	Prostrate forming a low-growing mat	10–50cm (4–20in)	1.5–2.5m (5–8ft)	Greenish-yellow all year round
J. horizontalis ICE BLUE ('Monber') AGM	Prostrate forming a low-growing mat	15cm (6in)	2m (6ft)	Silvery-blue, becoming plummy-bronze in cold winter weather

Juniperus communis 'Repanda'. A prostrate form suitable for ground cover.

Juniperus communis 'Repanda'. Detail of the foliage.

Juniperus × *pfitzeriana*

Hardiness: USDA 6b/7a. RHS H6.
A variable hybrid juniper, many selections of which are too fast-growing, spreading and, ultimately, too large to be considered slow-growing or compact. Only small or moderately sized cultivars are listed in Table 5.13. Whilst their main purpose in the garden or landscape is evergreen ground-cover, note their greater height and arching or ascending habit that makes them contrast against *J. horizontalis* cultivars.

TABLE 5.13 CULTIVARS OF *JUNIPERUS* × *PFITZERIANA*

Name	Habit	Height	Spread	Foliage
J. × *pfitzeriana* 'Carbery Gold' AGM	Slow-growing prostrate	0.5–1m (20–40in)	1.5–2m (5–6½ft)	Bright creamy-yellow. More golden in winter
J. × *pfitzeriana* 'Old Gold' AGM	Compact with ascending branches	0.5–1m (20–40in)	1.5–2.5m (5–8ft)	Bronze-gold foliage all year round all year round
J. × *pfitzeriana* 'Sulphur Spray' AGM	Ascending main branches and less compact than *J.* × *pfitzeriana* 'Old Gold'	1.5–2m (5–6½ft)	1.5–2.5m (5–8ft)	Pale sulphur-yellow

Juniperus × *pfitzeriana* 'Carbery Gold'. A brightly coloured, ground-covering cultivar.

Juniperus × *pfitzeriana* 'Carbery Gold'. Detail of the finely divided and coloured foliage.

Juniperus scopulorum (Rocky Mountain Juniper)

This is a small, conical tree with several main stems found in the Rocky Mountains of America. The slender branches have scale-like leaves tightly addressed to their surface. These leaves are glaucous and can vary from green to grey-green. In cultivation it has produced very fastigiate forms of great use in decorative horticulture.

Growing Requirements

Hardiness: USDA 6b/7a. RHS 6.
Aspect: Lack of light results in very poor, thin growth and where lighting is from one side this will result in unbalanced growth, growth being poor on the shaded side.
Sun: Full sun.
Soil: Under UK conditions, both cultivars have been found to require good drainage to grow well. Poor drainage makes them very vulnerable to root rot. They do grow well on poor but well-drained sandy, gravelly or chalky soils.
Longevity: They do not appear to be long-lived trees. *Juniperus scopulorum* 'Skyrocket' becomes thin and poor in appearance after approximately 20–25 years. The more recently introduced *Juniperus scopulorum* 'Blue Arrow' is considered an improvement on 'Skyrocket', making a smaller tree and less liable to weather damage. For the small or compact garden, *Juniperus scopulorum* 'Blue Arrow' is the best choice.

TABLE 5.14 CULTIVARS OF *JUNIPERUS SCOPULORUM*

Name	Habit	Height	Spread	Foliage	Cones	Remarks
J. scopulorum 'Blue Arrow' AGM	Narrow, upright habit	1.5–2.5m (5–8ft)	0.1–0.5m (4–20in)	Blue-grey	Blue-black, rounded	Slower growth rate makes it less liable to snow damage
J. scopulorum Skyrocket' (syn. *J. virginiana* 'Skyrocket')	Narrow, upright habit	4–8m (13–26ft)	0.5–1m (20–40in)	Blue-grey	Blue-black, rounded	Liable to snow damage. Branches may need tying up

Juniperus scopulorum 'Blue Arrow'. This very erect cultivar on sale at a garden centre. Ideal for creating a focal point or vertical contrast in a small- to medium-sized garden.

Juniperus scopulorum 'Skyrocket'. The vertical form and steely blue of the foliage look well against a modern building.

Juniperus scopulorum 'Skyrocket'. Detail of the foliage.

TABLE 5.15 CULTIVARS OF *JUNIPERUS SQUAMATA*

Name	Habit	Height	Spread	Foliage
J. squamata 'Blue Carpet' AGM	Undulating, slow-spreading ground-cover	0.1–0.5m (4–20in)	1.5–2.5m (5–8ft)	Juvenile, spikey, blue-green
J. squamata 'Blue Star' AGM	Dense, flat-topped dome	0.1–0.5m (4–20in)	0.5–1m (20–40in)	Juvenile, spikey, blue-green

Juniperus squamata (Flaky Juniper)

Hardiness: USDA 6a-1. RHS H7.
A shrubby juniper of very variable habit from Afghanistan eastwards to central China. The species is not grown as an ornamental plant but has produced several very good and deservedly popular garden plants.

Juniperus squamata 'Blue Star'. A well-grown specimen showing this cultivar's low-growing irregular, rounded, overall habit.

Juniperus squamata 'Blue Star'. Detail of the spikey, blue foliage.

MICROBIOTA (RUSSIAN ARBOR-VITAE OR SIBERIAN CARPET CYPRESS)

A monotypic genus, containing only *Microbiota decussata*.

Growing Requirements

Hardiness: USDA 6a-1. RHS H7.
Aspect: A site facing any direction, except due north, is acceptable.
Shelter: This may be exposed or sheltered.
Sun: Full sun is required.
Soil: Growth is good in soils of any texture. Soil drainage must be good, but some moisture should be retained in summer.
pH: The plant is not sensitive to soil pH and a very wide range from acid to alkaline is satisfactory.

Microbiota decussata AGM

Height: 0.1–0.5m (4–20in).
Spread: 1–1.5m (3–5ft)
Habit: Forms a low-growing mound.
Foliage: Finely divided.

This species, which grows in eastern Siberia, was only discovered in 1921. It is a small, evergreen shrub whose branches bear opposite pairs of scale-like leaves. The cones, both male and female, are small and inconspicuous.

The notable feature about this small conifer is its foliage, which is a light-green in summer but changes to reddish-brown in winter. Seen for the first time in winter, some might describe it as looking dead! However, when used as a feature plant with other much lighter-coloured, small shrubs, such as a white-flowered cultivar of *Erica carnea*, it takes on a much more interesting and contrasting appearance.

The species is also available as a top-worked plant. As such, it forms a small tree with a semi-weeping habit, which is distinctive both in summer and especially in winter.

PICEA (SPRUCES)

This genus consists of over forty species found widely distributed in the northern hemisphere. They are a genus of large, evergreen conifers, conical in overall shape with branches borne in whorls.

The leaves are short and needle-like, flattened or four-sided and arranged spirally or in two ranks. Characteristically, fallen leaves leave a short peg-like projection, which results in the branches being rough to the touch.

Male and female strobili are present on the same tree. Male cones are produced laterally, cone-shaped and yellow. Female cones are produced terminally and may be brightly coloured when young. Female cones are pendant, develop fully in one year but do not fall until the second year.

Growing Requirements

Hardiness: USDA 6a-1. RHS H7.
Aspect: The planting position can face in any direction.
Shelter: The various species do not require a sheltered site, but cultivars will do better with protection from cold, drying winds.
Sun: Full sun is required. Almost any degree of shade will result in poor growth and eventual death of branches on the shaded side, leading to a poor, one-sided specimen. Encroachment over the spruces must also be watched for since any other plant growth lying on the spruce foliage will result in its rapid death.
Soil: The soil can be of any texture, must be well drained but should retain moisture in the summer.
pH: The soil pH should be acidic or neutral and soils such as those derived from chalk or soft limestone are best avoided.

The spruces prefer cool, damp climates and do not grow well under hot, dry conditions.

Picea abies (Norway Spruce or Common Spruce)

The Norway spruce is a tree capable of rapid growth and will grow into large trees often over 30m (98ft) tall. It is known to many as the Christmas tree and there is often the temptation to plant out and try to grow on a tree after the Christmas celebrations. It is suggested that this is best avoided, as this species does not make a satisfactory small tree for where space is limited.

Picea glauca (White Spruce)

A large, conically shaped, evergreen conifer found in Canada and north-east USA.

Cultivars of *Picea glauca* are susceptible to adelgids, glasshouse red spider mite and aphids. To control red spider mite, water the tree and spray the foliage during dry spells of weather.

Picea pungens (Colorado Blue Spruce)

A medium-sized to large, conical, evergreen conifer found in western USA. Young needles are glaucous at first, but usually turn green as they age. Forms have been selected that retain the attractive blue-grey colour of the needles and are of smaller size. Amongst the best cultivars are those listed in Table 5.18.

Potted plant of *Microbiota decussata* showing the species' low-growing habit.

The reddish-brown foliage of *Microbiota decussata* after experiencing winter cold.

The green foliage of *Microbiota decussata* having experienced summer, just beginning to turn reddish as it feels the first autumnal cold.

TABLE 5.16 CULTIVARS OF *PICEA ABIES*

Name	Habit	Height	Spread	Foliage	Remarks
P. abies 'Acrocona' AGM	A small, evergreen tree with drooping branches	2.5–4m (8–13ft)	2.5–4m (8–13ft)	Dark-green	Bears cones even as a very young tree. The young female cones are dark-pink when young and pale-brown once mature
P. abies 'Little Gem' AGM	Very slow-growing, eventually forming a compact bun shape	0.1–0.5m (4–20in)	0.1–0.5m (4–20in)	Needles dark-green. Lighter green during spring flush of growth	
P. abies 'Nidiformis' AGM	A dwarf shrub that is wider than tall	1–1.5m (3–5ft)	2.5m –4m (8–13ft)	Dark-green, short needles	

TABLE 5.17 CULTIVARS OF *PICEA GLAUCA*

Name	Habit	Height	Spread	Foliage	Remarks
P. glauca var. *albertiana* 'Conica' AGM	Forms a regular cone shape	1.5–2.5m (5–8ft)	1.0–1.5m (3–5ft)	Short needles. New needles in spring are light-green turning darker green later	Very popular because of its 'classic' shape
P. glauca 'Echiniformis' AGM	Dwarf, dense, rounded shrub	0.1–0.5m (4–20in)	0.5–1m (4–20in)	Short grey-green needles	Very slow-growing

Picea glauca var. *albertiana* 'Conica'. This plant shows the typical, slightly irregular, conical shape developed by this cultivar.

Picea pungens (Glauca Group). This young tree is developing into a well-shaped specimen with a clear, straight, central leading shoot.

Picea pungens (Glauca Group) 'Koster'. An old tree displaying this cultivar's narrow, upright habit.

Picea glauca var. *albertiana* 'Conica'. The new foliage photographed in early summer will develop a darker green as the summer proceeds.

Picea pungens (Glauca Group) 'Koster'. New growth showing the intense blue coloration of the needles.

TABLE 5.18 CULTIVARS OF *PICEA PUNGENS*

Name	Habit	Height	Spread	Foliage	Remarks
P. pungens Glauca Group 'Globosa' AGM (syn. *P. pungens* 'Glauca Globosa')	Slow, dense-growing and spreading shrub, eventually producing a rounded shape	0.5–1m (1ft 6in–3ft)	0.5–1m (1ft 6in–3ft)	Prickly blue-grey needles	If vigorous, upright shoots are not curbed by pruning, a shrub with a pyramidal shape will be produced
P. pungens 'Koster' (Koster's blue spruce)	Small- to medium-sized tree with a conical shape	8–12m (26–40ft)	2.5–4m (8–15ft)	Intense blue-green.	To produce a good, regularly shaped specimen, training and pruning may be required when young. The leading shoot may need to be trained upwards and vigorous side-shoots pruned to keep them in check

PINUS (PINES)

This is a large genus of over 100 species of evergreen trees and shrubs. It occurs over a very wide part of the northern hemisphere, South and Central America and Indonesia. Economically, the genus is very important as timber trees. As a genus, it has produced many slow-growing or dwarf forms, some of which find an application in decorative horticulture. The best of those suited to where garden space is limited are mentioned in Table 5.19.

The leaves are long and needle-like. These are borne in bundles mainly of two, three or five, occasionally six, and in a single case as single needles. The number of needles in each bundle is a useful feature in beginning to determine which species an individual tree belongs to.

Male and female strobili are produced on the same tree. The female strobili develop into cones that can be a very attractive feature, as they vary considerably in size, shape and colour.

Pines make just one flush of growth at the beginning of their growing season. The resulting 'candles' of such growth often contrast strongly with older foliage and can be another very distinctive and attractive feature.

Growing Requirements

Hardiness: USDA 6a-1. RHS H7.
Aspect: A site of any orientation is acceptable.
Shelter: Shelter is not required.
Sun: Full sun is required, and they dislike any degree of shade.
Soil: A soil of any texture is acceptable, and many will ensure summer drought.
pH: Most are tolerant of a wide soil pH range, but the five-needled group do not like a calcareous soil.

Pinus contorta 'Chief Joseph' AGM

A cultivar of the lodgepole pine.

TABLE 5.19 *PINUS CONTORTA* 'CHIEF JOSEPH' AGM

Name	Habit	Height	Foliage	Remarks
P. contorta 'Chief Joseph' AGM	Broadly conical	2.5–4m (8–13ft)	A two-needled pine. Needles 4–5cm (1½–2in) long and twisted. Yellow-green in summer and bright-yellow in winter	Requires an acid or neutral soil. Avoid a north-facing site. Slow-growing

Pinus contorta 'Chief Joseph'. A slow-growing pine that develops a clear overall upright habit.

Pinus contorta 'Chief Joseph'. Showing the bright-yellow colour developed by the needles in winter.

Note the strong textural and colour contrast between *Pinus contorta* 'Chief Joseph' and *Nandina domestica* 'Seika'[PBR].

Pinus mugo (European Mountain Pine or Dwarf Mountain Pine)

A two-needled species that is very variable in habit and can be a large, spreading shrub or a small tree. Needles dark-green. Very hardy and drought-resistant. When raised from seed, it is not recommended for small gardens, since the eventual size and shape of the plants is so unpredictable.

TABLE 5.20 CULTIVARS OF *PINUS MUGO*

Name	Habit	Height	Spread	Foliage	Remarks
P. mugo 'Humpy' AGM	Very compact	0.5–1m (20–40in)	0.5–1m (20–40in)	Dark-green	Very short needles
P. mugo 'Mops' AGM	Slightly less compact	0.5–1.5m (20in–5ft)	1–1.5m (20in–5ft)	Dark-green	
P. mugo 'Winter Gold'	Spreading but much of the length branches upright	1.0–1.5m (40in–5ft)	1.5–2.5m (5–8ft)	Bright-green in spring. Mid-green in summer. Golden-yellow in winter	

Pinus mugo 'Mops'. A good dwarf, compact cultivar of the mountain pine.

Pinus mugo 'Mops'. Here, the compact nature and short needles of this cultivar can be clearly seen.

Pinus mugo 'Winter Gold'. This cultivar eventually makes a sizeable, spreading shrub.

Pinus mugo 'Winter Gold'. The needles develop yellow coloration in winter but change back to dark-green in the summer.

Pinus sylvestris (Scots Pine)

A large, evergreen conifer that is native to the UK. When young, a pyramidal tree, but lower branches slough off as the tree develops and the final shape is irregular, often consisting of a long, main trunk with a flattened head at the top, resulting, ultimately, in a very characteristic outline. Needles grey-green in pairs. Bark in large plates, often a reddish-brown colour. Ultimate height 12m+ (40ft+). Ultimate spread 8m+ (26ft+).

TABLE 5.21 CULTIVARS OF *PINUS SYLVESTRIS*

Name	Habit	Height	Spread	Foliage	Remarks
P. sylvestris Aurea Group 'Aurea' AGM	A tall, upright tree	12m (40ft)	4–8m (13–26ft)	Green in summer, yellow in winter	Included to show how careful cultivar selection is required!
P. sylvestris 'Gold Coin' AGM	Compact shrub	1–1.5m (40in–5ft)	1–1.5m (5–8ft)	Yellow all the year round	

This tree was labelled *Pinus sylvestris* 'Gold Coin' but is probably a good selection from *Pinus sylvestris* Aurea Group, since the foliage changes back to green in the summer, as shown in the following photograph.

Pinus sylvestris Aurea Group photographed in summer.

PODOCARPUS

A genus of evergreen conifers that may be trees or shrubs and are mainly confined to the southern hemisphere. Most species are not hardy and only those that have proved to be hardy are considered here. Unlike most conifers, instead of typical cones they produce a plum-like fruit and are commonly known as plum pines.

Most podocarps are dioecious with separate male and female plants. The female plant needs to be pollinated and fertilised by the male pollen before any fruits will be produced. It should be noted that some cultivars, being clones of an original plant, consist of only male or female plants. If fruit is required, a plant of the same species but different sex will need to be planted close to it.

Growing Requirements

Hardiness: USDA 7b/8a or 6b/7a. RHS 5 or 6.
Aspect: As a group they tolerate any aspect except one that is north-facing.
Shelter: Some shelter from the coldest drying winds is required.
Sun: Full sun will produce the best foliage colour, but part shade is tolerated.
Soil: The soil texture can be any, but it must be well drained.
pH: A wide pH range is accepted but avoid thin chalky soils with a high pH.

Garden Design Considerations

This small range of plants has a surprising range of useful applications.

The low-growing cultivars make effective, colourful, small-scale, ground-cover, providing any autumn leaf fall from trees above can be rapidly removed. They can make colourful contributions to plant associations designed to emphasise colour contrasts and make container plants with year-round interest. In a winter garden or border they will bring colour and, importantly, permanent structure.

The two low-growing cultivars mentioned in Table 5.21 would make very good candidates for a low-growing hedge. Both could be closely planed and then clipped once a year in spring to produce a neat, colourful, evergreen, dwarf hedge.

All three cultivars mentioned in Table 5.21 would be appropriate in a heather garden, or dwarf and slow-growing conifer collection or rock garden, bringing different colours and textures. The *Podocarpus totara* 'Aureus' would also provide a different habit, being boldly upright, and might be considered as an unusual substitute for the upright yellow-leaved *Taxus baccata* 'Standishii' AGM.

TABLE 5.22 CULTIVARS OF *PODOCARPUS*

Name	Habit	Height	Spread	Foliage	Remarks
P. 'Chocolate Box'	Low-growing and dome-shaped	40cm (1½ft)	60cm (2ft)	Evergreen. Dark reddish-green during growing season. Chocolate-coloured in winter	USDA 7b/8a. RHS H5. Responds well to light clipping
P. 'Country Park Fire'PBR AGM	Low-spreading, mounded shrub	60cm (2ft)	90cm (3ft)	Evergreen. Foliage consists of short, spirally arranged linear leaves. In spring, new growth is creamy-yellow from red shoots. This growth changes to pink through the growing season and later changes to greenish bronze. Finally, with winter cold the growth becomes reddish	USDA 6b/7a. RHS H6. Responds well to light clipping in spring. Red fruits are produced if a male plant is planted nearby. *Podocarpus* 'Flame' is a cultivar with coloured foliage and is male
P. totara 'Aureus'	Upright, slow-growing, small tree	2m (6½ft)	70cm (2¼ft)	Evergreen. Leaves yellowish-green, 2.5cm (1in) long and pointed	USDA 7b/8a. RHS H5

Podocarpus 'Chocolate Box'. Close-up of the foliage in summer. In winter, the foliage will become a darker chocolate colour.

Podocarpus 'Country Park Fire'. A mature plant showing the typical, spreading, irregular, mounded shape this cultivar eventually becomes.

Podocarpus 'Country Park Fire'. Photographed in summer. In winter, the foliage will assume more of a bronzed-red colour.

Podocarpus lawrencei 'Bluey' (syn. *P. alpinus* 'Bluey'). Here, the unusual blue-green foliage and low-mounded shape of this cultivar is seen. It could be useful for small-scale ground-cover and to provide a colour contrast with any other podocarp.

Podocarpus totara 'Aureus'. The photograph shows the sharply pointed, yellow-green foliage. In time this cultivar forms a distinctive, upright, small tree.

TAXUS

The yews are a genus of about ten species of evergreen trees and shrubs found in north temperate regions, North and South America and Indonesia. The leaves are linear and appear two-ranked with two indistinct bands of stomata on their undersides. Male and female strobili are usually produced on separate trees. The female strobilus develops into a fruit with a fleshy, usually red coloured cup called an aril, around a single poisonous seed.

Growing Requirements

Hardiness: USDA 6a-1. RHS H7.
Aspect: A site facing any direction.
Sun: Full sun to full shade.
Shelter: Sheltered or exposed is acceptable.
Soil: The soil can be of any texture but must be well drained.
pH: A very wide range of soil pH is tolerated.
Tolerance: Tolerant of drought and pollution.

Taxus baccata (Common Yew)

One of the only two conifer species native to the UK. When left to grow naturally it will eventually form a large tree with a massive trunk and is very long-lived. The Irish yew has been traditionally planted in the grounds of churches, supposedly to ensure farmers kept their cattle out, since when eaten the tree is toxic.

The species is considered outstandingly good when used to form an evergreen hedge. The hedge produced being a uniform, dense, fine-textured dark-green. It responds well to clipping and can be used to form low hedges and also up to 15m (49ft) tall, as seen at Powis Castle in Wales. *Taxus baccata* is excellent for use to form topiary of numerous different shapes and the topiaries so formed are capable of being very long-lived, as seen by the century's old species at Levens Hall in England.

The plant will tolerate drought and a high soil pH, growing well on thin, calcareous soils. It will grow well in full sun or full shade. It will not tolerate poor soil drainage, where it becomes susceptible to a range of diseases that attack the roots.

TABLE 5.23 CULTIVARS OF *TAXUS BACCATA*

Name	Habit	Height	Spread	Leaves	Remarks
T. baccata 'David'	Columnar, slow-growing	2.5–4m (8–13ft)	0.5–1m (20–40in)	Dark-green with yellow margins	A male clone
T. baccata 'Fastigiata' AGM (Irish yew)	Columnar, upright	8–12m (26–40ft)	2.5–4m (8–13ft)	Dark-green	A female clone
T. baccata 'Fastigiata Aureomarginata' AGM (golden Irish yew)	Columnar when young, becoming broadly columnar	2.5–4m (8–13ft)	1.5–2.5m (5–8ft)	Dark-green, yellow margined	
T. baccata 'Icicle' AGM	Columnar, slow-growing	0.5–1m (20–40in)	0.1–0.5m (4–20in)	Green edged yellow in summer. Green edged white in winter	To give best foliage colour, best in full sun or only part shade. Has the appearance of being covered in frost in winter
T. baccata 'Repens Aurea' AGM	Undulating and spreading	0.5–1m (20–40in)	1–1.5m (3–5ft)	Green edged yellow initially; as leaves age they become cream	Leaf colour best in full sun
T. baccata 'Standishii' AGM	Dense column	1–1.5m (3–5ft)	0.5–1m (20–40in)	Rich yellow	A female clone. Slow-growing, but very good leaf colour

Taxus baccata 'Standishii'. Several plants of this erect-growing cultivar planted together to form an upright feature.

Taxus baccata 'Standishii'. Showing the bright-yellow variegation of the foliage.

THUJA (ARBOR-VITAE)

A genus of five species of evergreen trees and shrubs widely dispersed in the northern hemisphere. Male and strobili are produced on the same tree, the female developing into a small, woody, oblong cone. The foliage consists of scale-like leaves arranged in four ranks. The distal branches hold the foliage in fan-like sprays. *Thuja* and *Chamaecyparis* are often confused, as the fan-like sprays of foliage are similar. The foliage is very pleasantly aromatic when brushed against or once held briefly in the hand to warm the foliage slightly and drive off the aroma. Clear points of difference are the characteristic smell of *Thuja* and the oblong shape of *Thuja* cones in contrast to the rounded cones of *Chamaecyparis*.

Growing Requirements

Aspect: The planting site can be of any orientation.
Shelter: Shelter does not have to be provided.
Sun: At least some direct sunlight is desirable.
Soil: A soil of any texture is acceptable, it should be well drained, retain some moisture in summer.
pH: A wide soil pH range is tolerated.

Thuja occidentalis

Hardiness: USDA RHS H7.

A large, columnar tree with reddish-brown bark from eastern North America where it is an important timber tree. It has produced many horticulturally important cultivars.

TABLE 5.24 CULTIVARS OF *THUJA OCCIDENTALIS*

Name	Habit	Height	Spread	Foliage	Remarks
T. occidentalis 'Danica' AGM	Dwarf, rounded tree	10–50cm (33–164ft)	10–50cm (33–164ft)	Upright sprays. Bright-green becoming bronzed in winter	Slow-growing
T. occidentalis 'Holmstrup' AGM	Dense, slightly irregular cone	2.5–4m (8–13ft)	1.5–2.5m (5–8ft)	Upright sprays of dark-green becoming bronzed in winter	
T. occidentalis 'Rheingold' AGM	Variable. Can be a regular pyramid or more irregular in shape	60cm–2m (23in–6½ft)	75cm–1.2m (29in–4ft)	Golden-yellow in summer, stronger colour in winter. May have both adult and juvenile foliage	Slow-growing
T. occidentalis 'Smaragd' AGM	Narrowly conical	1.5–5m (5–16ft)	60–75cm (29in–4ft)	Bright-green	
T. occidentalis 'Sunkist' AGM	Pyramidal or conical	1.5–2m (5–6½ft)	1–1.5m (3–5ft)	Flattened sprays. Lemon-yellow in summer, orange-yellow in winter	

Thuja occidentalis 'Amber Gold'. A good specimen of this dwarf cultivar.

Thuja occidentalis 'Amber Gold'. Showing the bright colour of the foliage.

Thuja occidentalis 'Rheingold'. Seen here contributing to the overall winter scene.

Thuja occidentalis 'Rheingold'. Photographed in winter when the colour of the foliage is exemplified.

Thuja orientalis 'Rosedalis'. A large, old specimen. This cultivar changes its foliage colour with the seasons. Cream-yellow in spring gradually changing to pale-green by autumn and then adopting a reddish-purple colour during winter.

Thuja orientalis 'Rosedalis'. The foliage in winter.

Thuja plicata (Western Red Cedar)

Hardiness: USDA 6b/7a. RHS H6.

A large, evergreen tree from western North America, where it is a very important timber tree. The species and its cultivars give off a pleasant fruity odour similar to pineapple when crushed.

Thuja plicata 'Atrovirens'. A well-maintained, evergreen hedge with the necessary batter to help avoid snow damage and lessen shade at the base of the hedge.

TABLE 5.25 CULTIVARS OF *THUJA PLICATA*

Name	Habit	Height	Spread	Foliage	Remarks
T. plicata 'Atrovirens' AGM	Tall, conical tree	12m+ (39ft+)	8m+ (26ft+)	Glossy dark-green, does not bronze in winter	Growth rate 40–60cm (15½–23in) per year. Makes an excellent hedge or screen. Tolerates clipping very well and can be maintained as a narrow very neat hedge. Slightly slower growing than the excessively vigorous × *Cuprocyparis leylandii* and much less liable to root-borne diseases
T. plicata 'Zebrina' AGM	Tall, conical tree	12m+ (39ft+)	4–8m (13–26ft)	Green with creamy-yellow stripes	Variegation stronger in full sun. Makes an excellent large specimen tree which stands out clearly. Also makes a bright coloured hedge
T. plicata 'Rogersii' AGM	Dwarf	30–40cm (11¾–15½in)	30cm (11¾in)	Yellow and bronze	Occasionally produces shoots which are stronger growing. These should be removed

× *CUPROCYPARIS LEYLANDII* (LEYLAND CYPRESS)

A bigeneric hybrid resulting from the cross *Cupressus macrocarpa* × *Xanthocyparis nootkatensis*. This is a very vigorous plant that grows into a large, dense, columnar tree. Although widely recommended for hedging and screening, it is definitely not for use in small gardens or where space is at all limited. If used in restricted areas, it will rapidly become a serious problem requiring frequent pruning to keep it within bounds.

Growing Requirements

Hardiness: USDA 6b/7a. RHS H6.
Aspect: A site of any orientation is acceptable.
Shelter: No special provision of shelter is necessary.
Sun: Sun or part shade.
Soil: The soil can be of any texture but must be well drained. Good drainage is essential since with poor drainage the plant becomes very susceptible to root pathogens.
pH: A wide range of soil pH is tolerated.

Garden Design Considerations

This evergreen conifer is often used to form a screen or hedge, principally because its growth is very rapid, and the desired effect

TABLE 5.26 × *CUPROCYPARIS LEYLANDII* AND CULTIVARS

Name	Habit	Height	Spread	Foliage	Remarks
leylandii	Tall, dark, upright, loose cone	12m+ (40ft+)	4–8m (13–28ft)	Dark-green sprays of scale-like leaves. Sprays less flattened than those of *Chamaecyparis lawsoniana*	To keep within bounds and tidy as a hedge, it needs frequent clipping and will not regrow from old wood if allowed to over-grow and then pruned
leylandii 'Castlewellan' (syn. 'Castlewellan Gold')	Tall, upright cone	12m+ (40ft+)	2.5–4m (8–13ft)	Light greenish-yellow	Same remarks apply to all cultivars
leylandii 'Gold Rider' AGM	Tall, upright, loose cone	12m+ (40ft+)	4–8m (13–28ft)	Bright-yellow, colour maintained throughout the year	
leyandii 'Haggerston Grey'	Tall, dark, upright, loose cone	12m+ (40ft+)	4–8m (13–28ft)	Greyish-green	
leylandii 'Naylor's Blue'	Tall, dark, upright, loose cone	12m+ (40ft+)	2.5–4m (8–13ft)	Glaucous grey colour more prominent in winter	
leylandii 'Silver Dust'	Tall, slightly less dark, upright loose cone	12m+ (40ft+)	4–8m (13–28ft)	Dark-grey with splashes of creamy-white	

× *Cuprocyparis leylandii*. Shown here producing a serviceable tall hedge on a working plant nursery.

× *Cuprocyparis leylandii* 'Gold Rider'. Once trained and clipped into a shape, regular clipping will be required to maintain a crisp outline. If this is not done, the rapid growth of the plant will soon cause the shape to be lost. Once lost, the shape will be difficult to restore, since this bigeneric hybrid will only regrow if clipped only into soft green growth. If harder pruning needs to be done to re-establish the shape, new growth will not occur and dead patches will appear in the shape.

× *Cuprocyparis leylandii*. A full-grown tree. This picture shows what a large tree will be produced if this bigeneric hybrid is allowed to grow to its full potential.

× *Cuprocyparis leylandii* 'Blue Jeans'. Here trained and clipped into a spiral shape. This hybrid is popular for this purpose, since its rapid growth allows the shape to be produced relatively quickly.

is quickly achieved. Rapid further growth will then occur. If the hedge or screen is to be kept tidy and within bounds, frequent further clipping will be required. If clipping is not kept up, a large amount of further growth will occur. The hedge owner then has a problem.

A light trim back into where the growth is still green, which is recommended for regular maintenance, will not suffice, since this will leave the hedge wider and taller than desired. If more severe pruning is done by cutting into brown-stemmed growth, the desired shape and height can be restored, but regrowth will not occur. The owner is left looking at a brown hedge that will eventually die where it has been pruned. A dead hedge is not a pretty sight!

A far better solution would be to plant another conifer such as a cultivar of *Thuja plicata*, which will be only slightly slower growing but produce a much lower maintenance and attractive hedge.

GARDEN DESIGN CONSIDERATIONS FOR ALL CONIFERS

Conifers have their own specific attributes and constraints when being considered in the context of garden design.

Attributes

Constancy

Almost all conifers planted in gardens are evergreen and, as such, provide a constant green background to other plantings. This constancy should not, however, be considered dull, since colour change will occur during the production of new growth and its subsequent maturing.

Texture

Conifers can provide a range of different textures that can be employed in conjunction with colour to provide a uniform background, as with a hedge of a cultivar of *Chamaecyparis lawsoniana* (Lawson's cypress) or to provide textural contrast as with the needle-like leaves of the small *Pinus mugo* 'Mops' planted adjacent to *Hosta* 'Big Daddy' with its comparatively huge, smooth, chordate leaves.

Colour

Again, conifers can provide intriguing colour combinations, e.g. the strong contrast between *Juniperus squamata* 'Blue Carpet' with its steel-blue foliage against the bright-yellow winter foliage of *Erica carnea* 'Foxhollow'. Colour harmonies can be easily achieved from within combinations within conifers or between conifers and any other plants.

Shapes and Forms

Conifers, especially the smaller and dwarf forms we are considering in this book, come in a vast range of different shapes and forms. The most extreme of these can be deliberately grouped together to give an interesting collection of plants or too great a diversity, a menagerie of different shapes, depending on the beholder's point of view.

Constraints on Use

Sun

Most conifers require planting in full sun or only very light shade. This is brought about by the fact that in nature they are normally the dominant vegetation and capture most, if not all, of the available light. Because of their dominance, they do not experience being covered by the fallen leaves of other deciduous trees and shrubs. The planting of conifers is, therefore, widely restricted to areas of full sun and the prompt removal of fallen leaves can make the use of conifers for ground cover a maintenance issue.

It is often said that it is difficult to combine conifers with other plants in a mixed planting. This may be true because of two main reasons.

Conifers, when found in nature, are often highly adapted to the often rather extreme conditions, they are growing in. This results in them often growing largely on their own and dominating an area without any other broadleaved plants. The eye is thus not accustomed to seeing conifers and broadleaved plants mixed together.

Conifers are not used to competition from broadleaved plants. In nature, competition is from their own kind and where two or more conifers grow together, the resultant shade often results in two or more one-sided plants, the shaded areas having only dead foliage. Consider conifers in a forestry planting where each tree often consists of a long, bare trunk and a small cone of growth where the top receives the sun. In the garden or planted landscape, the same occurs. Where overgrown by other plants, even the temporary growth of herbaceous plants, the conifer foliage will die, resulting in unsightly one-sided plants. Such sensitivity can make conifers difficult to incorporate into a mixed planting and where it is done, maintenance must be vigilant to prevent any degree of shading or overgrowth.

Drainage

With only very few exceptions, they require good drainage. Planting on badly drained ground is, therefore, not a viable option, unless effective measures are taken to improve the soil drainage prior to planting.

Growth Rate and Ultimate Size

Some conifers have a very fast rate of growth, e.g. the bigeneric hybrid × *Cuprocyparis leyandii* (Leyland cypress), and are difficult to place within a planting scheme. Others, especially wild-occurring species, naturally grow into large or very large trees and are not suitable for the small garden or restricted spaces that we are considering.

The difference between truly dwarf conifers that never become large trees and those called slow-growing must be fully appreciated. Although planted whilst small, slow-growing conifers will grow and may eventually become large. The question as to how slow-growing is a so-called slow-growing conifer, is often asked. To help answer this question, in this book the estimated size ten years after planting of dwarf and slow-growing conifers is given with an indication of when subsequent growth is liable to produce a much larger tree. This information must be taken into account when considering plant inclusion and spacing. A simple choice is to only plant true dwarf trees, but this severely limits plant choice. If slow-growing cultivars are chosen, then allowance must be made for their growth. This may involve the idea of removing certain plants after a period of time. Consider, will you still be in a position to see that this occurs? One too often sees a previously beautiful rock garden that is now dominated by one all-engulfing tree of *Chamaecyparis lawsoniana* 'Ellwoodii'.

Small Gardens and Where Space is Limited

Ground Cover

Prostrate cultivars of conifers, especially of junipers, can look outstanding as ground cover. Where space permits, a comparatively large area of one cultivar can have another of a different colour or height planted adjacent to it, giving a sculptural quality to the overall planting. To be effective, weed eradication must be effective before the conifers are planted and any fallen leaves from deciduous trees promptly removed.

Steep Banks

These are often very dry, sunny positions and potentially difficult to establish plants on. The choice of plants is restricted, but several conifers are adapted to these conditions, e.g. *Juniperus horizontalis* 'Blue Chip'.

Cryptomeria japonica 'Globosa Nana', *Hakonechloa macra* 'Aureola and *Acer griseum* forming an interesting plant association. The leaf colours and textures of the *Cryptomeria* and the *Hakonechloa* contrast strongly.

A collection of dwarf and slow-growing conifers in a rock garden. Each conifer is an interesting specimen and collectively they give structure and permanence to the planting, when the other alpine and rock garden plants are largely absent in the winter.

The Heather Garden

The combination of conifers with *Erica* and or *Calluna* has been advocated for many years. Unfortunately, some gardeners and professional horticulturists consider this has been rather overdone in the past and the suggestion is met with a yawn. The fact remains that these genera do favour the same conditions and grow together well. When planted imaginatively, with perhaps the inclusion of some deciduous trees that only cast light shade and produce little in the way of fallen leaves, such as the rowans (*Sorbus*), an extremely attractive and low-maintenance feature can be created.

Other Ideas

Dwarf and slow-growing conifers lend themselves very well to a variety of other situations, such as rock gardens, raised beds, dry stone walls, planters, troughs, bonsai and penjing. Their permanence gives structure to a planting of alpines and other rock garden plants. The flowering plants certainly give seasonal bright colours, but the conifers provide something of a background to such plantings and a continuity not provided by the others.

Another much valued property of conifers in such situations is their ability to create a miniature landscape, especially when used in conjunction with rocks, so placed as to give the impression of a natural rock exposure. This ability to produce a perfect miniature has long been appreciated by the devotees of both bonsai and penjing. In bonsai, meaning plant in a pot, the intention is to produce a perfect miniature tree. In penjing, the intention is to produce a miniature landscape. This may or may not include a plant or plants. Where a plant is a component of this landscape, miniature conifers fulfil this application to perfection.

Vertical Emphasis

Where a plant is needed to give vertical emphasis to a planting or to act as a vegetative exclamation mark, several conifers excel. Such a feature needs to be in scale with its surroundings and conifers provide a good choice. On a modern, small-garden scale, some of the fastigiate forms of *Taxus* (yew) would be appropriate, e.g. *Taxus baccata* 'Standishii', whereas on a very small scale, as in a trough garden, *Juniperus communis* 'Compressa' (Noah's ark juniper) would fulfil this role for many years before, if ever, becoming too large.

Formality

Taxus (yew) can be clipped into (and many other conifers naturally grow into) formal, regular shapes. These can be used to convey formality if required of a design, especially where space is limited, and other plants would be inappropriate because of their size.

Focal Points and Framing

A well-grown specimen conifer can be an arresting sight and functions well as a focal point in a design.

The dense, evergreen growth of conifers is very effective in limiting the viewer's field of view and directing sight to an intended view. This effect can be employed in gardens of any size. Even in the tiniest of spaces, a pair of vertical, pencil-like junipers can be employed to frame an intended feature.

Hedges and Screening

Plants within the conifer genera of *Chamaecyparis*, *Thuja* and, especially, *Taxus* are excellent for this purpose. Provided plants are selected from the tables provided, with the correct vigour and rate of growth to fit the space available, an entirely satisfactory hedge or screen can be created.

Thuja occidentalis 'Golden Smaragd', *Berberis thunbergii* 'Helmond Pillar' and two plants of a cultivar of *Festuca glauca* form an eye-catching combination.

Cupressus sempervirens 'Totem Pole'. Several trees of this cultivar creating points of vertical emphasis in an otherwise rather flat underplanting.

CHAPTER 6

DIRECTORY OF SUGGESTED *CALLUNA*, *DABOECIA* AND *ERICA*

CALLUNA (HEATHER)

A monotypic genus, the only species being *Calluna vulgaris*, which occurs in Europe, including the British Isles, and Asia Minor. A small, evergreen shrub never more than 1m (3ft) tall and usually 23–61cm (9–24in) tall. Leaves small, scale-like, oppositely arranged in four rows and pressed together. The flowers have four petals and four sepals. In contrast to *Erica*, it is the calyx and not the corolla that is coloured. Flowering occurs from late summer into autumn and is very attractive to bees. The flower does not drop, but gradually dries out, fades and eventually becomes light brown. 'Bud-blooming' cultivars have been developed where the flower develops colour normally, but never opens. This is claimed to confer a longer coloured flowering period.

Growing Requirements

Hardiness: USDA 6a-1. RHS H7.
Aspect: A site of any orientation is acceptable.
Shelter: No special degree of shelter is required.
Sun: Full sun is required.
Soil: The soil must be well drained and ideally moisture-retentive in summer. A light, sandy soil is favoured, clay and loam soils can be improved by the addition of organic matter prior to planting.
pH.: The soil pH should be acid or neutral and ideally 6.5 or lower.

Calluna vulgaris 'Athene'[PBR] (Garden Girl Series). A 'bud-blooming' cultivar in which the flower colours normally, but never opens. The flower never opening is claimed to make the flower last longer.

Calluna vulgaris 'Dart's Amethyst'.

Calluna vulgaris 'Ellen'.

Calluna vulgaris 'Robert Chapman'. This cultivar does produce some pink flowers but is mainly grown for its coloured foliage. The foliage is golden-yellow in the summer and then develops coppery-red colour in the winter.

Calluna vulgaris 'H. E. Beale'. A double-flowered cultivar.

Erica arborea var. *alpina* f. *aureifolia* 'Albert's Gold', providing bright foliage colour and a distinct habit in a frosty winter garden.

TABLE 6.1 CULTIVARS OF *CALLUNA VULGARIS*

Name	Height	Spread	Flower	Foliage	Remarks
'Anette'[PBR] (Garden Girl Series)	30cm (1ft)	30–45cm (1–1½ft)	Pink	Dark-green	Bud blooming
'Athene'[PBR] (Garden Girl Series)	30cm (1ft)	30–45cm (1–1½ft)	Crimson	Dark-green	Bud blooming
'Dart's Amethyst'	30cm (1ft)	30–45cm (1–1½ft)	Purple	Dark-green	
'Ellen'	30cm (1ft)	30–45cm (1–1½ft)	Dark-pink	Bright-green	
'Gold Spronk'	45cm (1½ft)	30–45cm (1–1½ft)	White	Lemon-yellow all the year	
'H. E. Beale' AGM	24cm (10in)	60cm (2ft)	Double silvery pink	Greyish-green	
'Peter Sparkes' AGM	20–50cm (8–20in)	60cm (2ft)	Double pink	Dark-green	
'Robert Chapman' AGM	25cm (10in)	10–50cm (4–20in)	Light purple	Golden-yellow becoming coppery-red in winter	

DABOECIA

A genus consisting of only two species, both of which are low-growing, evergreen shrubs. *Daboecia azorica* is found in the Azores. *Daboecia cantabrica* (Connemara heath or Saint Dabeoc's heath) is from Western Europe, including Ireland. *Daboecia × scotica* is the result of *Daboecia azorica* crossing with *Daboecia cantabrica*. These hybrids are smaller and more compact than *D. cantabrica*.

They are separated from the genus *Erica* by having broader, alternate, never whorled leaves. The flowers of *Daboecia* are nodding, with urn-shaped corollas. The flowers are larger than found in *Erica* and, in contrast with *Erica*, the corolla of *Daboecia* is deciduous, falling once the flower has finished. Flowering occurs from summer into autumn.

Growing Requirements

Hardiness: USDA 7b/8a. RHS H5.
Aspect: The orientation of the planting site can be any, except north-facing.
Sun: Full sun is required.
Shelter: No specific provision has to be made for shelter.
Soil: The soil, which must be well-drained, should be loam or sandy.
pH: The soil pH must be acid or neutral. Any alkalinity is not tolerated.

TABLE 6.2 CULTIVARS OF *DABOECIA CANTABRICA*

Name	Habit	Height	Spread	Flower colour	Flowering period
D. cantabrica 'Glamour'	Bushy	30cm (12in)	60cm (24in)	Light-pink	Summer into autumn
D. cantabrica 'Praegerae'	Bushy	30cm (12in)	60cm (24in)	Deep-pink	Summer into autumn
D. cantabrica 'Purpurea'	Bushy	45cm (18in)	60cm (24in)	Reddish-purple	Summer into autumn
D. cantabrica 'Tinkerbell'	Bushy	30cm (12in)	60cm (24in)	Rose-pink	Summer into autumn
D. cantabrica 'Waley's Red'	Bushy	35cm (14in)	60cm (24in)	Magenta	Summer into autumn
D. × scotica 'Jack Drake'	Bushy and compact	20cm (8in)	45cm (18in)	Dark-red	Summer into autumn
D. × scotica 'Katherine's Choice'	Bushy and compact	15cm (6in)	45cm (18in)	Pink	Summer into autumn
D. × scotica 'Silwerwells' AGM	Bushy and compact	15cm (6in)	45cm (18in)	White	Summer into autumn
D. × scotica 'William Buchanan' AGM	Bushy and compact	45cm	45cm (18in)	Deep-purple	Summer into autumn

Daboecia cantabrica 'Glamour'.

Daboecia cantabrica 'Tinkerbell'.

Daboecia × scotica 'Cora'. The *Daboecia × scotica* cultivars are shorter and spread less than the *Daboecia cantabrica* cultivars.

Daboecia cantabrica 'Tinkerbell'. A photograph to show the corolla that is expanded and coloured in *Daboecia*. After the flower finishes, the corolla drops off the plant. This is in contrast to *Erica* where it is the calyx that is coloured and this is retained on the plant once the flower is finished.

ERICA

A genus of evergreen shrubs and small trees whose main centre of diversity is in South Africa. Unfortunately, those from South Africa are unreliably or not hardy in north temperate regions. Other species are found in Europe and South-West Asia. Five species are native to the United Kingdom.

In *Erica*, it is the corolla that is conspicuously coloured and does not fall in the normal way, but withers and remains on the flower stalk for some time. The evergreen leaves are needle-like and produced in whorls up the stem.

Hardiness: South African species are not generally considered hardy, except in extremely favoured situations. Other species need to be considered on a species-by-species basis.

Erica arborea (Tree Heath)

A potentially large, evergreen shrub with dark-green, needle-like leaves from the Mediterranean basin, Ethiopia, Kenya and Tanzania. It produces honey-scented, white, bell-shaped flowers that are very attractive to bees in early spring. In the wild, it grows into a large shrub or small tree up to 7m (23ft) tall but does not achieve this in the garden situation.

Erica arborea var. *alpina* f. *aureifolium* 'Estrella Gold'. An upright-growing plant with distinctive yellow foliage.

Growing Requirements

Hardiness: USDA 8b/9a. RHS H4.
Aspect: The planting site can be of any orientation, except facing north.
Shelter: Avoid cold or windswept situations.
Sun: Full sun is required.
Soil: The soil can be of clay, loam or sandy texture, but must be well drained.
pH: The plant is calcifuge and soil pH must acidic or neutral.

Erica carnea (Syn. *E. herbacea*) (Winter Heath)

This is an important species for the gardener, especially where space is limited, since the species and many cultivars are dwarf, evergreen shrubs forming effective groundcover. Clipping over just after flowering to remove the old flower stems, but not into older growth, is recommended to encourage new, compact growth.

Collectively they are winter-flowering, the flowering season extending from late autumn to early spring. By careful selection of different cultivars, flower colour can be maintained over this long period. This is indicated in Table 6.4.

Growing Requirements

Hardiness: USDA 6b/7a. RHS H6.
Aspect: The growing site can face in any direction.
Shelter: No specific degree of shelter is necessary.
Sun: Full sun is best, but light shade is acceptable.
Soil: The soil can be of any texture but should be fine-structured to allow the very fine root system to penetrate. If the soil is heavy with a high clay content, organic matter should be added and incorporated to flocculate and break the clay into fine granules. Drainage should be good but retain adequate moisture in the summer.
pH: A wide pH range is tolerated, including mildly alkaline soils, but thin, calcareous soils are not recommended.

TABLE 6.3 *ERICA ARBOREA* AND CULTIVARS

Name	Habit	Height	Spread	Foliage	Flowers	Remarks
E. arborea	Upright	1–4m (3–13ft)	Variable can be tree or shrub-like	Dark-green	White in spring	Honey-scented and very attractive to bees. Susceptible to cold and wind damage
E. arborea var. *alpina*	Upright	1.5–2.5m (5–8ft)	0.5–1m (20–30in)	Dark-green	White in spring	Still rated USDA RHS 4, but in practice found to be hardier than *E. arborea*
E. arborea var. *alpina* f. *aureifolia* 'Albert's Gold' AGM	Upright	1.5–2.5m (5–8ft)	0.5–1m (20–30in)	Golden-yellow young foliage	White, but seldom blooms	Hardier than *E. arborea*
E. arborea var. *alpina* f. *aureifolia* 'Estrella Gold'	Upright	1m (40in)	75cm (30in)	Lime-green foliage, tipped yellow in spring/ early summer		Hardier than *E. arborea*

TABLE 6.4 CULTIVARS OF *ERICA CARNEA*

Name	Habit	Height	Spread	Foliage	Flowers	Flowering season *	Remarks
E. carnea 'Ann Sparkes' AGM	Low mat of growth	15cm (6in)	25cm (10in)	Golden-yellow in summer. Dark golden-yellow flushed red in winter	Rose pink	L	
E. carnea 'Eva' AGM	Low mat of growth	15cm (6in)	30cm (12in)	Dark-green becoming bronzed in winter	Pinkish-red	E	
E. carnea f. *aureifolia* 'Foxhollow' AGM	Low mat of growth	25cm (10in)	40cm (16in)	Yellow-green in summer. Reddish orange in winter	Pale-pink	L	Distinctive foliage colour in summer and winter
E. carnea 'Myretoun Ruby' AGM	Low mat of growth	15cm (6in)	45cm (18in)	Dark-green	Deep rose pink	L	
E. carnea f. *alba* 'Springwood White'	Low mat of growth	20cm (8in)	45cm (18in)	Bright-green	white	M	Very profuse flowering
E. carnea 'Vivellii' AGM	Low mat of growth	15cm (6in)	35cm (14in)	Green tipped purple in winter	Purple, pink, darken with age	M	Good combination of flower and foliage colour
E. carnea f. *alba* 'Winter Snow' AGM	Low mat of growth	20cm (8in)	45cm (18in)	Green	White	M	

* E =late autumn; M= mid-winter; L= late winter/early spring.

Erica carnea 'Ann Sparkes'. Although winter-flowering, this cultivar also provides some summer foliage colour.

Erica carnea f. *aureifolium* 'Foxhollow'. This cultivar provides foliage colour all year round.

Erica carnea 'Myretoun Ruby'. An excellent profuse-flowering cultivar.

Erica cinerea (Bell Heather)

A dwarf, evergreen shrub native to western Europe, including the British Isles. The leaves are linear dark-green, and the leaf margins are rolled with the edges rolled back. The flowers are small, urn-like and of a papery texture, remaining long after the flower colour has faded. Flowering occurs from early summer to early autumn.

The cultivars of this species are useful in that they provide summer colour, form effective low-growing ground-cover and combine well with *Erica carnea* cultivars to produce almost continuous flower colour throughout the year.

Clipping just after flowering, in the same manner and reason as for *E. carnea*, is recommended.

Growing Requirements

Hardiness: USDA 6a-1. RHS H7.
Aspect: The site can be of any orientation.
Shelter: No specific shelter is necessary.
Sun: Full sun is required.
Soil: The soil can be of any texture but should be fine structured to allow the very fine root system to penetrate. If the soil is heavy with a high clay content, organic matter should be added and incorporated to flocculate and break the clay into fine granules. Drainage should be good but retain adequate moisture in the summer.
pH: The species is ericaceous and requires a lime free or neutral soil.

TABLE 6.5 CULTIVARS OF *ERICA CINEREA*

Name	Habit	Height	Spread	Foliage	Flower
E. cinerea f. *alba* 'Alba Major'	Forms a mat-like carpet of growth	30cm (12in)	45–70cm (18–28in)	Dark-green	White
E. cinerea f. *aureifolia* 'Golden Hue'	Forms a mat-like carpet of growth	35cm (14in)	70cm (28in)	Lime-green	Small pale-pink
E. cinerea 'C. D. Eason' AGM	Forms a mat-like carpet of growth	25cm (10in)	45cm (18in)	Dark-green	Deep-pink
E. cinerea 'Pink Ice' AGM	Forms a mat-like carpet of growth	20 cm (8in)	40cm (16in)	Dark-green	Rich pink
E. cinerea 'Stephen Davis' AGM	Forms a mat-like carpet of growth	20 cm (8in)	45 cm (18in)	Dark-green	Purple
E. cinerea 'Velvet Night' AGM	Forms a mat-like carpet of growth	20cm (8in)	40–60cm (16–24in)	Dark-green	Blackish purple

Erica cinerea 'Nellie Dawson'. This cultivar flowers from late summer into autumn.

Erica × *darleyensis*

A hybrid between *Erica carnea* and *Erica erigena*. Cultivars of this cross are a little more vigorous and slightly taller than *Erica carnea*. In the garden and landscape, they make a natural combination with *Erica carnea* cultivars, providing some variation in height, extending the range of flower colour and adding to those providing foliage colour.

Growing Requirements

Hardiness: USDA 7b/8a. RHS H5.

Other requirements are the same as for *Erica carnea*.

TABLE 6.6 CULTIVARS OF *ERICA* × *DARLEYENSIS*

Name	Habit	Height	Spread	Foliage	Flowers
E. × *darleyensis* 'Arthur Johnson' AGM	Forms a mat-like carpet of growth	60cm (24in)	90cm (35in)	Dark-green	Pink-magenta. Winter into spring
E. × *darleyensis* 'Darley Dale' (syn. *E.* × *darleyensis* 'Pink Perfection')	Forms a mat-like carpet of growth	30–60cm (12–24in)	60–90cm (24–35in)	Dark-green	Pink. Over a long period
E. × *darleyensis* 'Furzey' AGM	Forms a mat-like carpet of growth	35–40cm !4–16in)	30–60cm (12–24in)	Dark-green	Rose-pink. From winter into spring
E. × *darleyensis* 'Ghost Hills' AGM	Forms a mat-like carpet of growth	30cm (12in)	50cm (20in)	Bright-green, tipped with cream in spring	Purple. Late winter into early spring
E. × *darleyensis* f. *aureifolia* 'Jack H. Brummage'	Forms a mat-like carpet of growth	30cm (12in)	60cm (24in)	Pale yellow foliage, becoming golden tinged red in winter	Deep-pink. Late winter into spring
E. × *darleyensis* f. *albiflora* 'Silberschmelze'	Forms a mat-like carpet of growth	35–40cm (14–16in)	100cm (40in)	Dark-green	White with a sweet scent. Early winter into late spring
E. × *darleyensis* f. *albiflora* 'Spring Surprise'[PBR]	Forms a mat-like carpet of growth.	30–45cm (12–18in)	45–60cm (18–24in)	Dark-green	Deep-pink. Early to late spring

Erica × *darleyensis* 'Arthur Johnson'. Showing the mat-like habit of these cultivars.

Erica × *darleyensis* 'Darley Dale'.

Erica × *darleyensis* 'Furzey'. In the picture, this cultivar, like the other cultivars in this group of hybrids, can be seen to be slightly taller than the *Erica carnea* cultivars.

Erica × *darleyensis* 'Furzey'. A good free-flowering cultivar.

Erica × *darleyensis* f. *albiflora* 'Silberschmelze'.

Erica × *darleyensis* f. *albiflora* 'Silberschmelze'. The flowers of this cultivar are sweetly scented.

Erica erigena (syn. *E. mediterranea, E. hibernica*) (Irish Heath)

A compact,evergreen shrub native to Ireland, south-west France, Spain, Portugal and Tangier. It has dark-green foliage and produces pink, scented flowers in late winter and early spring.

Growing Requirements

Hardiness: USDA 7b/8a. RHS H5.
Aspect: The planting site can be of any orientation, except facing north.
Sun: Full sun is required.
Soil: The soil can be of clay, loam or sandy texture, but must be well drained.
pH: The plant is calcifuge, and the soil pH must be acidic or neutral.

TABLE 6.7 CULTIVARS OF *ERICA ERIGENA*

Name	Habit	Height	Spread	Foliage	Flowers
E. erigena 'Irish Dusk AGM	Upright bush	0.5–1m (20–40in)	0.5m (20in)	Dark-green	Salmon-pink
E. erigena 'Superba'	Upright bush	1.5–2.5m (5–8ft)	0.5–1m (20–40in)	Dark-green	Shell-pink
E. erigena f. *alba* 'W. T. Rackliff' AGM	Rounded bush	60cm (24in)	55cm (22in)	Dark-green	White

Erica tetralix (Cross-Leaved Heath)

This species is native to Western Europe, including the British Isles as well as boggy regions of central Europe. It is found mainly in wet heaths, bogs and damp, coniferous woodland. A sub-shrub with needle-like leaves arranged in whorls of four, hence its common name. Pink, bell-shaped flowers are produced at the distal ends of shoots in summer and autumn.

Growing Requirements

Hardiness: USDA 6b/7a. RHS H6.
Aspect: The planting site can face in any direction.
Shelter: There is no necessity to provide shelter.
Sun: Full sun is required.
Soil: The soil must be well drained and can be of clay, loam or sandy texture.
pH: The plant is calcifuge and soil pH must be acid or neutral.

TABLE 6.8 CULTIVARS OF *ERICA TETRALIX*

Name	Habit	Height	Spread	Foliage	Flowers
E. tetralix f. *alba* 'Alba Mollis' AGM	Compact and forming a low mound	0.1–0.5m (4–20in)	0.1–0.5m (4–20in)	Dark grey-green	Nodding, bell-shaped, white flowers
E. tetralix 'Con Underwood'	Compact and forming a low mound	0.1–0.5m (4–20in)	0.1–0.5m (4–20in)	Grey-green	Magenta-crimson
E. tetralix 'Ken Underwood'	Compact and forming a low mound	0.1–0.5m (4–20in)	0.1–0.5m (4–20in)	Grey-green	Salmon-pink
E. tetralix f. *stellata* 'Pink Star' AGM	Compact and forming a low mound	0.1–0.5m (4–20in)	0.1–0.5m (4–20in)	Grey-green	Lilac-pink, less nodding than other cultivars

Erica vagans (Cornish Heath)

A dwarf, spreading, evergreen shrub native to south-west Europe and Cornwall. It has narrow, needle-like leaves arranged in whorls of four or five. Small, rounded flowers are produced on an erect, leafy, cylindrical raceme. Flowering occurs from late summer to early autumn.

Growing Requirements

Hardiness: USDA 7b/8a. RHS H5.
Aspect: The site can face any orientation.
Shelter: No specific shelter is required.
Sun: Full sun is required.
Soil: The texture of the soil may be clay, loam or sand and must be well drained**.**
pH: The soil must be acid or neutral, no alkalinity is tolerated.

TABLE 6.9 CULTIVARS OF *ERICA VAGANS*

Name	Habit	Height	Spread	Foliage	Flowers
E. vagans 'Birch Glow' AGM	Bushy mat	25–30cm (10–12in)	45–60cm (18–24in)	Dark-green	Rose-pink
E. vagans f. *alba* 'Cornish Cream' AGM	Bushy mat	35–45cm (14–18in)	60–75cm (24–30in)	Dark-green	Cream
E. vagans f. *alba* 'Lyonesse' AGM	Bushy mat	20–25cm (8–10in)	45–60cm (18–24in)	Bright-green	White with prominent brown anthers
E. vagans 'Hookstone Rose'	Bushy mat	40–45cm (16–18in)	60–75cm (24–30in)	Dark-green	Deep-pink
E. vagans 'Mrs D. F. Maxwell. AGM	Bushy mat	25–30cm (10–12in)	30–45cm (12–18in)	Dark-green	Rose-pink
E. vagans 'Pyrenees Pink'	Bushy mat	16–20cm (6.5–8in)	40–60cm (16–24in)	Dark-green	Pink, fading to white

Erica vagans 'Birch Glow'. A very good cultivar.

Erica vagans 'Pyrenees Pink'. A short-growing cultivar, but the same degree of spread as most others.

Erica vagans 'Hookstone Rose'. A tall and vigorous cultivar.

GARDEN DESIGN CONSIDERATIONS

This group of genera has a number of collective attributes and this is why they are all considered together in the context of garden design.

Attributes

Constancy

The plants in all three genera are evergreen. Thus, they cover the ground and provide backdrop to other plantings.

Texture

All three genera contain plants with a fine texture. In *Erica* and *Daboecia* the foliage consists of small, needle-like leaves; in *Calluna* the leaves are again small and appressed to the stems. As such, planted *en masse*, they give a textural uniformity to a planting. This may be what is desired, but if not, then plants from other genera will need to be deployed to give textural contrast.

Colour

This collective group should be highlighted here. Considering all three genera, flower colour can be provided all the year round.

The genus *Erica* can provide flowers for twelve months of the year. Cultivars of both *Erica carnea* and *Erica* × *darleyensis* will provide flowers throughout the winter. Late-flowering cultivars of *E. carnea*, *E.* × *darleyensis* and *E. arborea* plus its cultivars all flower in spring. *Erica cinerea* flowers from early summer into autumn and *E. vagans* a little after *E. cinerea*.

The genera *Daboecia* and *Calluna* both flower in the summer, adding even more colour.

Colour is provided also by foliage. Many shades of green are displayed, plus other cultivars deliberately selected for other foliage hues. *Erica arborea* var. *alpina* f. *aureifolium* 'Albert's Gold' has light-yellow foliage all the year round.

Several cultivars of *Erica carnea* and *E.* × *darleyensis* develop bronzing of normally green foliage in cold spells in the winter and *E. carnea* 'Ann Sparkes' has golden-yellow foliage in summer, which too becomes flushed with red during the winter.

Erica tetralix has grey-green foliage, which offers a slight contrast to the other *Erica* species.

Low-Maintenance Requirement

Once planted into weed-free soil, all three genera require little routine maintenance. After flowering, plants should be lightly clipped over to remove the old flower inflorescence, whilst taking care not to cut into older growth. If considered necessary to maintain vigour of the plants, they may be lightly top-dressed with an acidic organic compost, this being brushed into the plants so that it covers the roots at the soil surface. Any other fertiliser should not be required.

The need for the plants to receive maximum sunlight must be remembered. It is important that any fallen leaves from deciduous plants are quickly removed. If left on top of the plants, the covered areas will soon die back leaving unsightly areas.

The growth of trees in the heather planting or adjacent must be monitored and any growth causing shade to the plants beneath removed.

Constraints to Use

Soil

All plants from the genera *Calluna* and *Daboecia* require a soil with an acid pH or just neutral. Within *Erica*, of the commonly grown species, only *Erica carnea* and *Erica* × *darleyensis* will tolerate a mildly alkaline soil and even these prefer an acid soil. This tolerance of *E. carnea* and E. × *darleyensis* to neutrality or slight alkalinity is, however, very important, allowing much wider use of these species.

Plants from all three genera prefer a finely structured soil in which their fine roots can easily penetrate. Sandy and light soils are well liked, heavier clay soils are best improved by the addition of organic matter to flocculate the clay and bring the soil to a more crumbly, friable structure.

Sun

Although *E. carnea* and *E.* × *darleyensis* will tolerate a little light shade, the rest of the three genera require full sun. Obviously, this may preclude their use where this cannot be provided.

Uniformity of Habit and Texture

This uniformity of texture was mentioned with a caveat under Attributes. *Erica*, *Calluna* and *Daboecia* all tend to have a rounded habit and contribute little with regard to contrasting shapes within a planting. *Erica alpina* and its cultivars are an exception and can be used to give a marked variation in height and shape.

A heather bed in late winter. Much of the colour and interest is provided by cultivars of *Erica carnea* and *Erica* × *darleyensis*, plus a plant of *Helleborus* × *hybridus* in the foreground.

Potential Applications of *Erica*, *Calluna* and *Daboecia* in the Design of Small Gardens and Where Space is Limited

Heather Gardens with Dwarf and Slow-Growing Conifers Plus Other Small Trees
In heather gardens, *Erica*, *Calluna*, *Daboecia*, dwarf conifers and slow-growing conifers, plus other small trees, associate together very well. In the past this type of feature may have been over-emphasised to the detriment of its current popularity with gardeners.

Providing the choice of plants is carefully done, such a planting can be accommodated in a very small or larger space and be an excellent feature. Care must be taken not to plant conifers that will rapidly outgrow their allocated position. *Erica*, *Calluna* and *Daboecia* must also be chosen with care to match the soil present and to provide a long season of flower colour. Deciduous trees added to this association must be selected with regard to their size, and the density and duration of shade they cast. *Sorbus* (rowans) are often a good choice, not growing too fast or too large, producing only a light canopy of leaves and contributing to the design by adding height, plus seasonal flowers and fruit colour.

Rock Gardens, Raised Beds, Patio Gardens, Containers And Pots
Being small, relatively compact plants, *Erica*, *Calluna* and *Daboecia* are well suited to the rock garden, where their evergreen nature is in contrast to the often rather ephemeral appearance of many alpine and rock garden plants. The fact that they can provide a long flowering period is also very valuable. Rock garden plants and alpines are, in the main, spring-flowering, and when used exclusively leave a long gap in summer, and particularly winter, when they provide little in the way of flower colour.

Winter-flowering *Erica* is particularly useful for incorporating into small schemes for planters, and into containers of every sort, plus pots. The fact that its soil requirements are less demanding than *Calluna* and *Daboecia*, plus its small, compact nature, makes it very well-suited for this purpose.

Low-Maintenance Plantings

As explained under Attributes, the group does have a low-maintenance requirement and this is a distinct advantage where labour for maintenance is in short supply. Such plantings should also look very appealing, the choice available in flowering times and coloured foliage cultivars meaning that at no time during the gardening year should such a feature look dull or uninteresting.

Banks and Slopes

Banks and slopes exposed to full sun are often very hot and dry areas for planting. Many *Erica*, *Calluna* and *Daboecia* naturally grow in sunny, often excessively well-drained areas and are, therefore, well adapted to planting on banks and slopes. The elevation provided by the bank or slope can be used to display patterns due to differing foliage colours or flowering times, and coloured as created by the gardener or designer.

General Flower Borders and Beds

All three genera under consideration can be included in the mixed borders found today in most small gardens. Care must be taken to see that the soil is suitable and that they are not overgrown by other more vigorous plants. Given this, they can make a great contribution, especially to the winter garden scene.

A large heather garden, where *Stipa gigantea* acts as the background and *Yucca rigida* provides a focal point.

APPENDIX I

RHS HARDINESS RATINGS

Rating	Temperature ranges (°C)	Category	Definition	USDA
H1a	>15	Heated greenhouse – tropical	Under glass all year	13
H1b	10 to 15	Heated greenhouse – subtropical	Can be grown outside in the summer in hotter, sunny and sheltered locations (such as city centre areas), but generally perform better under glass all year round	12
H1c	5 to 10	Heated greenhouse – warm termperate	Can be grown outside in summer throughout most the UK while day-time temperatures are high enough to promote growth	11
H2	1 to 5	Tender – cool or frost-free greenhouse	Tolerant of low temperatures, but will not survive being frozen. Except in frost-free inner-city areas or coastal extremities, requires glasshouse conditions. Can be grown outside once risk of frost is over	10b
H3	1 to −5	Half-hardy – unheated greenhouse/mild winter	Hardy in coastal and relatively mild parts of the UK, except in severe winters and at risk of sudden (early) frosts. May be hardy elsewhere with wall shelter or good microclimate. Likely to be damaged or killed in cold winters, particularly with no snow cover or if pot grown. Can often survive with some artificial protection in winter	9b/10a
H4	−10 to −5	Hardy – average winter	Hardy through most parts of the UK apart from inland valleys, at altitude and central/northerly locations. May suffer foliage damage and stem dieback in harsh winters in cold gardens. Some normally hardy plants may not survive long, wet winters in heavy or poorly drained soil. Plants in pots are more vulnerable to harsh winters, particularly evergreens and many bulbs	8b/9a
H5	−15 to −10	Hardy – cold winter	Hardy in most places throughout the UK, even in severe winters. May not withstand open/exposed sites or central/northern locations. Many evergreens will suffer foliage damage, and plants in pots will be at increased risk	7b/8a
H6	–	Hardy – very cold winter	Hardy in all of UK and northern Europe. Many plants grown in containers will be damaged unless given protection	6b/7a
H7	<–20	Very hardy	Hardy in the severest European continental climates, including exposed upland locations in the UK	6a-1

Source: www.rhs.org.uk

USDA PLANT HARDINESS ZONES MAPS FOR THE US AND UK

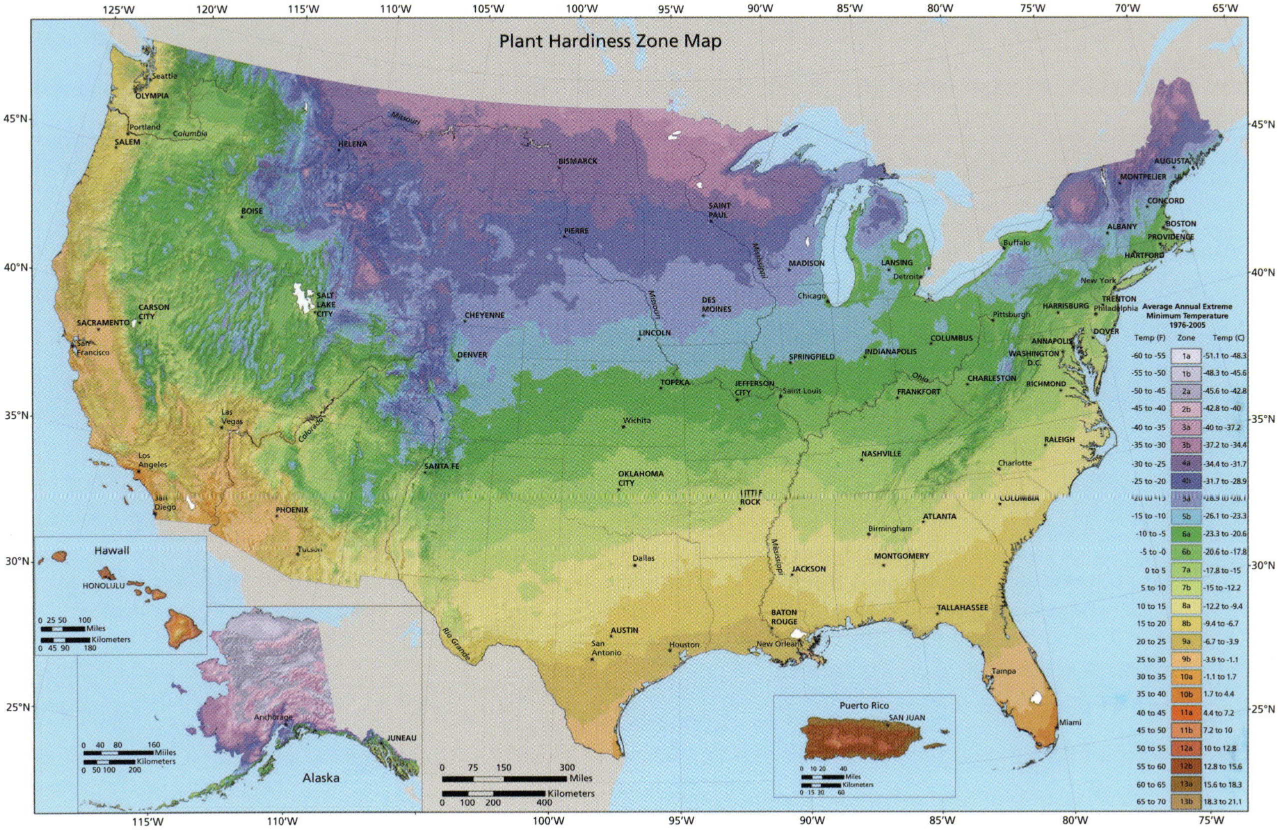

US hardiness zones map.

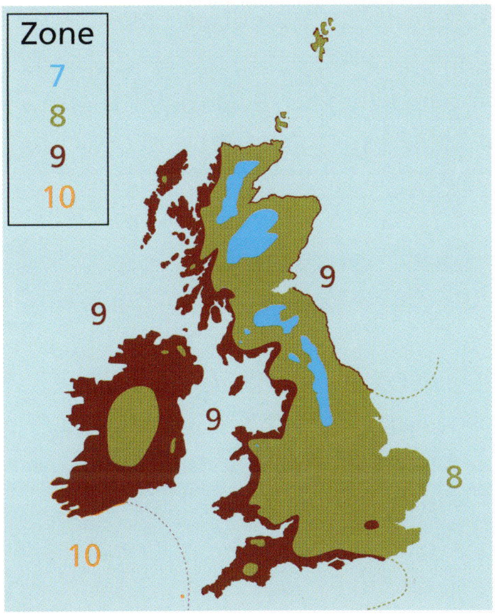

UK hardiness zones map.

Average Annual Extreme Minimum Temperature 1976-2005

Temp (F)	Zone	Temp (C)
-60 to -50	1	-51.1 to -45.6
-50 to -40	2	-45.6 to -40
-40 to -30	3	-40 to -34.4
-30 to -20	4	-34.4 to -28.9
-20 to -10	5	-28.9 to -23.3
-10 to 0	6	-23.3 to -17.8
0 to 10	7	-17.8 to -12.2
10 to 20	8	-12.2 to -6.7
20 to 30	9	-6.7 to -1.1
30 to 40	10	-1.1 to -4.4
40 to 50	'11	4.4 to 10
50 to 60	12	10 to 15.6
60 to 70	13	15.6 to 21.1

Actual temperatures recorded over thirty years.

APPENDIX III

SYMBOLS AND ABBREVIATIONS

Appearing to the left of the name	
+	Indicates a graft hybrid genus
Appearing to the right of the name	
AGM	The Award of Garden Merit is awarded by the Royal Horticultural Society to plants that, after assessment, are considered excellent garden plants.
PBR	Appears after the name of a plant covered by a grant of Plant Breeders' Rights. These are awarded by both the UK and EU Plant variety Rights offices. Plants awarded PBR must not be propagated for sale unless a licence has been awarded to the propagator by the rights holder.
Appearing as part of the name	
×	A hybrid of two or more species.
agg.	A name used to include a group of very similar plants.
cl.	A clone.
f.	A forma. A minor variant of a species.
Group	A collective name for a group of cultivars within a genus with similar characteristics. The word Group is always included and where mentioned with a cultivar name is enclosed in brackets ( ).
subsp.	A naturally occurring variant of a species.
Trade Designation	A trade designation, also known as Selling Name, is a name used to market a plant. Where a plant is sold in different countries its trade designation may vary from country to country. When ordering a plant both the cultivar name and trade designation should be given to ensure delivery of the correct plant. The trade designation is printed in A DIFFERENT FONT to distinguish it from the cultivar name.
var.	A varietas. A naturally occurring minor variant of a species.
'	Single quotation marks used to denote a cultivar.

BIBLIOGRAPHY

Bean, W. J. *Trees and Shrubs Hardy in the British Isles.* Vols 1–3, 7th edn (1950) John Murray, London.

Bitner, R. L. *Conifers for Gardens: an Illustrated Encyclopedia* (2007) Timber Press, Oregon.

Bloom, A. *Conifers for Your Garden.* (1972) Floraprint, Nottingham.

Bloom, A. *Gardening with Conifers* (2001) Frances Lincoln, London.

Bond, J. and Randal, L. *Dwarf and Slow-Growing Conifers. A Wisley Handbook.* (1987) The Royal Horticultural Society. Cassell, London.

Brickell, C. (ed.) *A–Z Encyclopedia of Garden Plants* (1997) The Royal Horticultural Society, Dorling Kindersley, London.

British Standards *BS3936-1-1992 Nursery Stock: Specification of Trees and Shrubs.*

British Containers Group. *Specification of Standards for the Production of Hardy Container Grown Plants.*

Cox, P. A. *Rhododendrons. A Wisley Handbook* (1985) The Royal Horticultural Society, Cassell, London.

Cubey, J. (ed in chief) *RHS Plant Finder*, 34th edn (2020) Royal Horticultural Society, London.

Dunn, N. *Trees for Your Garden* (2010) The Tree Council, London.

Edwards, D., Marshall, R. and Lancaster, R. (eds) *The Hillier Manual of Trees and Shrubs*, 9th edn (2019) Royal Horticultural Society, London.

Gardiner, J. *Magnolias Their Care and Cultivation* (1989) Cassell, London.

Gardiner, J. *Magnolias. A Gardener's Guide* (2000) Timber Press, Oregon.

Rushforth, K. *Trees for Small Gardens. A Wisley Handbook.* (1987) The Royal Horticultural Society, Cassell, London.

Tree Care Nursery Association (TCIA) *ANSI A300 Standards.*

Vertrees, J. D. *Japanese Maples*, 3rd edn (2002) Timber Press, Oregon.

GENERAL INDEX

garden design using
calluna, daboecia and erica 276
conifers 263–264
trees and shrubs 13–15
plant association 15–18
colour 16
fragrance, aroma, and odours 18
shape (or form) and habit 16
size at maturity 15
texture 17
sale of trees and shrubs 11
bare root 11
burlapped 11
containerised 11
container-grown 11
root-wrapped 11
shrubs production standards 9–11
ANSI A300 standards 10
British container grower group standards 11
British standard BS 3936 10
grading system for seedlings and very young trees 9
trees
environmental issues 7–8
native or exotic 19
tree production
tap roots 9
tree feathers 9
topworked 9
undercutting 9
types of tree 9–10
ANSI A300 standards 10
British Standard BS 3936 10
feathered 9
heavy feathered 10
light feathered 10
heavy standard 10
light standard 10
multistemmed 10
selected standard 10
semi-mature 10
shrubs converted into small trees 10
shrubs on a leg 10

PLANT INDEX

An entry in **bold** indicates a photograph.

Abelia 85
Abelia 'Edward Goucher' 86
Abelia chinensis 85,86
Abelia chinensis 'China Rose' 85
Abelia mosanensis 'Korean Spring' 85, 86
Abelia × 'Lake Maggiore' 86
Abelia × *grandiflora* 85, 86
Abelia × *grandiflora* 'Francis Mason' 86
Abelia × *grandiflora* 'Hopleys'PBR 86
Abelia × *grandiflora* 'Kaleidoscope'PBR 86
Abelia × *grandiflora* CONFETTI ('Conti'PBR) 86
Abies 240
Abies balsamea Hudsonia Group 240, **240**
Abies koreana 240
Abies koreana 'Icebreaker' **240**
Abies koreana 'Silberlocke' 240, **240**
Abies koreana 'Silver Shadow' **240**
Abies nordmanniana 240, **240**
Abies nordmanniana subsp. *nordmanniana* 'Golden Spreader' 240, **240**
Acer 21–28
Acer 'Serpentine' 27
Acer capillipes 27, **28**
Acer davidii 27
Acer davidii 'George Forrest' AGM 27, **28**
Acer davidii 27–28
Acer davidii VIPER ('Mindavi') 27, **28**
Acer griseum **21**–22, **22**
Acer griseum **264**
Acer grosseri 27
Acer grosseri var. *hersii* 27
Acer japonicum 21, 22, **22**
Acer japonicum 'Aconitifolium' 22, **22**
Acer japonicum 'Vitifolium' 22
Acer negundo 'Flamingo' 21, 22–23, **23**
Acer negundo 'Winter Lightning' **23**
Acer negundo var. *violaceum* 21
Acer palmatum 8, 21, 23–25, **24**
Acer palmatum 'Ariadne' AGM 25
Acer palmatum 'Bei-shidare' 25
Acer palmatum 'Burgundy Lace' AGM 25
Acer palmatum 'Dissectum Flavescens' 25
Acer palmatum 'Eddisbury' AGM 25
Acer palmatum 'Ginko-san' 25
Acer palmatum 'Inazuma' 25
Acer palmatum 'Koto-no-ito' 25, **26**
Acer palmatum 'Nuresagi' 25
Acer palmatum 'Orange Dream' 25, **26**
Acer palmatum 'Osakusuki' AGM 25
Acer palmatum 'Sangu-kaku' AGM 25, **26**
Acer palmatum 'Seiryu' AGM 25
Acer palmatum 'Skeeter's Broom' 25, **26**
Acer palmatum 'Wilson's Pink Dwarf' **12**
Acer palmatum var. *dissectum* Atropurpureum Group **26**
Acer pensylvanicum 'Erythrocladum' 27
Acer pseudoplatanus 'Brilliantissimum' 10
Acer rufinerve 27
Acer × *conspicuum* 'Phoenix' 27, **28**
Adiantum 248
Ajuga reptans 'Catlin's Giant' 217
Albizia jubilissin 29
Albizia jubilissin CHOCOLATE FOUNTAIN ('Ncaj1') 29
Albizia jubilissin f. *rosea* 'Summer Chocolate'PBR 29
Albizia jubilissin f. *rosea* AGM 29, **29**
Albizia jubilissin Ombrelle ('Boubri'PBR) 29
Amelanchier 30–31
Amelanchier 'La Paloma' AGM 31
Amelanchier alnifolia 'Obelisk'PBR 30, 31
Amelanchier canadensis 31
Amelanchier canadensis 'Glenn Form' 30, **30**, 31
Amelanchier laevis 'R.J. Hilton' AGM 31
Amelanchier laevis 'Snowflakes' 31
Amelanchier lamarckii 10, 30, **30**, 31
Amelanchier ovalis 30
Amelanchier ovalis 'Edelweiss' 31
Amelanchier × *grandiflora* 30
Amelanchier × grandiflora 'Ballerina' 31
Amelanchier × *grandiflora* 'Robin Hill' 30, 31
Anisodontea 86
Anisodontea 'El Rayo' 86, **86**, 87, **87**
Anisodontea 'Large Magenta' 86, 87
Anisodontea capensis 86, 87
Arbutus 32–33
Arbutus andrachne 33
Arbutus menziesii 32, 33
Arbutus unedo 32, **33**, 33
Arbutus unedo 'Atlantic' AGM 32, 33
Arbutus unedo f. *rubra* AGM 32, 33
Arbutus unedo ROSELILY ('Minilily'PBR) 33
Arbutus × *andrachnoides* AGM 33
Aucuba japonica 18
Aucuba japonica 'Variegata' 18
Azaleas 195
Azara microphylla 'Variegata' 32–33, **33**
Berberis 176
Berberis darwinii 'Compacta' 95
Berberis thunbergii 87
Berberis thunbergii 'Aurea' 88, 220
Berberis thunbergii 'Golden Torch' 88, **88**
Berberis thunbergii 'Helmond Pillar' **265**
Berberis thunbergii 'Orange Rocket' 88, **89**
Berberis thunbergii 'Orange Sunrise'PBR 88, **89**
Berberis thunbergii cvs. 95
Berberis thunbergii f. *atropurpurea* 87
Berberis thunbergii f. *atropurpurea* 'Atropurpurea Nana' 87, 89
Berberis thunbergii f. *atropurpurea* 'Helmond Pillar' 88, **89**
Berberis thunbergii f. *atropurpurea* 'Redtorch'PBR 88, **89**
Berberis thunbergii f. *atropurpurea* 'Rose Glow' 88, **89**
Berberis × *carminea* 88
Berberis × *carminea* 'Pirate King' **89**
Betula 34–37
Betula 'Fascination' AGM **34**
Betula 'Hergest' 35
Betula albosinensis 36
Betula albosinensis 'Bowling Green' 36
Betula gmelinii 35
Betula nigra **34**, 35
Betula nigra 'Little King' 35
Betula nigra 'Siloh Splash'
Betula pendula **34**, 35, 36
Betula pendula 'Laciniata' (syn. 'Dalecarlica') 35
Betula pendula 'Tristis' **36**
Betula pendula 'Youngii' 35, **36**
Betula pendula FASTIATA OES ('Jolep1'PBR) 35, **36**
Betula pendula subsp. *pendula* 'Schneverdinger Goldbirke' (syn. 'Golden Cloud') 35, **36**
Betula pubescens 35
Betula utilis subsp. *albosinensis* 13, 35
Betula utilis subsp. *albosinensis* 'China Rose' 36
Betula utilis subsp. *albosinensis* 'Kansu' AGM 36
Betula utilis subsp. *albosinensis* 'Red Panda' AGM 36
Betula utilis subsp. *albosinensis* 'Bowling Green' AGM 36, **37**
Betula utilis subsp. *albosinensis var. septentrionalis* 36
Betula utilis subsp. *jacquemontii* 35, 36, **37**
Betula utilis subsp. *jacquemontii* 'Dorenbos' (syn. Snow Queen') AGM 36
Betula utilis subsp. *jacquemontii* 'Jermyns' AGM **34**, 36
Betula utilis subsp. *jacquemontii* 'McBeath' 36
Betula utilis subsp. *jacquemontii* Silver Shadow' AGM 36
Brachyglottis 90
Brachyglottis (Dunedin Group) 'Drysdale' 90,
Brachyglottis (Dunedin Group) 'Sunshine' 90, **90**

Brachyglottis monroi 90, **90**
Brachyglottis WALBERTON'S SILVER DORMOUSE ('Walbrach') 90
Buddleja 91
Buddleja alternifolia 91, 93, **94**
Buddleja alternifolia 'Argentea' 93, 94
Buddleja davidii 91, 94
Buddleja davidii (Buzz Series) INDIGO 91
Buddleja davidii (Buzz Series) BUZZ CANDY PINK ('Tobudsopin') 91,
Buddleja davidii (Buzz Series) BUZZ HOT RASPBERRY 91
Buddleja davidii (Buzz Series) BUZZ IVORY ('Tobudivory'PBR) 91
Buddleja davidii (Buzz Series) BUZZ LILAC 91
Buddleja davidii (Buzz Series) BUZZ MAGENTA ('Tobudpipur'PBR) 92, **92**
Buddleja davidii (Buzz Series) BUZZ MAGENTA IMPROVED ('Tobudmagen') 91
Buddleja davidii (Buzz Series) BUZZ SKY BLUE ('Tobudskbi'PBR) 92
Buddleja davidii (Buzz Series) BUZZ VELVET ('Tobudvelve') 92
Buddleja davidii (Buzz Series) BUZZ VIOLET ('Tobudviole') 92
Buddleja davidii (English Butterfly Series) CAMBERWELL BEAUTY ('Camkeep') 93
Buddleja davidii (English Butterfly Series) MARBLED WHITE ('Markeep') 93
Buddleja davidii (English Butterfly Series) PEACOCK ('Paekeep'PBR) 93
Buddleja davidii (English Butterfly Series) PURPLE EMPEROR ('Pyrkeep') 93
Buddleja davidii 'Black Knight' 91
Buddleja davidii 'Dartmoor' 91
Buddleja davidii 'Empire Blue' **91**, 91
Buddleja davidii 'Harlequin' **91**, 91
Buddleja davidii 'Pink Delight' **91**, 91
Buddleja davidii NANHO BLUE 'Mongo' 91, **92**, 93
Buddleja davidii NANHO PURPLE ('Monum') 93
Buddleja davidii NANHO WHITE ('Monite') 93
Buddleja DREAMING LAVENDER ('Hindebud1'PBR) 93
Buddleja DREAMING PURPLE ('Hindbud4'PBR) 93
Buddleja DREAMING WHITE ('Hindbud3'PBR) 93
Buddleja globosa 91
Buddleja lindleyana 93, 94, **94**
Buddleja lindleyana 'Little Treasure' 93, 94
Buxus 94
Buxus sempervirens 18, 87, 95, 128, 169

Calluna 265, 267
Calluna vulgaris 267
Calluna vulgaris 'Anette'PBR (Garden Girl Series) 268, 267
Calluna vulgaris 'Athene'PBR (Garden Girl Series) 268
Calluna vulgaris 'Dart's Amethyst' 268, **267**
Calluna vulgaris 'Ellen' 268, 267
Calluna vulgaris 'Gold Spronk' 268
Calluna vulgaris 'H.E. Beale' 268, **268**
Calluna vulgaris 'Peter Sparkes' 268
Calluna vulgaris 'Robert Chapman' 268, **268**
Camellia 96
Camellia 'Black Lace' 101
Camellia 'Cornish Snow' 101, **102**, 102
Camellia 'Royalty' 101
Camellia 1001 SUMMER NIGHTS JASMINE ('Yulano1') 96, 100
Camellia azalea **96**
Camellia cultivars with complex or unknown parentage 100, 101
Camellia japonica 33
Camellia japonica 'Adolphe Audusson' 97, **97**
Camellia japonica 'Alba Simplex' 97, **97**
Camellia japonica 'Brushfield's Yellow' 97
Camellia japonica 'Dr Burnside' 97
Camellia japonica 'Dr King' 97
Camellia japonica 'Juno' **97**
Camellia japonica 'Kramer's Supreme' 97
Camellia japonica 'Margaret Davis' 97
Camellia japonica 'Nuccio's Gem' 97, **97**
Camellia japonica 'Nuccio's Pearl' 97
Camellia japonica 'Silver Anniversary' 97
Camellia japonica and *Camellia* × *williamsii* cultivars 98
Camellia japonica × *Camellia azalea* cultivars 96
Camellia sasanqua 'Cleopatra' 101
Camellia sasanqua 'Crimson King' **100**
Camellia sasanqua 'Dwarf Shishi' 101
Camellia sasanqua 'Jean May' 101
Camellia sasanqua 'Maiden's Blush' **100**
Camellia sasanqua 'Narumigata' 101
Camellia sasanqua 'Paradise Helen' 101
Camellia sasanqua 'Sekiyo' 101
Camellia sasanqua 'Shiro-wabisuke' **100**
Camellia sasanqua 'Survivor' 101
Camellia sasanqua 'Hugh Evans' 101
Camellia sasanqua cultivars 100, 102
Camellia transnokoensis 100, **100**, 101, 102
Camellia × *vernalis* 'Yuletide' 100, 101, 102
Camellia × *williamsii* 'Anticipation' 98, **98**
Camellia × *williamsii* 'Bow Bells' 98, **98**
Camellia × *williamsii* 'Brigadoon' 98, **99**
Camellia × *williamsii* 'Celebration' **99**
Camellia × *williamsii* 'Debbie' 98
Camellia × *williamsii* 'Donation' 98, **99**
Camellia × *williamsii* 'E.T.R. Carylon' 98
Camellia × *williamsii* 'Jury's Yellow' **99**
Camellia × *williamsii* 'Saint Ewe' **99**
Caragana 37
Caragana arborescens 37, 38
Caragana arborescens 'Lorbergii' 37, 38
Caragana arborescens 'Pendula' 37, 38, **38**
Caragana arborescens 'Walker' 37, 38,
Carpinus betulus 14
Caryopsis × *clandonensis* **84**
Caryopteris 102
Caryopteris incana 102, 104
Caryopteris × *clandonensis* 102
Caryopteris × *clandonensis* 'Arthur Simmons' 102, 103
Caryopteris × *clandonensis* 'Blue Cascade' 104, **104**
Caryopteris × *clandonensis* 'Dark Knight' 103, **103**
Caryopteris × *clandonensis* 'First Choice' 103
Caryopteris × *clandonensis* 'Grand Bleu Inoveris'PBR 103
Caryopteris × *clandonensis* 'Heavenly Blue' 103
Caryopteris × *clandonensis* 'Jason'PBR 104
Caryopteris × *clandonensis* 'Kew Blue' 103
Caryopteris × *clandonensis* 'Summer Sorbet'PBR 103
Caryopteris × *clandonensis* 'Sunshine Blue' 104
Caryopteris × *clandonensis* 'Worcester Gold' 103 **103**
Caryopteris × *clandonensis* HINT OF GOLD ('Lissaura'PBR) 103
Caryopteris × *clandonensis* STEPHI 'Lissteph'PBR 103
Caryopteris × *clandonensis* STERLING SILVER ('Lissilv'PBR) 103
Ceanothus 'Skylark' 48
Cedrus 241
Cedrus atlantica 'Glauca Pendular' **240**, 241
Cedrus deodara 'Feeling Blue' 241
Cedrus deodara 'Golden Horizon' 241
Ceratostigma 104
Ceratostigma griffithii 104, 105, **105**
Ceratostigma plumbaginiodes 104, 105, **105**
Ceratostigma willmottianum 104 105, **105**
Ceratostigma willmottianum 'Forest Blue' 105
Ceratostigma willmottianum DESERT SKIES 'Palmgold'PBR 105
Ceratostigma willmottianum SAPPHIRE RING ('Lissbrill'PBR) 105
Cercis 106
Cercis canadensis 106
Cercis canadensis LAVENDER TWIST ('Covey') 108
Cercis canadensis 'Alley Cat' **107**, 108
Cercis canadensis 'Forest Pansy' **107**, 108
Cercis canadensis 'Forest Pansy' 74
Cercis canadensis 'Ruby Falls' **107**, 108
Cercis chinensis 106
Cercis chinensis Avondale' 106, **106**
Cercis ETERNAL FLAME ('Nc2016–2') **107**, 108
Cercis siliquastrum 106, **106**
Cercis siliquastrum 'Bodnant' 106
Cercis siliquastrum 'White Swann' 106
Chaenomeles 108
Chaenomeles japonica 109
Chaenomeles japonica 'Orange Beauty' 109
Chaenomeles japonica MADAME BUTTERFLY ('Whitice') 109
Chaenomeles speciosa 'Nivalis' **108**, 109
Chaenomeles speciosa 'Scarlet Storm'PBR 109
Chaenomeles × *superba* 'Crimson and Gold' 109, **109**
Chaenomeles × *superba* 'Pink Lady' 109, **109**
Chaenomeles × *superba* 'Rowallane' 109, 109
Chamaecyparis 242–243, 261
Chamaecyparis lawsoniana 242, 263
Chamaecyparis lawsoniana 'Ellwood's Gold' 243
Chamaecyparis lawsoniana 'Ellwoodii' 243, **242**, 264
Chamaecyparis lawsoniana 'Gnome' 243
Chamaecyparis lawsoniana 'Green Globe' 243
Chamaecyparis lawsoniana 'Little Spire' **242**
Chamaecyparis lawsoniana 'Minima Aurea' 243
Chamaecyparis lawsoniana 'Tamariscifolia' 243
Chamaecyparis obtusa 243
Chamaecyparis obtusa 'Fernspray Gold' 243, **243**
Chamaecyparis obtusa 'Kosteri' 243
Chamaecyparis obtusa 'Nana Gracilis' 243, **243**
Chamaecyparis obtusa 'Nana Lutea' 243
Chamaecyparis pisifera 244
Chamaecyparis pisifera 'Boulevard' 244, **244**
Chamaecyparis pisifera 'Compacta' 244
Chamaecyparis pisifera 'Filifera Aurea' 244, **244**
Chamaecyparis pisifera 'Filifera Nana' 244
Chamaepericlymenum canadense 118
Choisya 110, 215
Choisya GREENFINGERS ('Lissfing') 111
Choisya ROYAL LACE ('Pmoore06'PBR) 111
Choisya ternata 18–19, 110, **110**, 111
Choisya ternata SUNDANCE ('Lich') 16, 110, **110**, 111
Choisya × *dewitteana* 110
Choisya × *dewitteana* 'Aztec Gold'PBR 111
Choisya × *dewitteana* 'Aztec Pearl' **101**, 111
Choisya × *dewitteana* APPLE BLOSSOM ('Pmoore09'PBR)
Choisya × *dewitteana* GOLDFINFERS ('Limbo'PBR) 111
Choisya × *dewitteana* GOLDFINGERS ('Limo'PBR) 220
Choisya × dewitteana WHITE DAZZLER ('Londaz'PBR) **101**, 111
Cistus 110, 152
Cistus 'Snow Fire' 112
Cistus corbariensis 112
Cistus ladanifer 112
Cistus laurifolius 112
Cistus pulverulentus 'Sunset' 112
Cistus × *argenteus* 'Peggy Sammons' 112
Cistus × *argenteus* 'Silver Pink' **111**, **112**
Cistus × *cyprius* 112
Cistus × *cyprius* var. *ellipticus* 'Elma' 112
Cistus × *dansereaui* 'Decumbens' 112
Cistus × *dansereaui* 'Jenkyn Place' 112
Cistus × *dansereaus* 'Decumbens' 111
Cistus × *hybridus* 112
Cistus × *laxus* 'Snow White' 111, 112
Cistus × *lenis* 'Grayswood Pink' 112
Cistus × *purpureus* **111**, 112
Clerodendrum 38, 112
Clerodendrum trichotomum var. *fargesii* 38, 39, **39**, 112, **112, 113**
Clerodendrum trichotomum var. *fargesii* 'Carnival' 38, 39, **39**, 113, **113**
Colchicum speciosum 104
Conus kousa 121, **121**, 122, 127, 128
Coprosma 113
Coprosma 'Beatson's Gold' 114
Coprosma 'Evening Glow' 114, **114**
Coprosma 'Fire Burst'PBR 114, **115**
Coprosma 'Rainbow Surprise'PBR 114, **115**
Coprosma 'Scarlet O'Hara' 114
Coprosma 'Inferno'PBR 114
Coprosma PACIFIC SUNSET ('Jwncopps') **115**
Coprosma repens 114
Coprosma repens 'Pacific Dawn' 114, **115**
Coprosma repens 'Pina Colada'PBR 114
Coprosma repens 'Tequila Sunrise' 114
Coprosma repens PACIFIC NIGHT 'Hutpac'PBR 114
Cornus 115, 127, 128
Cornus alba 115, 116, **116**, 124, 125, 128
Cornus alba 'Aurea' 116, **116**
Cornus alba 'Elegantissima' 116, **117**
Cornus alba 'Kesselring" 116
Cornus alba 'Sibirica Variegata' 116
Cornus alba 'Sibirica' 116, **117**, 127
Cornus alba 'Spaethii' 116, 127
Cornus alba 'Westonbirt' 116
Cornus alba BATON ROUGE ('Minbat'PBR) 116, **117**, **126**, 127
Cornus alba cvs. 216
Cornus alba IVORY HALO ('Baihalo'PBR) 116
Cornus alternifolia 115, 118, 127, 128
Cornus alternifolia 'Argentea' 118, **118**
Cornus canadensis 118, **118**, **127**
Cornus controversa 115, 119, **119**,127, 128
Cornus controversa 'Pagoda' 119
Cornus controversa 'Variegata' 82, 119, **119**
Cornus florida 120
Cornus florida 'Cherokee Chief' 120
Cornus florida 'Cherokee Princess' 120
Cornus florida 'Holman's Gold' 120
Cornus florida 'Rainbow' 120, 127
Cornus florida 'Rubra' 120
Cornus florida Sunset' 120, **120**
Cornus GOLDEN SHADOWS ('Wstackman'PBR) 118
Cornus kousa 'Madame Butterfly' **121**, 122
Cornus kousa 'Miss Satomi' **121**, 122, 127
Cornus kousa 'Schmetterling' 122
Cornus kousa 'Wolf Eyes' 122, **122**
Cornus kousa var. *chinensis* 121, **121**, 122, 127
Cornus kousa var. *chinensis* 'Wisley Cream' **121**, 122
Cornus mas 122, 123, **123**
Cornus mas 'Aurea' 123, **123**
Cornus mas 'Golden Glory' 123
Cornus mas 'Jolico' 123
Cornus mas 'Variegata' 123, **123**
Cornus sanguinea 124, **125**, 127
Cornus sanguinea 'Anny' 124
Cornus sanguinea 'Anny's Winter Orange' 124, **125**

Cornus sanguinea 'Magic Flame' 124
Cornus sanguinea 'Midwinter Fire' 124, **125**
Cornus sanguinea 'Winter Beauty' 124, **125**
Cornus sanguinea 'Winter Flame' 124
Cornus sericea 125, 126, 127, 128
Cornus sericea 'Budd's Yellow' 126, **126**
Cornus sericea 'Cardinal' 126, **126**
Cornus sericea 'Flaviramea' 117, 126, **126**
Cornus sericea 'Hedgerow's Gold' 126, 127
Cornus sericea 'White Gold' 126
Cornus sericea KELSEY'SGOLD 'Rosco' 126, 127
Corokia 128
Corokia cotoneaster 128, **128**, 129
Corokia × *virgata* 128, 129
Corokia × *virgata* 'Frosted Chocolate' 95, 129
Corokia × *virgata* 'Red Wonder' 129
Corokia × *virgata* 'Sunsplash' 129
Corokia × *virgata* 'Yellow Wonder' 129
Coronilla 129
Coronilla subsp. *glauca* 'Citrina' 129, 130
Coronilla valentina 129, 130
Coronilla valentina subsp. *glauca* 129, **129**, 130
Coronilla valentina subsp. *glauca* 'Variegata' 129, 130
Corus alba cvs. 218
Corylopsis 130
Corylopsis pauciflora 130, **130**
Cotinus 130
Cotinus 'Flame' **131**, 131, 132
Cotinus 'Grace' 74, **131**, 131, 132
Cotinus 'Ruby Glow' **131**, 132
Cotinus coggygria **131**, 132
Cotinus coggygria 'Lilia'PBR 132
Cotinus coggygria 'Royal Purple' 74, **131**, 132
Cotinus coggygria 'Young Lady'PBR 132
Cotinus coggygria GOLDEN LADY ('Mincojaug'PBR) 132
Cotinus coggygria GOLDEN SPIRIT ('Ancot'PBR) **131**, 132
Cotinus coggygria SMOKEY JOE ('Lisjo'PBR) 132
Cotinus DUSKY MAIDEN ('Londus'PBR) 132
Cotoneaster 132
Cotoneaster × *suecicus* 'Skogholm' 133
Cotoneaster 'Cornubia' 135
Cotoneaster 'Exburiensis' 136
Cotoneaster 'Gnom' 133
Cotoneaster 'Hybridus Pendulus' 8, 136, **136**
Cotoneaster adpressus 133, 134
Cotoneaster atropurpureus 'Variegatus' **132**, 133, **133**, 134
Cotoneaster conspicuus 'Decorus' 134, 134, **135**
Cotoneaster dammeri 133, **133**, 134
Cotoneaster franchetii 135, **135**, 135
Cotoneaster frigidus 'Cornubia' 135
Cotoneaster horizontalis 134, **135**
Cotoneaster humifusus 133
Cotoneaster lacteus 134, **135**, 135
Cotoneaster microphyllus 134
Cotoneaster nanshan 143
Cotoneaster procumbens 'Queen of Carpets' 133
Cotoneaster salicifolius 'Pendulus' 136
Cotoneaster simonsii 134, 135
Cotoneaster × *suecicus* 'Coral Beauty' 133, 136
Cotoneaster × *suecicus* 'Juliette' 133 **133**, 134, 136
Cotoneaster × *watereri* 'Cornubia' 135
Cotoneaster × *watereri* 'Pendulus' 136
Crataegus 39
Crataegus laevigata 40
Crataegus laevigata 'Crimson Cloud' 40
Crataegus laevigata 'Paul's Scarlet' 40, **41**
Crataegus laevigata 'Rosea Flore Pleno' 40
Crataegus monogyna 40, **41**
Crataegus orientalis 39, 40, **40**
Crateagus persimilis 'Prunifolia' 40, **41**
Crateagus × *lavalleei* 'Carrierei' 40
Cryptomeria japonica 'Elegans Viridis' 246
Cryptomeria japonica 245–246
Cryptomeria japonica 'Bandai-sugi' 246
Cryptomeria japonica 'Cristata' (syn. 'Sekkan-sugi') 245, **245**, 246
Cryptomeria japonica 'Elegans Compacta' 246
Cryptomeria japonica Elegans Group 'Elegans' 246
Cryptomeria japonica 'Globosa Nana' 245, **245**, 246, 264
Cryptomeria japonica 'Spiralis' 246
Cupressus 246–247
Cupressus macrocarpa 247
Cupressus macrocarpa 'Goldcrest' 247, **247**
Cupressus macrocarpa 'Wilma' 247, **247**
Cupressus sempervirens 246, 247
Cupressus sempervirens 'Swane's Gold' 247, **247**
Cupressus sempervirens 'Totem Pole' 247, **247**, 265
Cydalima perspectalis 95
Cylindrocladium buxicola 88, 95

Daboecia 268
Daboecia azorica 268–269
Daboecia cantabrica 268
Daboecia cantabrica 'Glamour' 268, **269**
Daboecia cantabrica 'Praegerae' 268
Daboecia cantabrica 'Purpurea' 268
Daboecia cantabrica 'Tinkerbell' 268, **269**
Daboecia cantabrica 'Waley's Red' 268
Daboecia × *scotica* 'Cora' **269**
Daboecia × *scotica* 'Jack Drake' 268
Daboecia × *scotica* 'Katherine's Choice' 268
Daboecia × *scotica* 'Silverwells' 268
Daboecia × *scotica* 'William Buchanan' 268
Danae 137
Danae racemosa 137, **137**
Daphne 137
Daphne bholua 138, **138**, 144
Daphne bholua var. *glacialis* 'Gurkha' 138, 144
Daphne bholua 'Jacqueline Postill' 138, **138**, 144
Daphne caucasica 144
Daphne cneorum 138
Daphne cneorum 'Exima' 138, 139, 144
Daphne gemmata 'Royal Crown' 139
Daphne laureola 137, 139, 144
Daphne laureola subsp. *philippi* 139, **139**, 144
Daphne mezereum 137, 140, 144
Daphne mezereum f. *alba* **140**, 140
Daphne mezereum var. *rubra* 140, 144
Daphne odora 140, 141, 144
Daphne odora 'Aureomarginata' 140, **141**
Daphne odora 'Mae-jima' 140, **141**
Daphne odora 'MARIANNI ('Rogbret') 140
Daphne PERFUME PRINCESS ('Dapjur01'PBR) 141, **141**, 144
Daphne petraea 143
Daphne pontica 142, **142**, 144
Daphne sericea Collina Group 143, 144
Daphne tangutica 144, 145
Daphne tangutica Retusa Group 144, 145
Daphne × *rollsdorfii* 143
Daphne × *rollsdorfii* 'Arnold Cihlaz' 143, **143**
Daphne × *rollsdorfii* 'Wilhelm Schacht' 143, **143**
Daphne × *rollsdorfii* cultivars 144
Daphne × *transatlantica* 144
Daphne × *transatlantica* ETERNAL FRAGRANCE ('Balfra'PBR) 144, 145, **145**
Daphne × *transatlantica* PINK FRAGRANCE ('Blapink'PBR) 144, 145, **145**
Elaeagnus 'Quicksilver' 146, **147**
Elaeagnus 146
Elaeagnus pungens 'Maculata' 146, **147**
Elaeagnus pungens 'Frederici' 146, **147**
Elaeagnus × *ebbingei* 146
Elaeagnus × *submacrophylla* 146, **147**
Elaeagnus × *submacrophylla* 'Guilt Edge' 146, **147**
Elaeagnus × *submacrophylla* 'Limelight' 146
Elaeagnus × *submacrophylla* 'Vivleg'PBR 146
Erica 265, 267, 270
Erica arborea 270
Erica arborea var. *alpina* 270
Erica arborea var. *alpina* f. *aureifolia* 'Albert's Gold' **266**, 270, 276
Erica arborea var. *alpina* f. *aureifolia* 'Estrella Gold' 270, **270**
Erica carnea 254
Erica carnea (syn. *E. herbacea*) 270
Erica carnea 'Ann Sparkes' 271, **271**, 276
Erica carnea 'Eva' 271
Erica carnea 'Foxhollow' 263
Erica carnea 'Myretoun Ruby' 271, **271**
Erica carnea 'Vivellii' 271
Erica carnea 271
Erica carnea f. *alba* 'Springwood White' 271
Erica carnea f. *alba* 'Winter Snow' 271
Erica carnea f. *aureifolia* 'Foxhollow' 271, **271**
Erica cinerea 272
Erica cinerea 'C.D. Eason' 272
Erica cinerea 'Nellie Dawson' **272**
Erica cinerea 'Pink Ice' 272
Erica cinerea 'Stephen Davis' 272
Erica cinerea 'Velvet Knight' 272
Erica cinerea f. *alba* 'Alba Major' 272
Erica cinerea f. *aureifolia* 'Golden Hue' 272
Erica erigena (syn. *E. mediterranea*, *E. hibernica*) 274
Erica erigena 'Irish dusk' 274
Erica erigena 'Superba' 274
Erica erigena f. *albiflora* 'W.T. Rackliff') 274
Erica tetralix 'Con Underwood' 274
Erica tetralix 274, 276
Erica tetralix f. *alba* 'Alba Mollis' 274
Erica tetralix f. *stellata* 'Pink Star' 274
Erica vagans 'Birch Glow' 275, **275**
Erica vagans 'Hookstone Rose' 275, **275**
Erica vagans 275
Erica vagans f. *alba* 'Cornish Cream' 275
Erica vagans f. *alba* 'Lyonesse' 275
Erica vagans 'Pyrenees Pink' 275, **275**
Erica × *darleyensis* 276
Erica × *darleyensis* 'Arthur Johnson' 273, **273**
Erica × *darleyensis* 'Darley Dale (syn. *E.* × *darleyensis* 'Pink Perfection') 273, **273**
Erica × *darleyensis* 'Furzey' 273, **273**
Erica × *darleyensis* 'Jack H. Brummage' 273
Erica × *darleyensis* f. 'Spring Surprise'PBR 273
Erica × *darleyensis* f. *albiflora* 'Silberschmelze' 273, **273**
Erwinia amylovora 40
Escallonia 148
Escallonia 'Apple Blossom' 148
Escallonia 'Edinensis' 148
Escallonia 'Iveyi' 148
Escallonia GLOWING EMBERS ('Lowat21'PBR) 148, **148**
Escallonia GOLDEN CARPET ('Alcaura'PBR) 148
Escallonia PINK ELLE ('Lades'PBR) 148, **148**
Euonymus 42
Euonymus alatus 44, **44**, 45
Euonymus alatus 'Blade Runner' 44, 45
Euonymus alatus 'Compactus' 44, **44**, 45
Euonymus europaeus 44, **44**, 45
Euonymus europaeus 'Red Cascade' 44, **44**, 45
Euonymus fortunei 'Emerald 'n' Gold' 42, **43**
Euonymus fortunei 'Emerald Gaiety' 42, **43**
Euonymus fortunei 'Silver Queen' 42, **43**
Euonymus fortunei cvs. 95
Euonymus fortunei GOLDEN HARLEQUIN ('Hoogi'PBR) **43**
Euonymus japonicus 'Bravo' **10**, 42, **43**
Euonymus japonicus 'Green Rocket' 42, **43**
Euonymus japonicus 'Green Spire' 42
Euonymus japonicus 'Kathy'PBR 42
Euonymus japonicus 'Microphyllus Aureovariegatus' 42
Euonymus japonicus 'Ovatus Aureus' 42
Euonymus japonicus EL DORADO ('Br2012'PBR) 42, **43**
Euonymus microphyllus cvs. 95
Exochorda 149
Exochorda racemosa MAGICAL SPRINGTIME ('Kolmasprit') 149, 150
Exochorda serratifolia 'Snow White' 149, **149**, 150
Exochorda × *macrantha* 'Niagra'PBR) 149, 150
Exochorda × *macrantha* 'The Bride' 149, **149**, 150

Fagus sylvatica 14
Fuchsia 150
Fuchsia 'Alice Hoffman' 152
Fuchsia 'David' **151**, 152
Fuchsia 'Display' **151**, 152
Fuchsia 'Dollar Princess' 152
Fuchsia 'Dollar Prinzessin' 152
Fuchsia 'Genii' 152
Fuchsia 'Hawkshead' 152
Fuchsia 'Lady Boothby' 152
Fuchsia 'Lady Thumb' **151**, **152**
Fuchsia 'Mrs. Popple' **151**, 152
Fuchsia 'Riccartonii' **151**, 152
Fuchsia 'Tom Thumb' 152
Fuchsia 'Delta's Sarah' **151**, 152
Fuchsia magellanica 152
Fuchsia magellanica var. *gracilis* 'Versicolor' **151**, 152

Galanthus 144, 230
Ginkgo biloba (syn. *Salisburia adiantifolia*) 248, 249
Ginkgo biloba 'Fastigiata Blagon' 248, **249**
Ginkgo biloba 'Princeton Sentinel' 248
Ginkgo biloba 'Saratoga' 248, **249**
Ginkgo biloba 'Troll' 248, **249**
Gleditsia 45
Gleditsia triacanthos 45
Gleditsia triacanthos f. *inermis* 45, 46
Gleditsia triacanthos f. *inermis* 'Rubylace' 45, 46, **46**
Gleditsia triacanthos f. *inermis* 'Spectrum' 45, 46
Gleditsia triacanthos f. *inermis* 'Sunburst' 45, 46, **46**

Hakonechloa macra 'Aureola' **264**
Halimium 152
Hamamelis 175
Heptacodium miconioides 153, **153**
Hibiscus 154–155
Hibiscus moscheutos 154
Hibiscus syriacus 154
Hibiscus syriacus 'Bredon Springs' 154, **155**
Hibiscus syriacus 'Diana' 154
Hibiscus syriacus 'Dorothy Crane' 154, **155**
Hibiscus syriacus 'Hamabo' 154, **155**
Hibiscus syriacus 'Oiseau Blue' 154, **155**
Hibiscus syriacus 'Red Heart' 154
Hibiscus syriacus 'Woodbridge' 154, **155**
Hibiscus syriacus LAVENDER CHIFFON ('Notwoodone'PBR) 154
Hibiscus syriacus MAGENTA CHIFFON ('Rwood5'PBR) 154
Hibiscus syriacus PINK CHIFFON ('Jwnwood4'PBR) 154
Hibiscus syriacus STARBURST CHIFFON ('Rwoods6') 154
Hyacinthus 144
Hydrangea 156–164
Hydrangea macrophylla 156–158

Hydrangea macrophylla 'All Summer Beauty' 156
Hydrangea macrophylla 'Ami Pasquier' 156, **157**
Hydrangea macrophylla 'Bela'PBR **157**
Hydrangea macrophylla 'Cardinal Red' 156
Hydrangea macrophylla 'Deutschland' 156, **157**
Hydrangea macrophylla 'Doctor Jean Varnier' 158
Hydrangea macrophylla 'Frau Nobuko'PBR (Lady Series) 156, **157**
Hydrangea macrophylla 'French Bolero Pink' 156, **157**
Hydrangea macrophylla 'French Cancan' (Rendezvous Series) 158, **158**
Hydrangea macrophylla 'Klavern' 158, **158**
Hydrangea macrophylla 'Lanarth White' 158, **158**
Hydrangea macrophylla 'Madame Emile Mouillère' 156
Hydrangea macrophylla RUNAWAY BRIDE SNOW WHITE ('Ushyd0405'PBR) 158, **158**
Hydrangea macrophylla 'Veitchii' 158, **158**
Hydrangea macrophylla 'Zoro'PBR 158, **158**
Hydrangea macrophylla MAGICAL NOBELESSE (Hokomano'PBR) 156, **157**
Hydrangea paniculata **159**,159–161
Hydrangea paniculata 'Floribunda' 160
Hydrangea paniculata 'Grandiflora' 160, **160**
Hydrangea paniculata 'Jane' (syn. *H. paniculata* LITTLE LIME) 160, **161**
Hydrangea paniculata 'Limelight' 160, **160**
Hydrangea paniculata 'Magical Moonlight' 160
Hydrangea paniculata 'Phantom' 160, **161**
Hydrangea paniculata 'Unique' **159**, 160
Hydrangea paniculata 'Wim's Red'PBR 160
Hydrangea paniculata BOBO ('Ilvobo'PBR) 160, **160**
Hydrangea paniculata DART'S LITTLE DOT ('Rendia'PBR) 160, **160**
Hydrangea paniculata FRAISE MELBA ('Renba'PBR) 160
Hydrangea paniculata PASTELGREEN ('Rencolor'PBR) 160
Hydrangea paniculata PINKY WINKY ("Dupplinky'PBR) 160, **161**
Hydrangea paniculata SUNDAE FRAISE ('Rensun'PBR)
Hydrangea paniculata VANILLE FRAISE ('Renhy'PBR) 160, **161**
Hydrangea quercifolia 161–164
Hydrangea quercifolia 'Alice' 162
Hydrangea quercifolia 'Burgundy' 162
Hydrangea quercifolia 'Little Honey'PBR 162
Hydrangea quercifolia 'Pee Wee' 161, 162
Hydrangea quercifolia 'Ruby Slippers' 161, 162
Hydrangea quercifolia SNOW FLAKE ('Brido') 162, **163**
Hydrangea quercifolia SNOW QUEEN ('Flemygea') 162, **162**
Hypericum 163
Hypericum 'Rowallane' 163, **164**, 165
Hypericum calycinum 163, 163, 165
Hypericum kouytchense 163, **164**, 165
Hypericum lancasteri 163, 165
Hypericum olympicum 163, 165
Hypericum olympicum 'Citrinum' 163, **164**, 165
Hypericum reptans 163, 165
Hypericum × *drummeri* 'Peter Drummer' 163,165
Hypericum × *hidcoteense* 'Hidcote' 163, **164**, 165
Hypericum × *inodorum* (Magical Series) MAGICAL BEAUTY ('Kolmbeau'PBR) **164**
Hypericum × *inodorum* (Magical Series) MAGICAL PUMPKIN ('Kolmapuki'PBR) **164**
Hypericum × *inodorum* (Magical Series) MAGICAL UNIVERSE ('Kolmundi'PBR) 165
Hypericum × *inodorum* (Miracle Series) MIRACLE MARVEL ('Allmarvel'PBR) **164**
Hypericum × *inodorum* 'Elstead' 165
Hypericum × *moserianum* 'Tricolor' 165

Ilex 166–170
Ilex aquifolium 166–167, 168
Ilex aquifolium 'Alaska' 166, **167**
Ilex aquifolium 'Argentea Marginata Pendula' 166, **167**
Ilex aquifolium 'Argentea Marginata' 166, **167**
Ilex aquifolium 'Bacciflava' 166, **167**
Ilex aquifolium 'Ferox Argentea' 166
Ilex aquifolium 'Ferox' 166
Ilex aquifolium 'J. C. van Tol' 166, **167**
Ilex aquifolium 'Pendula' 166, 167
Ilex aquifolium 'Silver Queen' 166
Ilex crenata 95, 168, 169
Ilex crenata 'Glory Gem' 169
Ilex crenata 'Convexa' 169
Ilex crenata 'Fastigiata' 169
Ilex crenata 'Green Hedger' 169, **169**
Ilex crenata 'Shiro-furkurin' **169**
Ilex verticillata 170
Ilex verticillata 'Jim Dandy' 170
Ilex verticillata 'Red Sprite' 170
Ilex verticillata 'Southern Gentleman' 170
Ilex verticillata 'Winter Red' 170
Ilex verticillata male 170
Ilex × *altaclerensis* 'Belgica Aurea' 168, **168**
Ilex × *altaclerensis* 'Golden King' 168, **168**
Ilex × *altaclerensis* 'Ripley Gold' 168
Ilex × *altaclerensis* 168
Ilex × *meserveae* 168–169
Ilex × *meserveae* BLUE ANGEL ('Conang') **168**, 169
Ilex × *meserveae* 'Casanova'PBR **168**, 168
Ilex × *meserveae* BLUE MAID ('MESID') 169
Ilex × *meserveae* BLUE PRINCESS ('Conapri') 169
Iris reticulata 230

Juniperus 250–253
Juniperus chinensis 'Aurea' 251
Juniperus chinensis 'Kaizuka' 251
Juniperus chinensis 250, 251
Juniperus communis 250, **250**, 251
Juniperus communis 'Compressa' 250, 251, 264
Juniperus communis 'Green Carpet' 251
Juniperus communis 'Hibernica' 251, **251**
Juniperus communis 'Repanda' 251, **251**
Juniperus horizontalis 250, 251
Juniperus horizontalis 'Bar Harbor' 251
Juniperus horizontalis 'Blue Chip' 251, 264
Juniperus horizontalis 'Golden Carpet' 251
Juniperus horizontalis ICE BLUE ('Monber') 251
Juniperus scopulorum 252–253
Juniperus scopulorum 'Blue Arrow' 253, **253**
Juniperus scopulorum 'Skyrocket' (syn. *J. virginiana* 'Skyrocket') 253, **253**
Juniperus squamata 253
Juniperus squamata 'Blue carpet' 253
Juniperus squamata 'Blue Star' 253, **253**
Juniperus × *pfitzeriana* 252
Juniperus × *pfitzeriana* 'Carbery Gold' 252, **252**
Juniperus × *pfitzeriana* 'Old Gold' 252
Juniperus × *pfitzeriana* 'Sulphur Spray' 252

Koelreuteria paniculata 47, **47**
Koelreuteria paniculata 'Coral Sun' 47

Laburnum 48
Laburnum alpinum 48, 49
Laburnum alpinum 'Pendulum' 48
Laburnum anagyroides 48, 49
Laburnum anagyroides 'Yellow Rocket' 48, 49
Laburnum × *watereri* 'Vossii' 48, **48**, 49, **49**
Lagerstroemia 170–172
Lagerstroemia indica 170, 171, **171**
Lagerstroemia indica 'Fuchsia d' 'Été' 171
Lagerstroemia indica 'Violet Fili'PBR 171
Lagerstroemia indica BERRY DAZZLE ('Gamad V1') 171
Lagerstroemia indica MIMIE FUCHSIA ('Dablage01'PBR) 171
Lagerstroemia indica PETITE PINKIE ('Monkie') 171
Lagerstroemia indica RAPSODY IN PINK ('WhitV11') 171, **172**
Lagerstroemia indica VIOLET D'ETE' ('Indyvio') 171, **172**
Lagerstroemia indica WITH LOVE BABE ('Milaperl'PBR) 171
Lagerstroemia indica WITH LOVE ETERNAL ('Milavio'PBR) 171
Lagerstroemia indica WITH LOVE KISS ('Milarosso') 171
Leycesteria 173–174
Leycesteria 'Smoldering Embers' 173, 174
Leycesteria formosa 173, 174, **173**
Leycesteria formosa 'Gold Leaf' 174
Leycesteria formosa 'Purple Rain' 174
Leycesteria formosa GOLDEN LANTERNS ('Notbruce'PBR) 173, **173**, 174, **174**
Leycesteria formosa LITTLE LANTERNS ('COD 01') 173, 174, **174**
Liquidambar 50
Liquidambar orientalis 50
Liquidambar stryraciflua 'Slender Silhouette' 50, 51
Liquidambar styraciflua **50**, 50
Liquidambar styraciflua 'Corky' 50, 51
Liquidambar styraciflua 'Gum Ball' 51
Liquidambar styraciflua 'Lane Roberts' **50**, 51, **51**
Liquidambar styraciflua 'Palo Alto' 50, 51, **51**
Liquidambar styraciflua 'Worplesdon' 51
Loropetalum 175
Loropetalum chinense 175
Loropetalum chinense CAROLINA MOONLIGHT ('Nci 002') 175
Loropetalum chinense var. *rubrum* 'Burgundy' 175
Loropetalum chinense var. *rubrum* 'Chang Nian Hong'PBR (syn. 'Ever Red') 175
Loropetalum chinense var. *rubrum* 'Pearl' 175
Loropetalum chinense var. *rubrum* BLACK PEARLPBR **175**
Loropetalum chinense var. *rubrum* RUBY SNOW ('Iwai'PBR) 175
Loropetalum chinense var. *rubrum* 'Blush' 175
Loropetalum chinense var. *rubrum* 'Fire Dance' 175
Luma 52
Luma apiculata 52, **52**
Luma apiculata 'Saint Hilary' 52
Magnolia 53
Magnolia 'Butterflies' 55, **55**
Magnolia 'Elizabeth' 55, **55**
Magnolia 'Genie'PBR 56, **56**
Magnolia 'Princess Margaret' 57, **57**
Magnolia 'Susan' 53, 59, **59**
Magnolia acuminata 53
Magnolia BLACK TULIP ('Jurmag'PBR) 54, **54**
Magnolia BURGUNDY STAR ('Jurmag4') 54
Magnolia FAIRY MAGNOLIA BLUSH ('Micjur01'PBR) **18**, 55, **55**
Magnolia FAIRY MAGNOLIA CREAM ('Murjur02'PBR) 55, **55**
Magnolia FAIRY MAGNOLIA WHITE ('Murjur05'PBR) 56, **56**
Magnolia grandiflora 59
Magnolia grandiflora 'Galissonniere' 56, **56**
Magnolia liliflora 'Nigra' 56
Magnolia seiboldii 59
Magnolia stellata 53, 58, **58**
Magnolia stellata 'Royal Star' 58, **58**
Magnolia wilsonii 59
Magnolia × *brooklyensis* 'Yellow Bird' 54, **54**
Magnolia × *loebneri* 'Leonard Messel' 57, **57**
Magnolia × *loebneri* 'Merrill' 57, **57**
Magnolia × *soulangeana* 58, **58**, 5 9
Magnolia × *soulangeana* 'Rustica Rubra' 58
Mahonia 176–178
Mahonia aquifolium **17**, 176, 177
Mahonia aquifolium 'Apollo' 177
Mahonia aquifolium 'Smaragd' 176, 177
Mahonia eurybractea subsp. *gonpinensis* 'Soft Caress' **176**, 177
Mahonia gracilipes 177
Mahonia japonica **177**, 178, **178**
Mahonia nervosa **176**, 177
Mahonia nitens 'Cararet'PBR 177
Mahonia oiwakensis subsp. *lomariifolia* **177**, 178
Mahonia × *media* 'Buckland' 178
Mahonia × *media* 'Charity' 178, **178**
Mahonia × *media* 'Lionel Fortescue' 178
Mahonia × *media* 'Winter Sun' 177, 178, **178**
Mahonia × *wagneri* 'Pinnacle' 177
Malus 60
Malus 'Adirondack' **60**, 62
Malus 'Evereste' **61**, 62
Malus 'Royal Beauty' 62
Malus CORALBURST ('Coracole') **60**, 62
Malus JELLY KING ('Mattfru') 62
Malus sargentii 'Tina' 62
Malus transitoria 62
Malus trilobata **61**, 62
Malus × *floribunda* **61**, 62
Malus × *moerlandsii* 'Liset' 62
Malus × *robusta* 'Red Sentinel' **6**
Malus × *robusta* 'Red Sentinel' 60, **61**, 62
Malus × *scheideckeri* 'Red Jade' 62
Malus × *zumi* 'Golden Hornet' 62, **62**
Manglietia 53
Melampsora hypericorum 165
Mespilus germanica 63, **63**, **64**
Mespilus germanica 'Flanders Giant' 64
Mespilus germanica 'Large Russian' 64
Mespilus germanica 'Nottingham' 64
Mespilus germanica 'Royal' 63, 64
Michelia 53
Microbiota 254
Microbiota decussata 254, **254**
Morus 64
Morus 'Matsunaga' 65
Morus 'Waisei-Kirishima-shikinari' 64, 65
Morus alba 64, **64**, 65
Morus alba 'Pendula' 64, **64**, 65
Morus Charlotte Russe 65
Morus Mojo Berry 65
Morus nigra 64, 65
Morus rotundiloba 65
Myrtus 179
Myrtus communis 179, **179**
Myrtus communis 'Variegata' 179, **179**
Myrtus communis subsp. *tarentina* 179

Nandina 179–181, 215
Nandina domestica 179, **180**, 181
Nandina domestica 'Fire Power' **180**, 181
Nandina domestica 'Sunset'PBR 181
Nandina domestica 'Gulf Stream' **180**, 181
Nandina domestica 'Richmond' 181
Nandina domestica 'Seika'PBR **180**, 181
Nandina domestica 'Twilight' **180**, 181
Nandina domestica BLUSH PINK ('Aka'PBR) 181
Nandina domestica BRIGHTLIGHT 'Shelton004') 181
Nandina domestica MAGICAL LEMON AND LIME ('Lemlin'PBR) **180**, 181
Nectria cinnarbarina 106

Osmanthus 182–183
Osmanthus delavayi **182**, 183
Osmanthus delavayi 'Pearly Gates' 183

Osmanthus fragrans 8, 182
Osmanthus heterophyllus 183
Osmanthus heterophyllus 'Goshiki' **182**, 183
Osmanthus heterophyllus 'Variegatus' **182**, 183
Osmanthus × burkwoodii **182**, 183
Otiorhynchus sulcatus 42

Pachysandra 184
Pachysandra procumbens 184, **184**
Pachysandra terminalis 184
Pachysandra terminalis 'Green Carpet' 184
Pachysandra terminalis 'Green Sheen' 184
Pachysandra terminalis 'Variegata' 184
Perovskia 185 All transferred to *Salvia yangii* 214
Perovskia 210
Philadelphus 186–187
Philadelphus 'Beauclerk' 186
Philadelphus 'Belle Etoile' 186, **187**
Philadelphus 'Monteau d' Hermine' 186
Philadelphus coronarius 'Aureus' 186, **187**
Philadelphus coronarius 'Variegatus' 186
Philadelphus maculatus 186
Philadelphus maculatus 'Sweet Clare' 186, **187**
Philadelphus Silberregen' (syn. 'Silver Showers') 186
Phillyrea 185
Phillyrea angustifolia 185
Phillyrea angustifolia f. *rosmarinifolia* 185, **185**
Phillyrea angustifolium f. *rosmarinifolia* 95
Phillyrea latifolia 185
Phormium 'Sundowner' **17**
Physocarpus 131
Physocarpus opulifolius 223
Physocarpus opulifolius 'Diabolo'PBR **16**
Physocarpus opulifolius LITTLE ANGEL ('Hoogi16'PBR) 131
Picea 254–256
Picea abies 254
Picea abies 'Acrocona' 255
Picea abies 'Little Gem' 255
Picea abies 'Nidiformis' 255
Picea glauca 254
Picea glauca 'Echiniformis' 255
Picea glauca var. *albertiana* 'Conica' 255, **255**
Picea pungens 254–256
Picea pungens (Glauca Group) 'Globosa' **256**
Picea pungens (Glauca Group) 'Koster' **255**, 256
Picea pungens Glauca Group 225
Pinus 256
Pinus contorta 'Chief Joseph' **180**, **238**, 256, **256**
Pinus mogo 'Humpy' 257
Pinus mugo 'Mops' 257, **257**, 263
Pinus mugo 'Winter Gold' 257, **257**
Pinus mugo 257
Pinus sylvestris 258
Pinus sylvestris (Aurea Group) 'Aurea' 258
Pinus sylvestris 'Gold Coin' 258, **258**
Pittosporum 65
Pittosporum tenuifolium 65, 66
Pittosporum tenuifolium 'Golden Ball', 66
Pittosporum tenuifolium 'Golf Ball'PBR 65, 66, **67**, 95
Pittosporum tenuifolium 'Irene Patterson' 65, 66, **67**
Pittosporum tenuifolium 'Silver Queen' 66
Pittosporum tenuifolium 'Tom Thumb' 65, 66, **67**
Pittosporum tenuifolium 'Variegatum' 66, **67**
Pittosporum tobira 'Nanum' 66, **67**
Pittosporum tobira 66
Podocarpus 258–259
Podocarpus 'Chocolate Box' 259, **259**
Podocarpus 'Country Park Fire'PBR 259, **259**
Podocarpus lawrencei 'Bluey' (syn. *P. alpinus* 'Bluey') **259**
Podocarpus totara 'Aurea' 258, 259, **259**
Potentilla 187
Potentilla fruticosa 'Abbotswood' 187
Potentilla fruticosa 'Goldfinger' 187, **188**
Potentilla fruticosa 'Goldstar' 187, **188**
Potentilla fruticosa 'Hopleys Orange' 187, **188**
Potentilla fruticosa 'Limelight' 187
Potentilla fruticosa 'Pink Beauty'PBR (syn. P. 'Lovely Pink') 187
Potentilla fruticosa 'Tangerine' 187
Potentilla fruticosa 187
Potentilla fruticosa BELLA SOL ('Hachdon') 187, **188**
Potentilla fruticosa PRINCESS ('Blink') 187
Potentilla fruticosa var. *pumila* 'Red Ace' 187
Prunus 68
Prunus incisa 'Kojo-no-mai' **69**, 70
Prunus 'Accolade' **68**, 70
Prunus 'Amanogawa' **68**, **69**, 70
Prunus 'Blushing Bride' 71
Prunus 'Chocolate Ice' 70, **72**
Prunus 'Fugenzo' **69**
Prunus 'Hally Jolivette' 70
Prunus 'Kursar' 70, **71**, **72**
Prunus 'Matsummae-fuki' 70, **72**
Prunus 'Pandora' 71, **73**
Prunus 'Shirofugen' **69**
Prunus 'Shogetsu' 71
Prunus 'Snow Showers' **20**
Prunus 'The Bride' 71, **73**
Prunus avium 68
Prunus domestica 68
Prunus FRILLY FROCK ('Fpmspi') 70
Prunus incisa 'Paean' 70, **70**
Prunus incisa 'Pendula' 70, **70**
Prunus incisa 'The Bride' 71
Prunus mume 68
Prunus mume 'Beni-chidori' 70, **72**
Prunus mume 'Omoi-no-mama' 70, **72**
Prunus nipponica var. *kurilensis* 'Brilliant' 68, 70, **72**
Prunus padus 68
Prunus serrula 68, 71, 73
Prunus serrula 'Branklyn' 71
Prunus serrula 'Dorothy Clive' 71, **73**
Prunus serrulata 'Longipes' 71
Prunus spinosa 68
Prunus × subhirtella 'Autumnalis Rosea' 71, **73**
Prunus × subhirtella 'Autumnalis' 71
Pyrus salicifolia 'Pendula' 146

Rheum palmatum **26**
Rhododendron 'Dopey' 194
Rhododendron 'Dreamland' 194, **194**
Rhododendron 'Irohayama' 197
Rhododendron 'Sapphire' 191, **192**
Rhododendron 189
Rhododendron ponticum 189
Rhododendron williamsianum 189
Rhododendron 'Anneke' 195, **195**
Rhododendron 'Azuma-kagami' 197
Rhododendron 'Baden-Baden' 191
Rhododendron 'Bashful' 194
Rhododendron 'Baslisk' 195, **195**
Rhododendron 'Bengal Fire' 197
Rhododendron 'Blue Danube' 196, **196**, 197
Rhododendron 'Blue Diamond' 191
Rhododendron 'Britannia' 193
Rhododendron 'Christmas Cheer' 193, **193**
Rhododendron 'Cilpinense' 191, **191**
Rhododendron 'Curlew' 191
Rhododendron 'Cynthia' 193
Rhododendron 'Doncaster' 193
Rhododendron 'Elsie Lee' 197
Rhododendron 'Emanuela' 194
Rhododendron 'Florida' 197
Rhododendron 'Gibraltar' 195, **195**
Rhododendron 'Golden Eagle' 195, **195**
Rhododendron 'Golden Lights' 195
Rhododendron 'Golden Torch' 194, **194**
Rhododendron 'Hatchman's Marlis' 194
Rhododendron 'Hatchman's Porzellan' 194
Rhododendron 'Hino-crimson' 197
Rhododendron 'Homebush' 195
Rhododendron 'Kermesinum' 197
Rhododendron 'Koster's Brilliant Red' 195, **195**
Rhododendron 'May Day' 191, **191**
Rhododendron 'Mother's Day' 197
Rhododendron 'Orange King' 197
Rhododendron 'Palestrina' 197
Rhododendron 'Percy Wiseman' 194, **194**
Rhododendron 'Persil' 195
Rhododendron 'Pink Pearl' 193, **193**
Rhododendron 'Praecox' 189
Rhododendron 'Ptarmigan' 191, **192**
Rhododendron 'Pucella' 195, **195**
Rhododendron 'Razorbill' 191, **192**
Rhododendron 'Sappho' 193
Rhododendron 'Snipe' 191
Rhododendron 'Squirrel' 197
Rhododendron 'Surrey Heath' 194, **194**
Rhododendron 'Torchlight' 197
Rhododendron 'Vulcan' 193, **193**
Rhododendron augustinii 190, **190**
Rhododendron Blue Tit Group 191, **191**
Rhododendron bureavii 17
Rhododendron catawbiense 193
Rhododendron ciliatum 189
Rhododendron Cunningham's White 189
Rhododendron dauricum 190
Rhododendron Elizabeth Group 191, **191**
Rhododendron forrestii 190
Rhododendron luteum (syn. *Azalea ponticum*) 190, **190**
Rhododendron yakushimanum 190, **190**, 194, 196
Rhus 130
Robinia 74
Robinia hispida 74, **74**, 75
Robinia pseudoacacia 74, 75
Robinia pseudoacacia 'Frisia' 46, 74, **74**, 75
Robinia pseudoacacia 'Lace Lady' 74, **74**, 75
Robinia × ambiqua 'Decaisneana' 75
Robinia × marginata 'Pink Cascade' 75
Rosa 197
Rosa gallica 'Versicolor' 199, **199**
Rosa SUN SEEKER ('Dicraker') 203, **203**
Rosa 'Bloomfield Abundance' 205
Rosa 'Cinco de Mayo'PBR 201
Rosa 'Comte de Chambord' 200
Rosa 'De Resht' 200
Rosa 'Duchess of Portland' 200
Rosa 'Flower Carpet Coral'PBR 206, **206**
Rosa 'Gruss an Aachen' 205
Rosa 'Helen Knight' 198, 207
Rosa 'Magic Carpet' 207
Rosa 'Nice Day' 207
Rosa 'Nozomi' 206
Rosa 'Raubritter' 206
Rosa 'Robin Redbreast' 207
Rosa 'ROTE MAX GRAF ('Kormax') (syn. *Rosa* 'Red Max Graf') 206
Rosa 'The Fairy' 205, **205**
Rosa 'White Pet' (syn. R. Little White Pet') 205
Rosa 'Yvonne Rabier' 205
Rosa A WHITER SHADE OF PALE (Peaanfare'PBR) 202, **202**
Rosa BERKSHIRE ('Korpinka') 206
Rosa CAREFREE DAYS ('Meirivoui'PBR) 203, **203**
Rosa DOUBLE DELIGHT ('Andeli') 202
Rosa FLOWER CARPET AMBER ('Noa97400a'PBR) 206
Rosa FLOWER CARPET RED VELVET (Noare'PBR) 206, **206**
Rosa FLOWER CARPET WHITE ('Noaschnne'PBR) 206
Rosa FLOWER POWER ('Frycassia'PBR) 203
Rosa FOR YOUR EYES ONLY ('Cheweyesup'PBR) 201, **201**
Rosa gallica var. *officinalis* 199, **199**
Rosa glauca 198, **189**, 207
Rosa HOT CHOCOLATE ('Wekpaltlez') 201
Rosa JOIE DE VIVRE ('Korfloci 01'PBR) 203, **203**
Rosa KENT ('Poulcov') 206
Rosa LANCASHIRE ('Korstegli'PBR) 206
Rosa LAURA FORD ('Chewarvel') 204, **204**
Rosa Little Bronze 204, **2 04**
Rosa LOVELY BRIDE ('Meiratcan') 203
Rosa MAGIC CARPET ('Jacklover'PBR) 206
Rosa MARGARET MERRIL ('Harkuly') 201,
Rosa MARJORIE FAIR ('Harhero') 205
Rosa moyesii 197
Rosa MY VALENTINE ('Mormyval') 204
Rosa ORANGE MEILOVE ('Meitongos') 201, **201**
Rosa PAPA MEILLAND ('Meisar') 202
Rosa PEARL ANNIVERSARY ('Whitson'PBR) 204
Rosa QUEEN MOTHER ('Korquemu') 203
Rosa RUBY ANNIVERSARY ('Habonny'PBR) 203
Rosa spinoisissima (syn. *R. pimpinellifolia*) 200
Rosa spinoisissima 'double white' 200
Rosa spinoisissima 'Grandiflora' 200
Rosa spinoisissima 'William 111' 200
Rosa SURREY ('Korlanum') 206
Rosa SWEET DREAM ('Fryminicot') 203, 207
Rosa SWEET MAGIC ('Dicmagic'PBR) 204
Rosa THANK YOU ('Chesdeep'PBR) 203
Rosa TROIKA 'Poumidor' 202, **202**
Rosa TRUMPETER ('Mactru') 201, **201**
Rosa virginiana 198, 207
Rosa × *jacksonii* 'Max Graf' 206
Rosa × odorata (Sanguinea Group) 'Bengal Crimson' 198, 207
Rosa × odorata 'Mutabilis' **111**
Rosa × odorata 198
Rosa xanthina 197
Rosa YELLOW MEILOVE ('Meitongos') 202
Rosmarinus 207 (All plants previously known as *Rosmarinus* have been transferred to the genus *Salvia*)
Rosmarinus 210
Rubus 127, 208–210
Rubus 'Betty Ashburner' 209
Rubus cockburnianus 'Goldenvale' 208, 209, **210**
Rubus fruticosus 208
Rubus spectabilis 'Olympic Double' 208, 209, **210**
Rubus thibetanus 208, 209, **210**
Rubus tricolor 209

Salix 127
Salix alba 'Kilmarnock' 10
Salvia 'Mirage Deep-Purple' 210
Salvia 210–215
Salvia 'Alba' 210
Salvia 'Amethyst Lips' 210
Salvia 'Amistad' 210
Salvia 'Desert Blaze' (syn. *Salvia* 'Sparkler') 210
Salvia 'Dyson's Joy' 210
Salvia 'Emperor' 210
Salvia 'Hot Lips' 213, **212**
Salvia 'Icing Sugar'PBR 210
Salvia 'Lipstick' 210
Salvia 'Nachtvlinder' (syn. *S. × jamensis* 'Nachtvlinder') **212**, 213
Salvia 'Peach' 210
Salvia 'Royal Bumble' 213, **212**
Salvia 'Silas Dyson' 213
Salvia 'Strawberries and Cream' 210
Salvia 'Sungold' (syn. *Salvia* 'Devon and Cream') 210
Salvia 'Tricolor' 213
Salvia AMETHYST LIPS ('Dyspurp') 213
Salvia argentea **17**
Salvia greggii 210
Salvia guarantica 'Blue and Black' 210
Salvia involucrata 210
Salvia lavandulifolia 213
Salvia microphylla (syn. *S. neurepia*, *S. macellaria*) 212
Salvia microphylla 'Aphrodite' 212
Salvia microphylla 'Blush Pink' 212

Salvia microphylla 'Cerro Potosi' 212
Salvia microphylla 'Delice Fiona' 212
Salvia microphylla 'Glitter' 212
Salvia microphylla 'Kew Red' 212, **212**
Salvia microphylla 'Newby Hall' 212
Salvia microphylla 'Wild Watermelon' 212
Salvia microphylla 'Wine and Roses' 212
Salvia microphylla var. *microphylla* 212
Salvia officinalis 213, 215
Salvia officinalis 'Berggarten' 213
Salvia officinalis 'Icterina' 213
Salvia officinalis 'Purpurascens' 213, 213
Salvia rosmarinus 19
Salvia rosmarinus Prostrata Group 215
Salvia rosmarinus (syn. *Rosmarinus officinalis)* 214, **214**
Salvia rosmarinus 'Benenden Blue' 214
Salvia rosmarinus 'Green Ginger' 214
Salvia rosmarinus 'Lady in White' 214
Salvia rosmarinus 'Miss Jessop's Upright' 214, 215
Salvia rosmarinus 'Severn Sea' 214
Salvia rosmarinus Prostrata Group 214
Salvia × *jamensis* 210, 211
Salvia × *jamensis* 'Californian Sunset' 211
Salvia × *jamensis* 'Cherry Lips' 211
Salvia × *jamensis* 'Dark Dancer' 211
Salvia × *jamensis* 'Flammenn'^{PBR} **211**
Salvia × *jamensis* 'James Compton' 211
Salvia × *jamensis* 'Javier' 211
Salvia × *jamensis* 'La Luna' 211
Salvia × *jamensis* 'La Siesta' 211
Salvia × *jamensis* 'Maraschino' 211, **211**
Salvia × *jamensis* 'Moonlight Serenade' 211
Salvia × *jamensis* 'Pat Vlasto' 211
Salvia × *jamensis* 'Peter Vidgeon' 211, **211**
Salvia × *jamensis* 'Pluenn'^{PBR} 211
Salvia × *jamensis* 'Raspberry Royale' 211
Salvia × *jamensis* 'Red Velvet' **211**
Salvia × *jamensis* 'Sierra San Antonio' 211
Salvia yangii (Previously classified as *Perovskia atriplicifolia*) 214, 215
Salvia yangii 'Blue Jean Baby' 215
Salvia yangii 'Blue Shadow' 215
Salvia yangii 'Blue Spire' 215, **215**
Salvia yangii 'Blue Steel' 215
Salvia yangii 'Filigran' 215
Salvia yangii 'Little Spire'^{PBR} 215
Salvia yangii LACEY BLUE ('Lisslitt'^{PBR}) 215
Salvia yangii SILVERY BLUE ('Lissvery'^{PBR}) 215
Santolina 216–217
Santolina chamaecyparissus (syn. *S. incana)* 216
Santolina chamaecyparissus 'Lambrook Silver' 216
Santolina chamaecyparissus 'Lemon Queen' 216
Santolina chamaecyparissus 'Nana' 215, 217
Santolina chamaecyparissus 'Pretty Caroll' 216
Santolina pinnata 'Sulphurea' **215**
Santolina pinnata subsp. *neopolitana* 'Edward Bowles' 216
Santolina pinnata subsp. *neopolitana* 216
Santolina rosmarinifolia (syn. *S. virens* and *S. viridis)* 216
Santolina rosmarinifolia 'Lemon fizz' 216, **216**, 217
Santolina rosmarinifolia subsp. *rosmarinifolia* 'Primrose Gem' 216
Santolina rosmarinifolia subsp. *rosmarinifolia* 216
Sarcococca 217–218
Sarcococca confusa **217**, 218
Sarcococca hookeriana 'Purple Stem' 218
Sarcococca hookeriana 218
Sarcococca hookeriana var. *digyna* **217**, 218
Sarcococca hookeriana var. *humilis* 218
Sarcococca hookeriana WINTER GEM ('Pmoore03'^{PBR}) 218
Sarcococca ruscifolia 218
Sarcococca ruscifolia var. *chinensis* 218
Sarcococca ruscifolia var. *chinensis* 'Dragon's Gate' 218
Senecio 90
Skimmia 218–220
Skimmia japonica 'Magic Marlow'^{PBR} 220
Skimmia japonica **219**, 220
Skimmia japonica **67**
Skimmia japonica 'Fragrans' **219**
Skimmia japonica 'Fragrans' (syn. *S. japonica* 'Fragrant Cloud') 220
Skimmia japonica 'Kew White' 220
Skimmia japonica 'Nyman's' 220
Skimmia japonica 'Rubella' **219**
Skimmia japonica OBSESSION ('Obsbolwi'^{PBR}) 220
Skimmia × *confusa* 218
Skimmia × *confusa* 'Kew Green' **219**, 220
Sophora japonica 82
Sophora japonica 'Pendulum' 82
Sorbaria 220–221
Sorbaria sorbifolia 'Sem'^{PBR} 221, **221**
Sorbaria sorbifolia PINK HOPI ('Cousorb05') 221
Sorbus 75, 264, 277
Sorbus 'Asplenifolia' 76
Sorbus 'Chinese Lace' **77**, 78
Sorbus 'Glendoick Spire' 78
Sorbus 'John Mitchell' **18**
Sorbus 'Joseph Rock' **77**, 78
Sorbus aria 76
Sorbus aria 'Lutescens' **77**, 78
Sorbus aucuparia **18**, 18, 76, **76**
Sorbus aucuparia CARDINAL ROYAL ('Michred') 76
Sorbus AUTUMN SPIRE (Fanrock') **77**, 78
Sorbus bissetii 'Pearls' 79
Sorbus cashmiriana 76, 79, **79**
Sorbus commixta 'Olympic Flame' 78
Sorbus dodong 'Olympic Flame' 78
Sorbus frutescens 79
Sorbus fruticosa 'Koehneana' 79
Sorbus hupehensis 'Rosea' 79
Sorbus koehneana 76, 79, 80
Sorbus pogonopetala 79
Sorbus pohuashanensis 79
Sorbus pseudohupehensis 79, **80**
Sorbus pseudohupehensis 'Pink Pagoda' 79
Sorbus pseudovilmorinii 76, 79
Sorbus reducta 79, **80**
Sorbus sargentiana 79, **81**
Sorbus ulleungensis 'Olympic Flame' **77**, 78
Sorbus vilmorinii 79, **81**
Sorbus × *kewensis* 79
Spiraea 220, 222
Spiraea 'Arguta' 222, **222**
Spiraea japonica 222, 223
Spiraea japonica 'Gold Mound' 223
Spiraea japonica 'Goldflame' **222**, 223
Spiraea japonica DOUBLE PLAY BIG BANG **222**, 223
Spiraea japonica DOUBLE PLAY GOLD ('Yan') **222**
Spiraea japonica PLAY ARTISAN ('Galen') **222**, 223
Spiraea nipponica 'Snowmound' 223
Stachys byzantina 17
Stipa gigantea **277**
Storax 83
Styphnolobium 82
Styphnolobium japonicum 82
Styphnolobium japonicum 'Pendulum' 82, **82**
Styrax 223 (All *Styrax* are considered in Chapter 3)
Styrax 83
Styrax japonicus 83
Styrax japonicus 'Sohuskan' 83
Styrax japonicus (Benibana Group) 'Fargesii' 83
Styrax japonicus (Benibana Group) 'Pink Chimes' 83, **83**
Styrax japonicus 'Emerald Pagoda' 83
Styrax japonicus 'Purple Dress' 83
Styrax obassia 83
Syringa 186, 223
Syringa meyeri 223, 224
Syringa meyeri (Flowerfesta Series) FLOWERFIESTA PINK (Anny 200817'^{PBR}) 224
Syringa meyeri (Flowerfesta Series) FLOWERFIESTA PURPLE (Anny200809'^{PBR}) 224
Syringa meyeri (Flowerfesta Series) FLOWERFIESTA WHITE (Anny 200810'^{PBR}) 224
Syringa meyeri 'Palibin' 224, **224**
Syringa pubescens microphylla 'Superba' 224
Syringa × *persica* 224
Syringa × *persica* 'Alba' 224

Taxus 260
Taxus baccata 95
Taxus baccata 'David' 260
Taxus baccata 'Fastigiata Aureomarginata' 260
Taxus baccata 'Fastigiata' 260
Taxus baccata 'Icicle' 260
Taxus baccata 'Repens Aurea' 260
Taxus baccata 'Standishii' 258, 260, **260**, 264
Theielaviopsis basicola 169
Thuja 261
Thuja occidental 'Danica' 261
Thuja occidentalis 'Amber Gold' **261**
Thuja occidentalis 'Degroot's Spire' **14**
Thuja occidentalis 'Golden Smaragd' **265**
Thuja occidentalis 'Holmstrup' 261
Thuja occidentalis 'Rheingold' 261, **261**
Thuja occidentalis 'Rosedalis' **261**
Thuja occidentalis 'Smaragd' 261
Thuja occidentalis 'Sunkist' 261
Thuja plicata 262
Thuja plicata 'Atrovirens' 13, 262, **262**
Thuja plicata 'Rogersii 262
Thuja plicata 'Zebrina' 262
Trillium 144

Veronica 225
Veronica verniculosa 229
Veronica 'Autumn Glory' 228
Veronica 'Blondie' **226**
Veronica 'Calcedonica' (syn. *Hebe* 'E.B.Anderson', *H.* 'Knightshayes') 228
Veronica 'Emerald Gem' (syn. *Hebe* 'Green Globe', *H.* 'Emerald Green') 228
Veronica 'Goldrush'^{PBR} 228
Veronica 'Great Orme' 228
Veronica 'Magic Summer'^{PBR} 228
Veronica 'Margaret' 228
Veronica 'Midsummer Beauty' 228
Veronica 'Mrs. Winder (syn. *Hebe vernicosa* 'Waikiki') 228
Veronica 'Neil's Choice' (syn. *Hebe* 'Purple Emperor') 228
Veronica 'Nicola's Blush' 228
Veronica 'Pascal' 229
Veronica 'Pewter Dome' 229
Veronica 'Pink Elephant' **226**
Veronica 'Red Edge' **227**
Veronica 'Red Edge' 229
Veronica 'Silver Queen' (syn. *Hebe* × *franciscana* 'Variegata', *H. elliptica* 'Variegata') 229
Veronica 'Wiri Cloud' 229
Veronica 'Youngii (syn. *Hebe* 'Carl Teschner') 229
Veronica albicans Recurva Group **227**, 228
Veronica hulkeana 228
Veronica macrantha 228
Veronica ochracea **16**
Veronica ochracea 'James Stirling' 229
Veronica odora 'Nana' (syn. *Hebe buxifolia* 'Nana') 229
Veronica odora **226**
Veronica pimeleoides 'Quicksilver' 227, 229
Veronica pinguifolia 'Pagei' **227**, 229
Veronica pinguifolia 'Sutherlandii' 227, 229
Veronica rakaiensis 18, 226, **227**, 229, 230
Veronica sempervirens 226
Veronica topiara 229
Veronica vernicosa **227**
Veronica × *andersonii* 'Variegata' 228
Veronica 'Goldrush' **226**
Viburnum 230–235
Viburnum 'Eskimo' 234
Viburnum davidii 33, 220
Viburnum farreri (syn. *fragrans*) 234, **232**
Viburnum farreri 'Candidissimum' 234, **232**
Viburnum farreri 'Nanum' 234, **232**
Viburnum lantana 234, **232**
Viburnum opulus 19
Viburnum opulus 'Compactum' 235, **232**
Viburnum plicatum f. *tomentosum* 'Mariesii' 82, 235, **232**
Viburnum plicatum f. *tomentosum* 'Pink Beauty' 235, **232**
Viburnum plicatum f. *tomentosum* KILIMANJARO ('Jww1'^{PBR}) 235
Viburnum plicatum f. *tomentosum* KILIMANJARO SUNRISE ('Jww5'^{PBR}) 235
Viburnum rhytidophyllum 17
Viburnum sargentii 'Onodaga' 235
Viburnum tinus 'Eve Price' 235
Viburnum tinus 'Gwenllian' 235, **232**
Viburnum × *bodnantense* 'Charles Lamont' 234, **231**
Viburnum × *bodnantense* 'Dawn' 234, **231**
Viburnum × *bodnantense* 'Deben' 234, **231**
Viburnum × *juddii* 234, **232**
Vinca major 'Variegata' 184
Volutella sp. 88, 95

Weigela 236–237
Weigela 'Bristol Ruby' 237
Weigela 'Florida Variegata' 237, **237**
Weigela ALL SUMMER PEACH ('Slingpink'^{PBR}) 237, **236**
Weigela ALL SUMMER RED ('Slingcol'^{PBR}) 237, **236**
Weigela BLACK AND WHITE ('Courtacad1'^{PBR}) 237, **236**
Weigela BRIANT RUBIDOR ('Olympiade') 237, **236**
Weigela CARNIVAL 'Courtalor' 237
Weigela florida 'Alexandra'^{PBR} 237
Weigela florida 'Folies Purpureis' 237, **236, 237**
Weigela florida MONET 'Verweig'^{PBR} 237
Weigela middendorffiana 237
Weigela middendorffiana 'Mango' 237

× *Cuprocyparis leylandii* 13, 262, 264
× *Cuprocyparis leylandii* 'Castlewellan' (syn. 'Castlewellan Gold') **263**, 262
× *Cuprocyparis leylandii* 'Silver Dust' 262
× *Cuprocyparis leylandii* 'Blue Jeans' **263**
× *Cuprocyparis leylandii* 'Gold Rider' 262, 263
× *Cuprocyparis leylandii* 'Naylor's Blue' 262
× *Cuprocyparis leylandii* 13
× *Cuprocyparis leylandii* Haggerston's Grey' 262
× Halimiocistus 152
× Halimiocistus 'Merrist Wood Cream' 153
× *Halimiocistus sahucii* 153
× *Halimiocistus sahucii* DANCER ('Ebhals'^{PBR}) 153
× *Halimiocistus wintonensis* 153

Yucca rigida **277**

COMMON NAME INDEX

Aaron's beard 165
African mallow 87
Alexandrian laurel 137
alpine laburnum 48
American box 129
apothecary's rose 199
arbor-vitae 239, 261
Atlantic cedar 241
autumn sage 210
azaleas 195

beech 14
bell heather 272
birches 34, 35
bird cherry 68
black locust 75
black mulberry 65
blackthorn 68
blue holly 168
box 94
box elder 22
box tree caterpillar
brambles 208
bridal wreath 222
bullace 68
Burmese plumbago 105
butterfly bush 91

camellia 96
Canadian bunchberry 118
Caucasian fir 239
cedars 241
cherries 68
Chilean myrtle 52
Chinese bramble 209
Chinese plumbago 105
Christmas box 217
Colorado blue spruce 254
common acacia 75
common box 18
common box 95
common crape myrtle 171
common dogwood 124
common gum 112
common hawthorn 40, 41
common holly 166
common juniper 250
common myrtle 179
common rosemary 214
common spruce 254
common yew 260
conifers 239
Connemara heath 268
coral bark maple 25, 26
cornel 115, 122
Cornelian cherry 122
Cornish heath 275
cotton lavender 216
crab apples 59
creeping cotoneaster 132
creeping dogwood 118
creeping juniper 250
cross-leaved heather 274
culinary apples 63
cypresses 246

dessert apples 63
dogwood 115
downy birch 35
downy Japanese maple 22
Dutch medlar 64
dwarf box 95

eastern thorn 40
edging box 95
edible dogwood 122
European cornel 122
European mountain pine 257

false acacia 74, 75
false cypresses 239, 242
fire blight 40
firs 239
flaky juniper 253
flowering cherries 16
flowering dogwood 120
flowering quince 108
foam of may 22
fragrant sweet box 218
fuchsia 150

garland flower 138
gean 68
golden Irish yew 260
golden rain tree 47
Grecian strawberry tree 33
green osier 118
greengages 68
grey snake bark maple 27
guelder rose 19

hardy plumbago 105
harlequin glory bower 38, 112
heavenly bamboo 179
Her's maple 27
herring bone cotoneaster 134
Highclere holly 167
Himalayan birch 35
Himalayan honeysuckle 174
Himalayan sweet box 218
honey locust 45
hornbeam 14
hybrid wire-netting bush 130
hydrangea 155

Irish heather 274
Irish yew 260

Japanese holly 168
Japanese maple 23
Japanese meadowsweet 222
Japanese pagoda tree 82
Japanese quince 108, 109
Japanese snowbell 83
Japanese spurge 184
japonica 108
Judas tree 106
juneberry 30
junipers 239, 250

Killarney strawberry tree 32, 33

laburnum 48
lace-cap hydrangea 157
lady of the woods 35
lamb's ears 17
laurel-leaf cistus 112
Lawson cypress 242
Lawson's cypress 263
Leyland cypress 262, 264
looking glass plant 114

madrone 33
magnolia 52
maidenhair tree 248
maple 7, 21
may 40
mazzard 68
medlar 62
Mexican orange 18
Mexican orange blossom 110, 111
mezereon 140
midland hawthorn 40
mirror plant 114
mock orange 186
monkshood 22
Monro's ragwort 90
mop-headed hydrangea 156
mountain ash 76

Nordmann fir 239
North American redbud 106
Norway spruce 254

oak-leaved hydrangea 161
old red damask 199
Oregon grape 17, 177
Oriental sweet gum 50
ornamental cherries 16
ornamental quince 108

Pacific madrone 33
pagoda dogwood 118
paper bark maple 15
pearl bush 149
Persian lilac 224
Persian silk tree 29
pheasant berry 173
pines 239, 256
plum 68
plum pines 258
Pontic daphne 142
Portland rose 200
pride of India 47

quick thorn 40

red birch 35
red osier dogwood 125
red rose of Lancaster 199
red vein maple 27
red-barked dogwood 115
rhododendron 189
river birch 35
rock roses 110
rocky mountain juniper 252
rose acacia 75
rose daphne 138
rose of Sharon 154, 165
rosemary-leaved cotton lavender 216
roses 197
rowan 76, 265
Russian arbor-vitae 254
Russian sage 210

sacred bamboo 179
sages 18
Saint Dabeoc's heath 268
salvia 210
sawara cypress 244
Scots pine 258
Scots rose 200
Scottish laburnum 48
seven son flower of Zhejiang 153
Siberian carpet cypress 254
Siberian pea tree 37
silver birch 35
silver firs 239
single-seeded hawthorn 40
sloe 68
small-leaved cotoneaster 134
smoke bushes 130
smooth service tree 29, 30
snake bark maples 27
snowbell 83
snowy mespilus
sorbet 122
spindle tree 45
spruces 239, 254
St. John's wort 163
star magnolia 58
storax 83
sweet box 218
sweet gum 51
syringa 186
Szechuan strawberry 121

tailored cherry 68
tawhiwhi 65
temu 52
thorns 39
Tibetan cherry 15
tree cotoneasters 135
tree heath 270
twin-flowered daphne 142

veronica 225 (including those previously called Hebe)

wall cotoneaster 134
wayfaring tree 234
wedding cake tree 119
western cedar 15
western red cedar 262
white mulberry 65
whitebeam 76
winged spindle 45
winter-flowering cherry 71
winter berry 170
winter heath 270
wire-netting bush 129
wolfsbane 22

yellow azalea 190
yew 265
Young's weeping birch 35

ACKNOWLEDGEMENTS

All photographs are by the author; those on pages 13 and 14 are reproduced by courtesy of the National Trust.

I would like to thank all the following individuals, organizations and businesses who very kindly allowed me to carry out photography on their premises and include the images in his book, these being:

Batsford Arboretum and Garden Centre, Batsford, Moreton-in-Marsh. GL56 9AT

Bourton House Garden, Bourton-on-the Hill, Moreton-in-Marsh. GL56 9AE

Cambridge University Botanic Garden, 1 Brookside, Cambridge. CB2 1JE

Grange Farm Nursery, Rectory Lane, Guarlford, Malvern. WR13 6NT

Hillier Nurseries Limited, Ampfield House, Ampfield, Romsey, Hampshire. SO51 9PA

John Richards Nurseries, Mathon Road, Malvern. WR13 6EW

Mr Adrian Bloom VMH, Foggy Bottom Garden, Bressingham, Norfolk

Mr John Massey VMH, at John's Garden at Ashwood Nurseries, Kingswinford. DY6 0AE

Pershore College, part of Warwickshire College, Avonbank, Pershore, Worcestershire. WR10 3JP

RHS Garden Rosemoor, Torrington. EX38 8PH

RHS Garden Wisley, Wisley Lane, Woking. GU23 6QB

Spetchley Park gardens, The Estate Office, Spetchley, Worcester. WR5 1RS

Wyevale Nurseries, Wyevale Way, Hereford. HR4 7AY

I would also thank to thank those persons listed below who helped me considerably in conversations about what plants to include and what to omit. Without their help several new plant introductions would have been missed.

Mr Adam Dunnett, Amenity Sales Director and Mr Charles Carr, Head of Wholesale Nurseries, at Hillier Nurseries Limited, Ampfield House, Ampfield, Romsey, Hampshire. SO51 9PA.

Mr Kyle Ross, Production Director, Wyevale Nurseries, Wyevale Way, Kings Acre, Hereford. HR4 7AY

Mr Nick Dunn, Frank P. Matthews Ltd. Berrington Court, Tenbury Wells, Worcestershire. WR15 8TH

First published in 2023 by
The Crowood Press Ltd
Ramsbury, Marlborough
Wiltshire SN8 2HR

enquiries@crowood.com
www.crowood.com

British Library Cataloguing-in-Publication Data
A catalogue record for this book is available from the British Library.

ISBN 978 0 7198 4219 1

Cover design by Sergey Tsvetkov

Graphic design and typesetting by
Peggy & Co. Design
Printed and bound in India by Parksons Graphics